GLASSY DISORDERED SYSTEMS

Glass Formation and Universal Anomalous Low-Energy Properties

GLASSY DISORDERED SYSTEMS

Glass Formation and Universal Anomalous Low-Energy Properties

Michael I. Klinger

Bar-Ilan University, Israel

World Scientific

NEW JERSEY • LONDON • SINGAPORE • BEIJING • SHANGHAI • HONG KONG • TAIPEI • CHENNAI

Published by

World Scientific Publishing Co. Pte. Ltd.
5 Toh Tuck Link, Singapore 596224
USA office: 27 Warren Street, Suite 401-402, Hackensack, NJ 07601
UK office: 57 Shelton Street, Covent Garden, London WC2H 9HE

British Library Cataloguing-in-Publication Data
A catalogue record for this book is available from the British Library.

GLASSY DISORDERED SYSTEMS
Glass Formation and Universal Anomalous Low-Energy Properties (Soft Modes)

ISBN 978-981-4407-47-2

Typeset by Stallion Press
Email: enquiries@stallionpress.com

Printed in Singapore by World Scientific Printers.

PREFACE

During the last fourty years, new experimental techniques and theoretical concepts have been suggested for understanding fundamental properties of glassy disordered systems, both at high temperatures close to the liquid-to-glass transition (glass formation) and in more detail, the so-called anomalous, including universal properties of glasses at low energies, i.e., low temperatures and/or frequencies which are lower than the Debye values. The present book discusses the fundamental properties of glassy disordered systems, both briefly at "high" temperatures close to the liquid-to-glass transition (i.e., glass formation) and in more detail, anomalous, including so-called universal, properties of glasses at low energies, in fact, at low temperatures and/or frequencies lower than the Debye values. Some earlier theoretical models for the glass formation and several more recent models for the anomalous, including universal properties of glasses are described and discussed in basic aspects.

In particular, the origin and main features of so-called *soft atomic-motion modes and excitations* and their role in the anomalous low energy properties of glasses are considered in detail. Basic concepts and results of the so-called *soft-mode model* of the anomalous properties are discussed. The analysis shows that the model is able to give a consistent description of the glassy properties at low energies. In accordance with the basic concepts of solid state theory, soft-mode (non-Debye) excitations of low energy determine the anomalous, including universal properties.

In fact, the present book demonstrates that the *soft-mode model* describes the origin of excess, non-Debye excitations at temperatures $T \ll T_D$ and frequencies $\nu \ll \nu_D$, including very low (T, ν), $T \lesssim T_m = h\nu_m/k_B \sim 1\,\text{K}$ or $\nu \lesssim \nu_m$ and moderately low ones, $T_m < (T, \nu) < T_M \sim 10^2\,\text{K}$.

The contribution of soft-mode excitations to the glassy properties is related either to soft-mode harmonic vibrational excitations (e.g. experimentally found boson peak typical of glasses) at moderately low (T, ν) or to soft-mode non-vibrational excitations like two-level systems at very low ones. In other words, in the model, the excess excitations can be soft-mode excitations mainly described by one of two fundamental quantum-mechanical systems: harmonic oscillators or strongly anharmonic two-level systems. Then an important *correlation* can be predicted in the model between the anomalous properties at very low (T, ν) and at moderately low (T, ν). This prediction naturally explains a recently found empirical *correlation*: universal anomalous dynamic and thermal properties of glasses at very low temperatures are found to be missing in amorphous materials where no boson peak is found. Additional manifestations of soft modes in glassy phenomena are also described which concern the origin of so called negative-U centres in electron properties of semiconducting glasses.

Other models of the anomalous properties of glasses (not in terms of soft modes) can be considered as the limit case of *soft-mode model* for either very low (T, ν) or moderately low ones.

The present book consists of two parts: the first part (I), Chapters 1–4, and the second part (II), Chapters 5–16, followed by Chapter 17 that represents **Summary, Conclusions and Problems**.

For brevity, "Chapter" and "Chapters" are often denoted as Ch. and Chs. respectively, while "Section" is denoted as Sec and "Sections" as Secs.

The first part is devoted to a review of basic general properties of glasses, including the basic process of glass formation from a supercooled liquid, as well as the thermal-equilibrium (thermodynamic) properties and non-equilibrium (relaxation) phenomena in glasses. This field has a rather long history, probably started from the 19th century. The properties and processes which appear to be fundamental for glass formation are mainly discussed.

The second part represents theoretical description of mostly universal dynamic and thermal properties of glasses, experimentally discovered (Zeller, Pohl, 1971) and then investigated in the following decades, at low temperatures $T \ll T_D$ and/or low external field frequencies $\nu \ll \nu_D$, which are anomalous in the sense that they are essentially different from the properties of corresponding crystalline materials (T_D and ν_D denote the Debye temperature and frequency characteristic of the materials respectively.). This part describes mostly the experimental background (Ch. 6) and the theoretical *soft-mode model* (SMM) of excitations (Chs. 7, 8, 10), probably with non-localized states in atomic dynamics of glasses (Klinger, 1979–1980) which appears to be able to describe at least qualitatively and scalewise the anomalous dynamic and thermal properties of glasses at low temperatures and frequencies. Several other theoretical models of low-energy atomic dynamics of glasses are also briefly described (Chs. 12, 13, 14).

The soft-mode model is a generalization to all low $T \ll T_D$ and $\nu \ll \nu_D$ of an earlier, *standard tunneling model* (P.W. Anderson, Halperin, Varma, 1972, W.A. Phillips, 1972) that characterizes the anomalous properties only at very low $T \lesssim T_m \approx 1\,\mathrm{K}$ and $\nu \ll \nu_m = k_B T/h \equiv T/h$ (in what follows the Boltzmann constant $k_B \equiv 1$). Both fundamentals and applications of the SMM, including low energy atomic dynamics and low temperature thermal properties, are described and discussed in Chs. 7, 8, 10. Other recent models of the dynamic and thermal properties are considered in Chs. 11–13 and compared to the soft-mode model. Moreover, anomalous electronic properties of semiconducting glasses, experimentally discovered in semiconducting glasses in the 1950s (Kolomiets *et al*) are theoretically described first in terms of postulates by Anderson and by Mott *et al.* in 1975 (Ch. 14) and as the soft-mode model of electron dynamics by Klinger and Taraskin in 1993–1995 (Ch. 15). Anomalous electronic properties, "photostructural changes" of semiconducting glasses under light of frequency related to the mobility-gap width were experimentally discovered in the 1980s (Kolomiets *et al*) and theoretically explained in the soft-mode model (Chs. 15, 16). Chapter 17 includes the **Summary, Conclusions and Problems**.

I dedicate this book to Irina.

Michael I. Klinger
Rehovot, Israel, Spring 2012.

CONTENTS

PART I

FUNDAMENTAL PROPERTIES OF GLASSES

1

GENERAL DESCRIPTION OF GLASSES AND GLASS TRANSITION

1.1 Metastability and disorder. Types of glasses

Glasses and glassy materials are macroscopic condensed systems constituting a very important variety of amorphous solids of which the formation is related to the so-called *glass transition* briefly described in what follows. Many materials are practically prepared as glasses and their number rapidly increases. It is important to understand how glasses differ from stable crystals in thermal equilibrium and from other types of amorphous solids and what the nature of phenomena characteristic of glasses is, including glass formation. Actually, as with other types of amorphous solids; glasses are metastable systems with a macroscopic life time t_g. For many glasses, t_g is very large, e.g., t_g can exceed centuries for window glasses, unlike perfect crystals with infinite life time, $t_g = \infty$ (see, e.g., [1, 2, 3, 4]).

A central problem of glass physics is the clarification of the origin of this metastability. It is often supposed that the metastability is related to the system nonergodicity [1]. As well known, conventional statistical mechanics deals with ergodic systems, like crystals, which uniformly move through all possible states in the phase space as time progresses in such a way that for a physical quantity, averaging over the observation time t is equivalent to averaging over an appropriate ensemble of systems (the Gibbs ensemble for thermal equilibrium). The resulting average value describes the behaviour of the ergodic system, for which the ensemble's most probable value of the quantity in question equals its average value. For non-ergodic systems' the two types of averaging do not necessarily give the same value, so the latter, generally speaking, differs from the most probable value of the quantity in the ensemble of systems. Nonergodicity is usually related to a macroscopic degeneracy of the system "ground states" whose motion in the phase space is nonuniform. Then, ensemble averaging embraces the whole phase space while time averaging embraces only the phase space region accessible for the particular system. The deviations from ergodicity depend on the relation between the observation time t and the times τ of transitions between the different "ground states" (with almost the same free energies), so the system becomes again ergodic for $t \gg \tau_{max}$, as long as the maximum transition time τ_{max} is finite. Unlike crystals, amorphous systems are not necessarily ergodic. For a non-ergodic system, it is important to also find the probability distribution of its characteristics, not only the average values.

Another important problem of glass physics is the description of the disorder of amorphous systems which, unlike crystals, do not have long range order. In this case, relaxation processes in viscous liquids and glasses, generally speaking, are different from those in crystals (see, e.g., [5]). Moreover, topological properties of the systems turn out to be more essential than metric ones, so topology becomes the relevant geometry of glasses [1, 2, 6].

Principal glass-forming elements belong to the fourth (Si, Ge, Sn), fifth (P, As, Sb) and sixth (O, chalcogenides {S, Se, Te}) groups of the periodic table. Of these elements, mainly sulfur and selenium form usual glasses, the majority of glasses being amorphous (a-) alloys of these elements with each other and with some other elements. The three main types of glasses can be distinguished by the characterization of their electrical and optical properties, in particular, the fundamental optical absorption gap width E_{opt}. The types are dielectric, semiconducting and metallic glasses (also called glassy dielectrics, semiconductors and metals).

Dielectric glasses having broad optical gaps (with typical $E_{opt} \approx 5$–10 eV) contain materials with basic covalent bonding (e.g., oxide glasses, in particular silicate glasses based on a-SiO_2), with largely ionic bonding (e.g., a-BeF_2, a-K-Ca-H_2O) and with hydrogen and/or van der Waals bonding (amorphous water, a-H_2O and polymers). Adding ions of the first three groups of the periodic table (Na, B, etc.) by introducing relatively low concentrations ($\ll 1$) of appropriate molecules (Na_2O, B_2O_3, etc.) to a metastable supercooled liquid generally favours glass formation at liquid cooling (see, e.g., [2, 7]). Such additions can significantly change the characteristic temperature of the liquid-to-glass transition (or, often, glass transition). These ions can also noticeably modify the random atomic structure of the so-called "pure network glasses" (e.g., SiO_2, GeO_2, BeF_2) of which the structure "consists of a three-dimensional random network of AB_4 (e.g., SiO_4) tetrahedra linked at the vertices with random angles and relative orientations". Therefore, such ions are often called "network modifiers" while the resulting glasses are "modified network glasses". Some useful comments on "pure" and "modified" network glasses, including splat-cooled oxide glasses, can be found in the mentioned early reviews.

Semiconducting glasses with moderately broad optical gaps ($E_{opt} \approx 1$–3 eV) include chalcogenide glasses, amorphous binary alloys a-A_xCl_{1-x} of a chalcogen Cl and an element A of the fourth or fifth groups with $0 \leq x \lesssim 0.4 - 0.6$ (and ternary alloys), and some oxide glasses and amorphous alloys of transition metal elements like a-V_2O_5-P_2O_5-CaO (see, e.g., [2, 7]). Covalent bonding may predominate in the glasses with a low average atomic coordination number $\bar{z}$, at $2 \leq \bar{z} \lesssim 2.5-2.7$ (cf. non-glassy amorphous silicon, a-Si, with $\bar{z} = 4$).

Finally, metallic glasses, with no optical gap, mainly reduce to amorphous alloys a-M_xQ_{1-x} of transition metal elements (M=Pd, Co, Ni. etc.) and "metalloid" elements Q of the third, fourth and fifth groups, with $x \approx 0.7 - 0.8$. In these materials, metallic bonding related to the electronic Fermi liquid is important (although quasi-covalent, directed bonds of metalloid atoms with metal ones can play a noticeable part to some extent); see, e.g., [1, 8, 9].

1.2 Qualitative description of glass (liquid-to-glass) transition

A brief consideration of the phenomenology of glass formation is presented here (for more details, see some reviews quoted in Ch.3). Traditionally, a glass is considered as an amorphous solid produced by rapid cooling of a melt (classical liquid) at a constant (hydrostatic) pressure $p = const$, at the average cooling rate R_c exceeding a minimum rate R_{min} characteristic of the liquid. A significant change of some essential properties of a liquid occurs around a characteristic temperature $T = T_g$, the so-called calorimetric glass transition temperature, in a relatively small temperature range

$\delta T \ll T_g$. This empirical fact is considered to correspond to a liquid-to-glass transition or a glass transition for brevity, a process whereby a solid glassy system is formed in a normal liquid as temperature is lowered rapidly enough from the melt temperature T_m to T_g (see, e.g., [10]).

Actually, the glass formation from a supercooled liquid is solidification without crystallization. Since the thermal-equilibrium phase at $T < T_m$ is a crystal, a supercooled liquid is a thermodynamically metastable system with a finite, macroscopically large, lifetime t_l in the temperature range $T_g < T < T_m$. Practically, the time t_l is close, at least in scale, to the cooling time t_f which determines the cooling rate $R_c \approx (T_m - T_g)/t_f$ (as noted above, at $T < T_g$ the glass is certainly a metastable system with lifetime t_g much larger then t_f). The metastability of glass-forming liquids or glass formers is shown in particular in the dependence of both T_g and other properties on the cooling rate R_c, $T_g(R'_c) > T_g(R''_c)$ for $R'_c > R''_c$, so $T_m > T_g \geq T_g^0 \equiv T_g(R_{\min})$. It is empirically found that $T_m/2 \lesssim T_g \lesssim 2T_m/3$, often $T_g \approx 2T_m/3$, and

$$T_g(R_c) \propto [\ln(const./R_c)]^{-1} \propto [\ln(const. \cdot t_f)]^{-1}. \qquad (1.1)$$

It is worth adding that a glassy material can also be obtained as a film by applying techniques with much higher cooling rates, e.g., by deposition on cold substrates by sputtering or evaporation. However, the important properties of the (annealed) films are practically the same as those of bulk glass samples obtained by cooling the melt. On the other hand, films of amorphous silicon, hydrogenated (a-Si : H) or not (a-Si), do not show some of the properties of typical semiconducting glasses, though both the latter and a-Si are covalently bonded amorphous semiconductors. These facts can be considered as an indication that the glass-forming ability of a material is essentially due to fundamental properties of the interatomic interactions and resulting local atomic configurations, though it also depends on the cooling rate. Finding the basic criteria for the glass-forming ability is a fundamental problem of glass physics even though only some "empirical" criteria are found (see, e.g., [2, 7] and references therein). With the above remarks in mind, the problem seems to be whether or not a liquid is an effective glass former, rather than what the minimum cooling rate is for a given liquid to be a glass former.

Both the empirical or calorimetric, glass-transition point T_g and the minimum cooling rate $R_{\min}$ show large variations for different types of glasses mentioned above. In glassy dielectrics, T_g and $R_{\min}$ vary over significant ranges, from $T_g \simeq 140K$ (amorphous water) to $T_g \simeq 1500K$ (a-SiO_2) and from $R_{\min} \approx 10^{-4} - 10^{-5} K/s$ to $10^{-1} K/s$ while in glassy semiconductors, $R_{\min} \approx 1 - 10^{-2} K/s$ and $200K \lesssim T_g \lesssim 700-800K$ for glass formation from supercooled liquids. Moreover, $R_{\min}$ is very high for metallic glasses when applying rapid cooling techniques, e.g., $R_{\min} \approx 10^5 - 10^6 K/s$ (except for some materials, e.g., a-$Pd_{77}Cu_6S_{17}$ with $R_{\min} \sim 10^3 K/s$) or even larger, $R_c \sim 10^{10} K/s$ when using laser pulses in a material layer with width $\approx$ 100 Å. Much larger values of $R_c (\gtrsim 10^{12} K/s)$ might be simulated on a computer for a small system of typically $10^3 - 10^4$ atoms in a simple liquid like liquid argon which would behave as if a glass transition occurs. If such computer simulations correspond to processes in actual materials, glass formation would be practically a universal phenomenon and the problem of glass-forming ability would be what $R_{\min}$ is required for the liquid-glass transition. However, this viewpoint does not seem to have sufficient empirical ground. The liquid cooling can be performed either at a constant rate, $R_c = -dT/dt = const$, or in discrete steps ΔT at time intervals Δt with average rate $R_c = dT/dt$. Moreover, the glass transition can be detected by measuring the response of the liquid to external forces oscillating with a finite frequency ν, in particular the complex shear modulus $\mu(\eta)$ for the response to shear stress. The temperature dependence of the response is noticeable at a given R_c when ν is high enough, e.g., at $\nu > R_c/\Delta T$, with ΔT, the minimal range where this dependence is observable. For a constant external force, the response time τ_r takes the part of ν, so it should be $\tau_r < \Delta t = \Delta T/R_c$ when the liquid cooling is performed in discrete steps ΔT. Changes of the volume $V(T)$ and enthalpy $H(T)$ (the extracted heat) can also be measured near the "glass transition" point (e.g., [2, 4]).

These and other properties of the system in question may demonstrate that the glass transition is a very complex phenomenon determined by many factors. Indeed, if a simple glass-forming liquid is cooled from the melt temperature T_m, a strong increase of the viscosity is observed and a simple criterion of the phenomenon is a sharp suppression of a normal liquid viscous flow by increasing the shear (dynamic) viscosity η and an appearance of the glass shear elasticity at significant values of the static shear modulus μ_0 characteristic of a solid. In fact, if a glass-forming liquid is cooled rapidly enough from the melt temperature T_m, a strong increase of the viscosity is observed from $\eta \approx 10^{-2}$–10^{-1} poise, a typical value for a simple liquid in its normal state, to η that may exceed 10^{10} poise within the supercooled liquid phase. This process is often called a "viscoelastic transition" from a liquid to a glass, which is accompanied in particular by an increase of acoustic absorption (viscoelastic damping). On the other hand, since the time scale τ of atomic motions is actually related to the viscosity scale as $\eta \propto \tau^{-1}$, the time scale of supercooled liquid dynamics closes the gap between microscopic oscillation times and macroscopic times, the latter being typical for slow processes usually described as structural relaxation processes. Thus, the basic phenomenon observed in a supercooled liquid is the formation of a glass as a nonequilibrium amorphous solid. The system is a solid in the sense that it is experimentally characterized, in particular, by static shear stresses but does not have a long range spatial order typical of crystalline solids. At the same time, there is no clear experimental evidence for assuming that the general origin of glass formation is due to a conventional, thermodynamical phase transition from a liquid state to a solid one. In particular, there are no well-defined differences between thermodynamical properties of the glass and supercooled liquid at the glass transition (see below). Moreover, there are reasons for suggesting that the glass formation is rather associated with a dynamic transition from a liquid to a solid. The related theoretical, mode-coupling, model (see, e.g., reviews [11, 12] and also Sec. 3.3 and Sec. 12.1 below) is mathematically linked to some works (e.g., [13]) suggested earlier for a different research field.

Actually, the glass becomes solid in the sense that the time scale τ for liquid-like flow phenomena becomes infinite on normal empirical time scales. Hence, it was concluded in many works (see, e.g., the above-mentioned reviews) that there is nothing like a precise liquid-to-glass transition temperature but a rather small transformation region, so the term "glass" may mean just a high viscosity supercooled liquid state. Such a conclusion, however, has been found in some works, including the above-mentioned ones, to be premature. The characteristic time scale τ for typical atomic motions in simple glass-forming liquids is the same as in normal liquids, $\tau \approx 10^{-12}$–10^{-14} s. Since in normal classical liquids [10] there are also low-frequency viscoelastic effects (see below) related to structural relaxation phenomena, a rather large frequency interval, e.g., from 1 MHz to 1 THz, exists, in which the phenomena must change with cooling of the liquid and can become precursors of glass transition. In fact, the main result of the abovementioned theoretical model appears to be that a finite temperature T_c can exist, characterizing a dynamic crossover between a normal liquid phase and a glass-forming phase. A question is what the relation between the calorimetric glass transition temperature T_g and the temperature T_c is.

1.3 Kinetic and thermodynamic properties

For most glass-forming liquids, the increase of the viscosity η shows considerable deviations from Arrhenius behaviour (the Vogel–Fulcher law) [14, 15] and can be mainly described as:

$$\eta(T) \approx \eta^{(0)} \exp(\varepsilon_a/T), \quad \text{with } \varepsilon_a \equiv \varepsilon_a(T) = \varepsilon_0 T (T - T_0)^{-1}$$
$$\text{at } \varepsilon_0 \approx 0.1 - 1 \text{ eV}, \quad T > T_g > T_0. \tag{1.2}$$

(in what follows $k_B \equiv 1$). A strong increase of the apparent activation energy $\varepsilon_a(T)$ with decreasing temperature for $T_g - T_0 \ll T_g$ is characteristic of this law (e.g., ε_a increases about 10 times for typical values $T_g - T_0 \sim 50$ K and $T_g \simeq 500$ K). However, in the nearest neighbourhood of T_g, the behaviour of $\eta(T)$ is close to the Arrhenius law $\eta^{(1)} \exp(\varepsilon_1/T)$ with the constant activation energy ε_a saturated at a value ε_1 of the order of $100T_g$, $\varepsilon_1 \gg \varepsilon_0$ (see, e.g., [4, 16, 17]). There are not so many glass-forming liquids like SiO_2 (fused quartz) and BeF_2, which do not show large deviations of $\eta(T)$ from the Arrhenius law (as if $T_0 \ll T_g$) [17], and the respective glasses are classified as "strong" ones whereas in the opposite limit case the glasses are classified rather as "weak" ones.

Note that for usual viscous liquid, the "viscoelastic" transition introduced by Maxwell [18], from viscous flow for $\nu\tau_s \ll 1$ to shear elasticity for $\nu\tau_s \gg 1$, occurs at

$$2\pi\nu\tau_s(T_\nu) = 1, \tag{1.3}$$

describing the response to shear stress oscillating with frequency ν. Here, τ_s is the average shear stress relaxation time, $\tau_s \approx \mu_\infty^{-1}$ for a Maxwell viscoelastic medium whereby μ_∞ is the high-frequency shear stiffness. This "viscoelastic" transition can be detected by varying either ν or T as it is a "relaxation driven" transition. It is often supposed that the glass transition may be essentially related to such a viscoelastic transition at the lowest frequency $\nu = \nu_{\min} = R_{\min}/\Delta T$, with $1/\nu_{\min}$ playing the role of a typical time t, $\nu_{\min}^{-1} = t = \alpha(\varepsilon_0/T_g)T_g/R_{\min}$; the function $\alpha(x)$ characterizes a concrete model of the system [4, 19]. If this is valid, the equation

$$2\pi\nu_{\min}\tau_s(T_g) = 1 \tag{1.4}$$

can determine T_g, so $\tau_s \gg t$ and macroscopic relaxation is largely suppressed at $T < T_g$ and usual experimental times t. The glass looks rather as a highly viscous liquid with characteristic $\eta(T_g) \approx 10^{13}–10^{14}$ poise at typical times $t \approx 10^3–10^4$ s. Very large changes of η are also typical as well as other kinetic characteristics like atomic relaxation times and the (self-)diffusion coefficient D in large regions with size $L \gg a_1$ when the temperature varies near T_g (a_1 is the average atomic separation, $a_1 \approx 3\text{Å}$). For instance, with decreasing T, η varies from $10^6 - 10^{10}$ poise for $T_g < T \lesssim T_m$ to $\eta \approx 10^{13} - 10^{14}$ poise at $T \approx T_g$, while τ_s increases from microscopic $\tau_s \approx 10^{-12}$ s to macroscopic $\tau_s \approx 10^3 - 10^4$ s.

The kinetic effects under discussion can be treated as related to constraints imposed by interatomic forces on the atomic motion and relaxation which prevent the formation of critical-size crystallization nuclei and thus the crystallization of the liquid. Some features which favour the glass-transition associated with the kinetic effects mentioned have been empirically revealed: proximity of the melt to the eutectic (the melting point T_g and the supercooling range $T_m - T_g$ decrease); occurrence of polymorphic crystalline phases; complex structure of the crystalline cell and specific bonding in the crystal. As noted above, kinetic effects in the glass transition are also manifested in the dependence of the transition point T_g on the cooling rate or the cooling time, the transition being in this sense time dependent. The empirical relation (1.1), $T_g \simeq \varepsilon_1/\ln(const\ t_f)$, can be obtained from Eqs. (1.2)–(1.4) and $\eta \propto \exp(\varepsilon_1/T)$ for temperatures close to T_g. However, as noted below, the real glass transition does not reduce to the "viscoelastic" transition and the related kinetic effects and thus, the transition is much more complicated.

For a few glass-forming liquids (SiO_2, BeF_2, GeO_2) for which $\eta(T)$ does not show large deviations from the Arrhenius law, one can suppose that a developed network structure over the whole range $T_g \lesssim T \lesssim T_m$ is formed ("strong" network liquids, see Ch. 3 and [4]); the thermodynamical properties change smoothly upon cooling through the glass-transition region. However, noticeable step-like changes of thermodynamical properties (although less pronounced than the kinetic properties like $\eta(T)$) are observed near T_g in the majority of glass formers. For instance, the thermal expansion coefficient $\varsigma(T)$ and isothermal compressibility $\varkappa(T)$ decrease with decreasing T from ς_l and $\varkappa_l$ of the liquid to smaller values ς_g and $\varkappa_g$ for the glass, whereas the static shear modulus $\mu_0(T)$ and specific heat $C_p(T)$ (at pressure $p = const.$) increase more or less in a similar way.

The fact that elastic stiffening at the viscoelastic transition and discontinuities of the thermodynamical properties occur near T_g was assumed to indicate that both kinds of features are due to the arrest of diffusive atomic motion and relaxation on macroscopic space and time scales in glasses; see, e.g., [4] (actually, reconstructions of local atomic configurations still occur in glasses. Then, the observed changes in thermodynamical properties can be interpreted by supposing that at $T = T_g$, a liquid stops behaving like an ergodic system in a phase space and a transition to non-ergodic phase space motion may occur [21, 22, 23]).

The existence of discontinuities $\Delta C_p(T_g) \equiv [C_p^{(l)}(T_g + \frac{1}{2}\delta T) - C_p^{(g)}(T_g - \frac{1}{2}\delta T)]$, $\Delta\varsigma(T_g)$, $\Delta\varkappa(T_g)$ and $\Delta\mu_0(T_g)$ was supposed in some works to indicate a similarity of the glass transition to a thermodynamic second-order phase transition along a line $T = T_c(p)$ in the temperature-pressure plane. In this respect, the validity of the well-known Ehrenfest relations for a second-order phase transition was discussed (see, e.g., [4] and [24, 25]):

$$dT_c/dp = VT_c\Delta\varkappa(T_c)/\Delta C_p(T_c)$$
$$dT_c/dp = \lambda\Delta\varkappa(T_c)/\Delta\varsigma(T_c) \quad \text{for } \lambda = 1, \tag{1.5}$$

so the Prigogine-Defay ratio are as follows:

$$\frac{\Delta C_p(T_c)\Delta\varkappa(T_c)}{VT_c[\Delta\varsigma(T_c)]^2} = \frac{1}{\lambda} = 1 \quad \text{at } T = T_c(p).$$

Applying the relations to such glass transition (at a typical value of R_c), with $T_g(p)$ substituted for $T_c(p)$, one can find that only the first Ehrenfest relation (1.5) is exactly valid in the majority of glasses studies. This means that the entropies of liquid and glass are the same on either side of the transition line $T = T_c(p)$, independently of the pressure p applied. The second and third Ehrenfest relations (1.5) are violated for the glass transition since $\lambda < 1$ (typical values of parameter λ are at $\lambda^{-1} \approx 2$–4. Moreover, the viscosity η is experimentally found to be approximately constant along the glass transition line $T = T_g(p)$.

The situation described above shows that both thermodynamic and nonequilibrium factors may contribute to real glass transition. The significant role of nonequilibrium factors is supported, besides the cooling rate dependence of T_g, also by the occurrence of hysteresis effects in the behaviour of the material physical properties like $C_p(T)$, $\varsigma(T)$, $\varkappa(T)$ and $\mu_0(T)$ at cooling and heating through this glass transition. In particular, as the glass is heated near T_g, a peak of heat absorption was observed, looking as a peak of the specific heat which was sometimes interpreted as a manifestation of a thermodynamic second-order phase transition. However, the specific heat measured near T_g on cooling shows only a smooth decrease, and there is no entropy release characteristic of a configurational rearrangement. Thus the process does not appear essentially similar to a second-order phase transition (Sec. 3.3).

1.4 Slow relaxation processes

An important feature of the supercooled liquid near the glass-transition point is related to the existence of slow relaxation processes for small deviations of macroscopic properties $P(t, T)$ from their thermal-equilibrium values $P_{\rm eq}$. The universal expression describing the relaxation processes is the Kohlrausch–Williams–Watts (KWW) law or the "stretched-exponential" formula [26, 27] (see also, e.g., [4, 5] and references therein):

$$\delta P(t) \equiv P(t) - P_{\rm eq} \propto \exp\left[-(t/\tau)^{\upsilon}\right], \quad 0 < \upsilon < 1, \qquad (1.6)$$

for time intervals t much larger than the typical local relaxation time $\approx 10^{-13} - 10^{-12}$ s. The original Kohlrausch relaxation and the simplest

Maxwell–Debye relaxation are characterized respectively by $\upsilon = 1/2$ and $\upsilon = 1$. The parameters τ and υ practically do not depend on t over many orders of magnitude and υ does not noticeably depend on T, both being hardly sensitive to the particular property P (the original Kohlrausch relaxation, discovered in the 19th century, corresponds to the value of $\upsilon = 1/2$). In general, the KWW relaxation differs significantly from the Maxwell–Debye exponential relaxation at $\upsilon = 1$, and is also observed in a wide variety of materials, for relaxation of various properties after a sudden change of temperature in supercooled liquids (see, e.g., [5, 29, 30]). The dependence of τ on T in Eq. (1.6) is often described by the Vogel–Fulcher law (1.2), in particular by the Arrhenius law for strong liquids, at least in the nearest neighbourhood of the glass-transition temperature. Usually, the KWW formula is found to be valid in the empirically essential range of t, from $\delta P(t) \simeq \delta P(0)$ to $\delta P(t) \ll \delta P(0)$, say, from $\delta P(t) = 0.99\delta P(0)$ to $\delta P(t) = 0.01\delta P(0)$.

The non-standard character of the relaxation can also be associated with nonequilibrium features of the glass transition. As noted above, it does not seem obvious, however, that the glass transition reduces to the viscoelastic transition determined by purely kinetic factors. In fact, the empirical viscoelastic behaviour of metastable supercooled liquid seems to differ significantly from that of a viscous Maxwell liquid. For the latter, the real part of the complex shear modulus appears to be [31]

$$\mathrm{Re}\,\mu(\nu) = (2\pi\nu)^2\tau_s^2\mu_0\left[1 + (2\pi\nu)^2\tau_s^2\right]^{-1}, \tag{1.7}$$

while for a supercooled liquid (see, e.g., [32, 33])

$$\mathrm{Re}\,\mu(\nu) \propto \mathrm{Re}\left(\frac{i\nu\tau_s}{[1 + (2\pi i\nu\tau_s)^{1/2}]^2}\right). \tag{1.8}$$

To conclude, the observed transition appears to be a very complex phenomenon containing contributions of both a viscoelastic phenomenon and an "ideal" glass transition from a metastable, supercooled liquid to a metastable glass in which either an "ideal" glass transition is a conventional first- or second-order phase transition or a glass is an essentially nonergodic, metastable phase, so the glass transition is a dynamic transition between an ergodic metastable phase (supercooled liquid) and a metastable non-ergodic phase, a glass (see Ch. 3). It is worth adding that very slow atomic relaxation processes due to local atomic rearrangements are also observed in some glasses even at low temperatures ($T \ll T_g$), indicating that the relaxation is considerably slower than the KWW one.

2

MODELS OF GLASSY (TOPOLOGICALLY DISORDERED) STRUCTURES

2.1 Characteristics of glassy structures

A most important feature of the glass structure is its topological disorder. Each atom A_i of type $i(=1, 2, \dots)$ in a glass is characterized on a length scale $a_1 \approx 2$–3 Å, the average interatomic separation, by a short-range order with coordination number z_i and pair (radial) distribution function $\rho_i(R)$ for nearest neighbours (see, e.g., [1, 2]). This short-range order turns out to be in important aspects incompatible with crystalline periodicity (long-range order) at the characteristic length scale a_1. On the other hand, an intermediate-range order on a length scale $a_2 \approx 10$–30 Å is observed in glasses which corresponds to spatial correlations of the short-range order atomic configurations; see, e.g., [3]. In addition, glasses do not exhibit a crystalline long-range order, but also a topological order (see, e.g., [6]).

Actually, a long-range order exists in a perfect d-dimensional crystal ($d = 3$ or for a surface or interface $d = 2$). The long-range order determines both the metric properties (conventional geometry) of a perfect crystal lattice and its topological properties (topology) related to the connectivity in this regular network of atoms (see, e.g., [32, 33]). For a real (imperfect) crystal lattice containing point defects (interstitials, vacancies, etc.) with concentration $x^* \ll 1$, the long-range order can be violated but the topological order can be preserved. However, dislocations and other non-point defects, e.g., dislocations and disclinations, can violate the topological order (see, e.g., [34]). In particular, a disclination can be created by cutting the regular structure and removing (or adding) a wedge of material between two edges of the cut. The relevant symmetry operation is a rotation with characteristic Frank angle α between $\alpha = 0$ and $\alpha = 2\pi$, unlike a translation for a conventional dislocation with a characteristic Burgers vector B. For the sake of simplicity, a glass as a topologically disordered structure can sometimes be roughly modelled as a real crystal with concentration $x^* \sim 1$ of the non-point defects, although the latter assertion should not be understood too literally (the concept of defects hardly is meaningful for $x^* \sim 1$). Actually, an amorphous solid can be defined through its topological properties, which do not reduce to those of a crystalline lattice. In other words, an amorphous material cannot be produced by continuously distorting a crystalline lattice; see also [35]. This definition is invariant, taking into account that for an amorphous system, unlike a crystal, its topological properties, rather than the metric ones, are essential, the topology being the relevant geometry (see, e.g., [6, 36]).

A glass is a topologically disordered structure, as if $B \sim L$ and $x^* \sim 1$ in real crystals, although the latter assertion should not be understood too literally (the concept of defects is hardly meaningful for $x^* \sim 1$, see also below). In fact, an amorphous solid can be defined through its topological properties which do not reduce to those of a crystalline lattice. In other words, an amorphous material cannot be produced by continuously distorting a crystalline lattice; see also [35]. This definition is invariant, taking into account that for an amorphous system, unlike a crystal, its topological properties, rather than the metric ones, are essential, the topology being the relevant geometry as for a rubber medium (see, e.g., [6, 36]).

There are a series of useful experimental methods for investigating amorphous structures (see, e.g., [2]). One of the most effective is related to diffraction methods based on studies of X-ray and neutron scattering. These methods, however, are not as effective for amorphous structures as for crystals, only giving rise to the radial distribution function $\rho(R)$ averaged over the amorphous alloy composition, for instance, $\rho(R) = x\rho_A(R) + (1-x)\rho_B(R)$ for an alloy a-A_xB_{1-x} ($0 \lesssim x \lesssim 1$) with an average coordination number $\bar{z} = xz_A + (1-x)z_B$. For a given type of atom (A or B in A_xB_{1-x}), the radial distribution function $\rho_i(R)$ can be obtained by the well-known EXAFS method which considers the fine structure of the soft X-ray absorption band for a definite shell of the atom with coordination number z_i (see, e.g., [2, 3, 37]).

As a rule, the distribution function $\rho_i(R)$ or $\rho(R)$ shows two, or possibly three peaks corresponding to the first ($j = 1$), second ($j = 2$) and possibly third nearest neighbours. The peak widths Δ_j change with increasing j; for instance, for SiO_2 glass, it is found that $\Delta_1 a_1^{-1} \sim 10^{-2}$ and the relative r.m.s. fluctuation of the interatomic bond length $\Delta a_1/a_1 \sim 10^{-2}$, while $\Delta_2 a_1^{-1} \sim 10^{-1}$ due to much larger fluctuations of the interbond angle α_2(SiOSi), $\Delta\alpha_2/\alpha_2 \sim 10^{-1}$, near the average angle $\overline{\alpha_2} \simeq 150°$.

In any case, these methods give effectively a "linearized" information about the short-range order or the three-dimensional amorphous structure, but not about the intermediate-range order in length of the order of $a_2 \approx 10$–30 Å. However, it is empirically found that amorphous structures with a similar short-range order do not always have similar physical properties [2, 38]. This fact indicates that a description of a topologically disordered system through its short-range order and radial distribution function is hardly sufficient. In other words, a description of intermediate-range structures with size ≈ 10–30 Å can be essential, allowing one to discriminate between different amorphous systems with a similar short-range order. The description can contain some important features of the global three-dimensional amorphous structure, although the scale of the intermediate-range order can depend to some extent on the physical property under consideration and vary from ≈ 10 Å to ≈ 30 Å. The problem of the description of intermediate-range order is rather difficult, since it is not easy to choose a small number (two or three) of most essential structural parameters in their large variety without losing the basic information after averaging over the other parameters.

Some modern empirical methods are able to give useful information about the intermediate-range order. One of these methods seems to be based on studying the first sharp diffraction peak (FSDP) and the structural factor $I(p)$, the Fourier component of the radial distribution function at a characteristic value of the momentum transfer $|p|_0 \approx 0.1$–0.03 Å^{-1}, corresponding to the scale of the intermediate-range order $a_2 \approx |p|_0^{-1}$. The FSDP method has been applied to the glasses a-SiO_2, a-As_2Se_3 and a-$GeSe_2$. Methods using small-angle X-ray scattering, electron beam scattering, ESR and NMR effects, IR absorption, Raman scattering and the Mössbauer effect seem to be useful for studying structures of scale $a_2 \approx 10$–30 Å and of larger scale (e.g., voids, large atomic clusters) by revealing dangling bonds, vibrational modes of the structures, variations of the interatomic bond type, etc. Nevertheless, at present, the empirical methods applied are hardly able to give an unambiguous description of the topological disorder in glasses and other amorphous solids. The problem of revealing and describing elementary excitations of an amorphous system is to some extent associated with the difficulty mentioned.

Therefore, it becomes important to use theoretical models which are able to describe most essential features of the microscopic structure of an amorphous system and for glasses, to understand the relations between the material structure and the glass-forming ability. However, before considering the models, one can summarize the principal features of the glass structure as follows:

1. there is no long-range and topological order;
2. there is a short-range order (z_i, $\rho_i(R)$) on a length $a_1 \approx 3$ Å;
3. there is an intermediate-range order on a length scale $a_2 \approx 10$–30 Å at least;
4. the glass formation from supercooled liquids is determined by intrinsic atomic interactions and related constraints in atomic dynamics and relaxation;
5. the metastability of the glass structure is related to nonequilibrium features and constraints in atomic dynamics which determine very large lifetimes with respect to crystallization.

2.2 Homogeneous (ideal) models

A popular approach to the glass structure problem is based on constructing "ideal" models by directly using balls (atoms) and sticks (bonds) or by computer simulations under some empirically justified rules. Two types of such models (see, e.g., [2, 7]) are

(1) a continuous random network (CRN) in interatomic bonds;
(2) random dense packing (RDP) of "atoms", e.g., hard spheres or "incompressible" structure units chosen by taking into account the chemical composition of the glass.

The CRN model, for a two-dimensional (2D) amorphous system at least, corresponds to a topological graph or a set of sites (vertices) connected by bonds (edges) which divide the Euclidean space in an unambiguous way in polygons for 2D systems (more or less in unambiguous way, in polyhedra for 3D systems). Such a model is assumed to characterize in a simple way a covalent amorphous structure in which atoms with coordination numbers $z \leq 4$ are connected by directed covalent bonds in a united ideal network, as first proposed in [39] (see also [38]).

On the other hand, the RDP model first presented in [40] describes such materials like amorphous metal-metalloid alloys and metallic glasses. A random cellular structure is in fact implied which fills the space and consists of polyhedra (polygons), Voronoi cells, and are an analogue of well-known Wigner-Seitz cells in crystals; see, e.g., [41]. Such a cell around each atom (or around a structural unit, e.g., a quasimolecular group) contains all the points which are closer to this atom than to any other. The cellular structure filling the space is often named a "froth". The mathematical consideration of the RDP model is based on an analysis of topological constraints occurring for a possible most dense filling of the Euclidean space by Voronoi cells. Each face of a Voronoi cell belongs to two three-dimensional (3D) cells, each edge to three 3D (or two 2D) cells and each vertex to four 3D (or three 2D) cells [42]. The topological constraints are first of all the well-known Euler relations, conservation laws for numbers of vertices (V), edges (E), faces (F) and 3D cells (C):

$$F - E + V = \chi, \quad \text{(2D)} \tag{2.1}$$

or

$$-C + F - E + V = \chi. \quad \text{(3D)} \tag{2.2}$$

For a space tiled by a random structure, the Euler-Poincaré invariant χ is a number of the order of 1, so for the whole structure $\chi \ll \{F, E, V, C\}$ one can approximate $\chi \simeq 0$. A 3D froth is characterized by two topological random parameters, the number f of faces per one 3D cell and the number p of edges per one 2D face while a 2D froth is only characterized by p. The Euler relation and the vertex coordination numbers (number of edges having a common vertex) determine the following topological identities for a froth (see, e.g., [2, 42]):

$$\bar{p} = 6, \quad \text{(2D)} \tag{2.3}$$

or (the Coxeter identity)

$$\bar{f} = 12/(6 - \bar{p}), \quad (3D) \tag{2.4}$$

where $\bar{p}$ and $\bar{f}$ are the average values of the random parameters p and f.

The edges and vertices of the Voronoi cells in the RDP model also form a graph, so the CRN and RDP models are related to each other in an unambiguous way, for 2D systems at least. With this in mind, one of the models, the CRN model, is considered in more detail here. In the approach under consideration, in which the CRN (and RDP) models are constructed, the principal problems seem to be as follows:

(I) to find the essential characteristics of the topological disorder and to describe in detail the amorphous structure;

(II) to establish interrelations between the glass structure and the glass-forming ability of the material, i.e., criteria for the glass-forming ability.

The important role of the ideal models under discussion is associated with the progress in describing in detail the topological disorder of structures. In fact, it is possible to determine in these models, with any given accuracy by computer simulations at least, the probability distribution for the atomic site positions, which can hardly be found experimentally in present.

The principal rules for constructing ideal models are as follows (see, e.g., [2, 42, 43]):

(1) A short-range order ($z_i, \rho_i(P)$) is given for each atom (A_i). Perfect chemical order and chemical stability is required. The requirement is satisfied for the CRN models when all covalent bonds of each atom are saturated, the Mott rule being largely valid, i.e., $z_i = 8 - n_i$ (or $z_i = n_i$) with the number of valence electrons $n_i \geq 4$, or $n_i \leq 4$ [44] (there are a few exceptions, e.g., for Te with $n = 6$ and both $z = 2$ and $z = 3$). The short-range order, in the RDP models at least, seems to be found incompatible with crystalline periodicity.

(2) Bond lengths are fixed, in accordance with their small fluctuations ($\Delta a_1/a_1 \lesssim 10^{-2}$) corresponding to the first (narrow) peak of the radial distribution function $\rho(R)$.

(3) A noticeable dispersion of interbond angles occurs, which corresponds to the significantly broader second peak of $\rho(R)$.
(4) There are no dangling bonds.
(5) Topological order does not occur.

Such CRN models are constructed, in particular, for amorphous silicon (a-Si) with $z = 4$ and a-As with $z = 3$, for glassy alloys a-A_xB_{1-x}, oxide and chalcogenide glasses like a-SiO_2 and $GeSe_2$ with layer-like chalcogenide glasses like a-As_2Se_3 and $z_A = 3$ and $z_B = 2$.

Before considering some results concerning the problem (I) in the CRN (or RDP) models, it is worth noting that the simplest parameter of an amorphous system is its average density, $\bar{\rho}$, which as a rule is lower than that (ρ_0) of the corresponding crystal. The value of the density deficit $|\delta\rho|$ is practically largest in glasses, $|\delta\rho|/\rho_0 \sim 0.1$. A quantitative characteristic of an amorphous structure explicitly found in the CRN (or RDP) models with fairly high accuracy is the radial (pair) distribution function $\rho(R)$ or its Fourier transform $I(p)$ which shows gross features similar to those found empirically. However, this gives only partial information on the actual three- or two-dimensional structure since the contributions of many parameters of the system are statistically averaged. The information for the CRN model concerns first of all the average value and r.m.s. fluctuation of the interbond angles. Moreover, this important information does not yet seem to be sufficient for obtaining unambiguous conclusions of the quantitative validity of the CRN or RDP models and for revealing the intermediate-range order experimentally observed in glasses.

An important characteristic of the topological disorder of amorphous structures is the statistical distribution $\varphi(p)$ of the number p of bonds in individual rings in a CRN model or of edges in polygons (for 2D systems) and faces in polyhedra (for 3D systems) in a RDP model. This distribution $\varphi(p)$ is centred at $\bar{p} = 6$ (2D) or $\bar{p} = (6\bar{f} - 12)/\bar{f}$ with $\bar{f} > 2$ (3D) and demonstrates an important feature of the CRN (and RDP) models of a topologically disordered system. In fact, they have large fractions $c(p)$ of "odd rings", i.e., rings having an odd number of bonds $p = 2k + 1$ ($k = 1, 2, 3, \ldots$) and with bond lengths fixed, a symmetry axis of $(2k + 1)$-th order which does not occur in crystals for well-known symmetry reasons (see, e.g., [45]). For instance, this concerns rings with

$p = 5$ and $p = 7$, while $p = 6$ only in corresponding crystals; in particular in two-dimensional CRN models, it is found that $c(p = 6) \approx 0.4$ while $c(p = 5) \approx 0.3$ and $c(p = 7) \approx 0.3$; analogous estimates are given for glasses like a-SiO_2 and a-As_2Se_3.

To some extent, a similar situation seems to be characteristic of RDP models which gives evidence of the importance of icosahedral short-range order structures incompatible with crystalline periodicity. In particular, in a computer simulation of the supercooling of Lennard-Jones fluid, it was found [46] that far below T_m, the range of icosahedral orientational order for bond orientations significantly exceeds the radial correlation length. In fact, an icosahedron, a polyhedron having 20 trigonal faces for 12 atoms, shows a symmetry of fifth order not permitted in crystals (the spacing between an atom and the centre of the icosahedron is $1.05a_1$, where a_1 is the interatomic separation). Therefore, the icosahedron cannot be an elementary cell of a usual crystal (although it is a basic structural element of a quasicrystal).

The energetic and geometric reasons for the essential role of imperfect icosahedral orientational order in RDP models of supercooled liquids and glasses are related to the fact that it trends towards close packing and high binding energy and are globally counteracting each other (see, e.g., [47]). In fact, to construct large regions of a RDP model, the icosahedra should be noticeably distorted since regular icosahedra cannot fill the Euclidean space. Some ideas related to "odd rings" in CRN models [32, 33, 36] and to imperfect icosahedral orientational order in RDP models [48, 49, 50] have also been developed.

At first glance, it seems that an atom with coordination number z' which is different from its normal value z can be defined as a point defect in models like the CRN model, e.g., $z' = z - 1$ ("dangling bonds") or $z' = z + 1$ ("overcoordinated atoms"). However, the problem of what a point defect is, at least in a glassy structure, appears to be a difficult problem. The issue is that an actual amorphous structure, unlike a crystalline lattice, is not unique but is only a (most probable) representative of an ensemble of topologically disordered structures.

It is worth adding some remarks concerning the mentioned problem (I). The ideal models appear to have maximal topological disorder, in the sense at least that the fractions of anomalous "odd rings" and similar noncrystalline structures are large. At the same time, these models are microscopically homogeneous and isotropic (for lengths $L_0 \gtrsim 5-10$ Å at least), though the homogeneity and isotropy corresponds to topological equivalence of vertices, edges and faces rather than the usual invariance with respect to translation and rotation (see, e.g., [50, 51]). Inclusion of "defects" like dangling bonds into ideal models does not seem to change fundamental topological properties of the models, for actual low concentrations at least.

As to the problem (II) from above, the criterion for the glass-forming ability in ideal CRN models seems to reduce the condition of global (perfect) chemical order or chemical stability which corresponds to saturation of the covalent bonds to each atom (the Mott rule) in network glasses. Essential features of the chemical bonding at short-range order and the global chemical order in the CRN (and RDP) models are taken into account by an appropriate choice of the structural units like $Si(O_{1/2})_4$ tetrahedra in SiO_2 glass. Such an approach to the criterion for the glass-forming ability can be insufficient to understand some recent experimental data, concerning in particular the value of the minimal cooling rate which are described in the next section. It appears, therefore, that the ideal models do not yet give the best description of the actual complex structure of glass.

2.3 Inhomogeneous (cluster) models

A different type of model for glassy structures has been recently developed in a series of papers, in order to explain some experimental data on network glasses like a-SiO_2 and chalcogenide glasses.

These models, named "cluster models", are used for network glasses at least, which are often considered as consisting of covalently bonded quasi-molecular groups of atoms held together by weak intermolecular forces like van der Waals or hydrogen bonding. This concerns, in particular, oxide glasses (a-SiO_2, etc.) and chalcogenide glasses (a-As_2Se_3, a-$GeSe_2$, etc.) with low average coordination number $\bar{z}$ in the range $2 \leq \bar{z} \lesssim 2.4-2.6$. The cluster models can be a generalization not only of the ideal (homogeneous) models but also of some earlier "crystalline" models of glass (see, e.g., [51, 52, 53] and [55, 194]). In the latter, the structure was believed to be formed by some weakly interconnected atomic clusters which have a crystalline topology but are both too small and too strained to produce the Laue diffraction patterns which are characteristic of polycrystalline materials.

A qualitative topological approach is sometimes used to see what can happen in a good glass-former when its melt is slowly cooled. For a glass-former containing N atoms in a a-dimensional space ($d = 3$ or $d = 2$), the motion of the system in the Nd-dimensional configuration space is expected to slow down with decreasing temperature and the viscosity η of the supercooled liquid increases much more rapidly than in lightly crystallizing liquids. Qualitatively the situation can be understood by assuming that atoms form larger clusters at lower temperature. As the average size or the average number N_0 or atoms of such a cluster increases, the cluster motion is expected to slow down. If these clusters were periodic with a crystalline topology, they could serve as crystallization nuclei. Then the avoidance of crystallization, due to kinetic effects, seems to be related to a structural problem: can large clusters with non-crystalline topology be formed in a supercooled liquid? Such large non-crystalline clusters can lead to glass formation insofar as their transformation to crystallization nuclei is frustrated.

In the cluster models, unlike the CRN models, the basic criterion of the glass-forming ability is related to the condition of mechanical stability of the glass-former rather than to the condition of chemical stability (or perfect chemical ordering). It was assumed that the criterion of mechanical stability is as follows:

$$\bar{N}_c(d) = N_d = d \tag{2.5}$$

for the actual d-dimensional space with $d = 3$ or $d = 2$. Here, real interatomic forces are approximated by mechanical constraints. $\bar{N}_c(d)$ is the average number of constraints per atom while N_d is the number of degrees of freedom per atom. A qualitative interpretation of the criterion (2.5) is given as follows. The topology of the cluster model can be characterized by counting the number of morphologically distinguishable clusters of N_0 atoms interacting via N_0N_c strong interatomic forces (constraints), the clusters being weakly coupled with each other (the cluster surface effects are neglected if N_0 is large enough). Two cases can be distinguished: $N_0\bar{N}_c < N_0N_d$ and $N_0\bar{N}_c > N_0N_d$. The constraints are ineffective at $\bar{N}_c < N_d$ for the rest of the degrees of freedom, so such clusters, e.g., diatomic molecules ($N_c = 1$), can easily crystallize with the glass-forming ability suppressed. In extreme examples of the opposite limit case, when $\bar{N}_c \gg N_d$, the large excess strain energy is expended by producing a large number of broken bonds. This effect destabilizes the structure and favours a rapid ("explosive") crystallization in quenched samples. Then, the glass-forming criteria are assumed here to coincide with the mechanical-stability criteria (2.5), so the glass-forming ability is essentially related to the mechanical stability of the material. However, generally speaking, the criteria of mechanical stability do not coincide with those of perfect chemical ordering (chemical stability). Some mathematical problems appear in the cluster model, concerning the derivation of Eq. (2.5) and the expression for $\bar{N}_c(d)$. These are discussed in a series of papers (see, e.g., [56, 57, 58, 59]) and reviews (see, e.g., [60]).

3

SOME THEORETICAL MODELS OF GLASS TRANSITION

3.1 Vogel–Fulcher relation and "entropy crisis"

As noted in Sec. 1.2, the liquid-to-glass transition, or glass transition, can be considered as a very complex phenomenon, containing changes in dynamic, kinetic and thermodynamic properties. The changes in thermodynamic properties at the glass transition are weakest in the so-called "strong-network" glass-forming liquids (SiO_2, BeF_2 and, GeO_2), in which an extensive network of directed bonds exists even at $T \approx T_m$ and small changes of the liquid structure occur at $T_g \lesssim T < T_m$ with decreasing T (see, e.g., [1, 4]). An unusual sharp though continuous decrease of kinetic coefficients $\mathcal{X}(T)$ like the fluidity η^{-1}, diffusion coefficient $D(T)$ and inverse relaxation times τ^{-1} is observed as the glass-forming liquid is cooled. This is a fundamental property of most glass-forming liquids near the glass transition point T_g [14, 15]. This behaviour can be described by the following relations:

$$\mathcal{X}(t) \simeq \mathcal{X}_{VF}(T)\phi(T - T_0) + \mathcal{X}_A(T), \tag{3.1}$$

$$\mathcal{X}_{VF}(T) = \mathcal{X}^{(0)} \exp[-\varepsilon_0/(T - T_0)], \tag{3.2}$$

$$\mathcal{X}_A(T) = \mathcal{X}^{(1)} \exp(-\varepsilon_1/T), \tag{3.3}$$

$$\phi(x) \equiv 1, \quad \text{for } x > 0, \quad \phi(x) \equiv 0, \quad \text{for } x < 0. \tag{3.4}$$

Here, $\mathcal{X}_A(T)$ corresponds to the common Arrhenius thermal activation mechanism and $\mathcal{X}_{VF}(T)$ to the unusual Vogel–Fulcher relations for which (3.2) can be valid in a rather wide range of temperatures, $T_m > T > T_g$, while the pre-exponential factors depends weakly on T. As also noted in Sec. 1.2, for the majority of liquids, the temperature T_0 characteristic of the material is rather close to T_g, $T_g - T_0 \ll T_0$, with typical $T_g - T_0 \sim 50\,K$. Moreover, the energies are material parameters and ε_1 considerably exceeds ε_0. Typical values of parameters are available, $\varepsilon_0 \approx 0.1-1$ eV $\ll \varepsilon_1 \sim 50T_g$, so the behaviour of $X(T)$ is determined by the Vogel–Fulcher relation (3.2) at T not very close to T_0, in agreement with the experimental data.

The following fundamental features of relationships (3.1)–(3.4) are noted. The properties are universal in the sense that for a particular glass-former, all the kinetic quantities $X(T)$ are characterized by the same energy ε_0 and temperature T_0. There is an essential singularity for $T \to T_0 + 0$, so that $X(T)$ cannot be reduced to a superposition $\int d\varepsilon g(\varepsilon)\exp(-\varepsilon/T)$ of Arrhenius-like contributions with a regular distribution $g(\varepsilon)$ for the activation energy. Obviously, the singularity is unattainable for the liquid at reasonable, finite apparent activation energies $\varepsilon_a(T)$ and experimental times $[\varepsilon_a(T) \equiv \varepsilon_0 T(T-T_0)^{-1} \to \infty$ for $T \to T_0 + 0]$. A crossover from the Vogel–Fulcher law (3.2) to the Arrhenius behaviour (3.3) occurs at temperatures close enough to T_0 (or to T_g), in accordance with experimental data. Moreover, the pre-exponential factor $X^{(0)}$ can be different, generally speaking, for diverse quantities. The viscosity $\eta^{(0)}$ should correspond to the empirically found values of the shear viscosity, $\eta(T_g) \approx 10^{13}-10^{14}P$, in accordance with the empirical criterion (1.4). For many network glass-forming liquids, $\eta(T)$ abruptly increases by 8–10 orders of magnitude in the range $T_m \gtrsim T \gtrsim T_g$. There are, however, a few glass-forming "strong-network" liquids mentioned above (SiO_2, GeO_2, BeF_2), for which the Arrhenius behaviour predominates in the whole range $T_m > T > T_g$, $X(T) \simeq X_A(T)$ with fairly high ε_1, e.g., $\varepsilon_1 \simeq 7.5eV \simeq 50\,T_g$ for a-SiO_2. This fact can be interpreted by applying formulae (3.1)–(3.4) when either $T_0 \ll T_g$ or $X^{(0)}/X^{(1)}$ is anomalously small.

The coexistence in the majority of supercooled liquids of Vogel–Fulcher behaviour for kinetic coefficients and large excess in thermodynamical properties is considered to indicate that the Vogel–Fulcher mechanism is associated with significant large-scale structural reconstructions. These reconstructions could mainly correspond to formation of atomic clusters whose size increases as the liquid cools, in agreement with some earlier ideas on a cooperative origin of the viscosity in supercooled liquids and a cluster-like structure of network glasses (see, e.g., [62, 63, 64, 65, 66]). However, clusters could also be characteristic of "strong-network" liquids in which the large-scale structural reconstructions at $T \approx T_g$ are considered to be relatively small.

The Vogel–Fulcher relation can be transformed to the so-called Doolittle formula [67],

$$\chi_D(T) = \chi^{(0)} \exp[-V^*/(V - V_0)], \tag{3.5}$$

by assuming that the relation $V_1 - V_2 \propto T_1 - T_2$ for the volume variation with temperature is valid down to around T_g in a supercooled liquid. In Eq. (3.5), $V - V_0 = \alpha_l(T - T_0)$ and $V^* \equiv \alpha_T \varepsilon_0$, with α_T the coefficient of thermal expansion; V_0 and V^* are two characteristic volume-like parameters of the liquid. It seems reasonable, however, that α_T is close to the glass coefficient α_0 typical of solids for $T < T_0(\lesssim T_g)$, while it is close to α_l for liquids for $T > T_g$, with α_l significantly larger than α_0. If a relation $V - V_0 \propto (T - T_0)^q$ with $q > 1$ is then introduced, $\chi_D(T)$ noticeably deviates from the generalized Vogel–Fulcher law (3.1)–(3.4). Moreover, the well-known equations of state for condensed systems like the van der Waals equation, can hardly give rise to the relation $V - V_0 \propto T - T_0$. Finally, the Doolittle formula is not expected to be more fundamental than the Vogel–Fulcher law, since formula (3.5) can fit some experimental data with a temperature dependent V_0 (see, e.g., [68, 69]).

One can conclude that the Vogel–Fulcher law is a fundamental property of the majority of glass-forming liquids. Two problems arise here [1] which concern the origin of the Vogel–Fulcher mechanism and its relation to the glass transition, and the meaning of T_0 and its relation to T_g. If T_0 corresponds to a true phase transition between different metastable phases (supercooled liquid and glass), then T_0 could be interpreted as $T_g(R_c)$ extrapolated to infinitely small cooling rates $R_c \to 0$. Such a phase transition (of a thermodynamic nature) has been assumed to underlie the real glass transition and was also called "ideal" glass transition. The latter is expected to be a transition from an ergodic phase (a liquid) to a non-ergodic phase (a glass) which does not reduce to a first- or second-order transition.

Another interesting property of the glass transition, suggesting the occurrence of a similar "ideal" glass transition at least for good glass-formers (e.g., for oxide and chalcogenide glasses), is associated with an experimentally found feature in the behaviour of entropy near T_g. As was observed many years ago [70], the excess entropy

$$\Delta S(T) \equiv S_l(T) - S_{cr}(T) = \int_{T^*}^{T} dT' \Delta C_p(T')/T' \quad (\text{at } T^* \ll T_g < T < T_m), \qquad (3.6)$$

of a supercooled liquid (S_l) relative to that of the corresponding crystal (S_{cr}) rather rapidly decreases as the liquid cools (and the excess specific heat $\Delta C_p \equiv C_{p,l} - C_{p,cr}$ decreases) and extrapolates to zero at a finite temperature T_0^*, which becomes negative for $T_0^* > T \geq 0$. However, S_{cr} for a crystalline material in thermal equilibrium should certainly be lower than S_l at all $T < T_m$, particularly at low $T(\rightarrow 0)$. This situation was called "entropy crisis". The temperature T_0^* is surprisingly close to both the Vogel–Fulcher temperature T_0 and to the glass transition point T_g, with typically $T_g - T^* \sim 50$ K for good glass-formers (e.g., [2]). The "entropy crisis" was considered to indicate that a true phase transition underlying the real glass transition can occur near $T = T^*$, so that a liquid cannot exist as a metastable phase at $T < T^*$ (see, e.g., [71]). The remarkable fact that T^* is close to T_0 in the Vogel–Fulcher law (3.2) may have a profound meaning. In this respect, it seems important to reveal the actual behaviour of the excess entropy $\Delta S(T)$ of a glass-former at low temperatures $T \ll T^*(\lesssim T_g)$. Using some experimental data, it was found that the glass entropy $S_g(T)$ decreases as the glass cools at $T < T_g$ and tends to a finite limiting value $S_g(0)$ at low $T \ll T_g$ (see, e.g., [72]). This means that the Nernst law is violated for a glass, the glass entropy $S_g(0) \equiv S_0$ being finite at zero temperature while for a crystal, certainly $S_{cr}(T = 0) = 0$. Moreover, the residual entropy S_0 was estimated to be large for good glass formers like a-SiO_2 and a-As_2S_3,

$$S_0 \equiv N\sigma_0 \sim N, \qquad (3.7)$$

with the residual entropy per atom $\sigma_0 \sim 1$ (e.g., $\sigma_0 \simeq 0.83$ for a-As_2S_3, $\sigma_0 \simeq 0.70$ for a-As_2Se_3 [61], $\sigma_0 \sim \ln 1.6$ for a-SiO_2 [4]). It is worth emphasizing that for glasses, not only is the Nernst law violated (this can actually occur in non-glassy amorphous solids as well) but the residual entropy per atom σ_0 is large, $\sigma_0 \sim 1$. This means that much of the configurational disorder of the liquid relative to the crystal remains in the glass and, thus, there is no "ideal" glass with a unique structure. This can be interpreted as follows. The potential energy of the system as a function of the variables in multi-dimensional configuration space has a large number [$\sim \exp(\sigma_0 N)$] of more or less equivalent valleys (local minima) separated by barriers whose height is variable within a certain range. Almost all of the barriers correspond to reconstructions of macroscopically large regions of the material ($\sim N$) and become essentially insurmountable at $T \ll T_g$, their height being macroscopically large ($\sim N^{1/2}$). The result is that the time evolution of a random glass configuration in phase space is non-uniform, taking place only in a single valley. In this sense, a glass really is a non-ergodic, non-equilibrium, system.

3.2 Role of configurational entropy, free-volume effects and "defects" diffusion

There are some thermodynamic models of the "ideal" glass transition from a glass-forming liquid to a glass which account for both the Vogel–Fulcher mechanism and the "entropy crisis" in the system under discussion. Three such models related respectively to either configurational entropy or free-volume of the system or to some mobile "defects" occurring in the system, are briefly considered below.

In the configurational entropy model, the "ideal" glass transition is directly related to the "entropy crisis" for the configurational entropy of a supercooled liquid [71, 72]. It is assumed that the change of vibrational entropy for a certain temperature variation is nearly the same for the liquid and the glass and that the glass entropy reduces mainly to vibrational entropy. On the one hand, the validity of the Ehrenfest relations (1.5) for the glass transition means that the liquid and the glass on either side of the transition line $T = T_c(p)$ have identical entropy. On the other hand, the viscosity η is experimentally found to be practically constant along the line $T = T_c(p)$, so one can expect that there exists a simple relation between the configurational entropy S_{conf} and η of the liquid which is supposed to be the Vogel–Fulcher law. In accordance with a derivation presented [4], the liquid is divided in N_0 identical blocks, each containing a macroscopically large number $n = N/N_0 \gg 1$ of atoms and occurring in two different configurations. The configurational entropy is approximated by the formula

$$S_{\text{conf}}(T) = \Delta C_p(T_g)\frac{T - T_c}{T_g}, \tag{3.8}$$

since the excess specific heat of the liquid is practically constant near T_g, $\Delta C_p(T) \simeq \Delta C_p(T_g)$. On the other hand,

$$S_{\text{conf}} = N_0 \ln \upsilon = N \ln \upsilon / n, \tag{3.9}$$

so that the block size is

$$n(T) = N_0 \ln \upsilon / S_{\text{conf}}(T), \text{ at } \upsilon = 2, \tag{3.10}$$

and $n(T) \to \infty$ for $T \to T_c$. It is argued that the time τ of a transition between the alternative configurations of the block is proportional to the volume of the phase space of the initial configuration, thus giving rise to $\tau \propto \exp[\gamma n(T)]$ with γ treated as a fit parameter. However, the time τ characterizes structural relaxation, so τ is proportional to the viscosity η. The Vogel–Fulcher law (3.2) with the correct pressure dependence of η [$\eta = const.$ along the transition line $T = T_c(p)$] follows from the above and formulae (3.8) and (3.10) at

$$T_0 = T_c (\equiv T^*) \text{ and } \varepsilon_0 = \gamma N T_g \ln 2/\Delta C_p(T_g) \tag{3.11}$$

if it is assumed additionally that $\gamma = const.$ A microscopic theory of the phase transition related to the "entropy crisis" was developed in a series of papers (e.g., [73]). The theory gives rise to a vanishing configurational entropy of the liquid at a temperature T_c at which a second-order phase transition occurs. The model used in the theory represents N identical flexible polymer chain molecules, each containing a large number $l \gg 1$ of segments, on a regular lattice of N_s sites. Only one segment can be placed on a site at any time. Each chain thus corresponds to a self-avoiding walk of l steps on a lattice. The configurational entropy S_{conf} of the system is calculated as the logarithm of the total number of distinguishable microscopic configurations at a given total energy E and is assumed to give the predominant contribution to the total entropy $S(E; N_s, N)$ of the liquid.

Applying the mean-field approximation in the theory of polymer systems, it is derived that the configurational entropy is

$$S_{\text{conf}}(E; N_s, N) = NS_{\text{intra}}(E, N_s) + \delta S_{\text{conf}}, \tag{3.12}$$

with a positive contribution of entropies of individual chain molecules ($NS_{\text{intra}} > 0$) and a negative contribution describing the relative configurations of the molecules ($\delta S_{\text{conf}} < 0$). The result is that the configurational entropy vanishes at a finite energy $E = E_c$ which corresponds to a positive finite temperature T_c of Gibbs ensemble, $T_c^{-1} = \partial S_{\text{intra}} | \partial E |_{E=E_c}$. It is argued that the mean-field approximation used to calculate δS_{conf} gives the correct result for $T > T_c$ with $S_{\text{conf}} > 0$ and $S_{\text{conf}} \propto (T - T_c)$, while S_{conf} is assumed to be zero rather than negative for $T < T_c$ and this removes the "entropy crisis". The supercooled liquid is identified with the phase at $T > T_c$ and the glass with the phase at $T < T_c$. The related phase transition, "ideal" glass transition, at $T = T_c$ is considered to be of second-order since it is found that $S_{\text{conf}}(T)$ continuously changes near T_c. It is supposed for the real glass transition, T_g varies proportionally to T_c with varying thermodynamic parameters of the system like the pressure.

The model under consideration was able to take into account qualitatively both the "entropy crisis" and for $\gamma = const.$, the Vogel–Fulcher behaviour of the viscosity η with its pressure dependence near the transition line, $\eta(T, p) = const$ at $T = T_c(p)$. Moreover, the microscopic model of the "ideal" glass transition may correctly reveal some trends of T_c with changing polymer chain length. The assumption of two ($\upsilon = 2$) alternative configurations of a microscopic block of supercooled liquid does not seem to be essentially restrictive, so that the conclusions mentioned above appear to also be relevant for the case with more than two block configurations with $2 \leq \nu \ll N$. On the other hand, some assumptions introduced in the model seem to restrict its applicability to the glass transition and related phenomena. In fact, in contrast to the assumption of a macroscopically large block size, $n \gg 1$, a block actually contains a few atoms for $T_g > T_c$, e.g., $T_g - T_c \simeq 50\text{K}$, with $S_{\text{conf}}(T_g) \sim S_0 \sim N$, and formula (3.7) above. Then it is not clear how such large relaxation times $\tau(\sim \eta/\mu_0)$ arise for such small blocks, which should correspond to huge $\eta(T_g) \approx 10^{13}$–$10^{14}P$. As noted in [4], the large τ required can be due to high free-energy barriers, separating two blocks configurations in configuration space which are not taken into account in the model under discussion. Generally speaking, this means that in (3.10), a noticeable temperature dependence of γ may occur which gives rise to a deviation of $\eta(T)$ from Vogel–Fulcher law (3.2). In other words, the behaviour of $\eta(T)$ in the supercooled liquid can be determined not only by the configurational entropy but also by the free-energy barriers and the transition paths between the block configurations. In other words, the model can hardly reproduce both the Arrhenius-like behaviour (3.3) of $\eta(T)$ close to T_g in the liquid and the non-ergodicity of the glass with $S_{\text{conf}} \approx S_0 \sim N$ in Eq. (3.7), below the transition, $T < T_c$ (see for some additional remarks also concerning this model, e.g., in [83, 84]).

Although the thermodynamic configurational entropy model reflects an important aspect of the "ideal" glass transition, the model does not seem to take into account at least two other basic factors: the non-ergodicity of glass and free-energy barriers in the liquid.

On the other hand, the free-volume model (see, e.g., [68, 74]) of the glass transition was considered as an alternative thermodynamic model, when applied to early models of liquids [75, 76, 77]. The basic ideas of the free-volume model can be presented as follows.

(i) A random local volume υ (e.g., a cell) of a molecular scale υ_m is associated with each "molecular" structural unit, both in a solid and in a liquid ($\upsilon = \upsilon_{cr} = const.$ in a crystal); an extra volume, a "free volume", $V_f \equiv V - V_0 \equiv N\overline{\upsilon}_f \equiv N(\overline{\upsilon} - \upsilon_0)$ is characteristic of a liquid where υ_0 is a material parameter ($\upsilon = const. \approx \upsilon_{\text{cr}}$).

(ii) The free volume is redistributed between the "molecules" without an appreciable activation energy as compared to T, with non-zero free volume $\upsilon_f \equiv \upsilon - \upsilon_0$ for a "molecule" only when $\upsilon - \upsilon_0 > 0$.

(iii) "Molecular" transport is due to movements of "molecules" into voids (formed by free volume redistribution) of which the free volume υ_f exceeds a critical volume $\upsilon_c - \upsilon_m$.

In other words, it is assumed that "molecular" transport in a glass-forming (very viscous) liquid is slow due to a deficit of free volume rather than to free-energy barriers. The free-volume model is supposed to be plausible for van der Waals liquids and simple metallic liquids. The Doolittle formula (3.5) with $V_0 = const.$ and $V^* = const.$ can be straightforwardly derived in this model, by taking into account that the fraction of "molecules" with a free volume is small, $\overline{v}_f/v_0 \ll 1$, in a glass-forming, very viscous liquid. It is plausible to assume that the random values of the free volume $v_f(>0)$ for different "molecules" are statistically independent. The distribution density $\rho(v_f)$ for the free volume of a "molecule" is of the usual exponential form

$$\rho(v_f) = (1/\overline{v}_f)\exp(-v_f/\overline{v}_f). \tag{3.13}$$

Since the molecular transport characteristics are determined by the probability $P(v_f > v_c)$ of a void with free volume $v_f > v_c$, the fluidity η^{-1} is described by the expression

$$\eta^{-1} \propto P(v_f > v_c) = \int_{v_c}^{\infty} \rho(v_f)dv_f = \exp(-v_c/v_f), \quad v_c \approx v_m, \tag{3.14}$$

which is actually similar to the Doolittle formula.

An attempt has been made to apply the free-volume model to treat the glass transition as a thermodynamic phase transition accompanied by a viscoelastic transition. The glass-forming material was considered to consist of two types of randomly distributed cells and clusters of cells: liquid-like cells and clusters with finite free volume $v_f \equiv v - v_0 > 0$, and solid-like cells and clusters, with no free volume ($v_f \equiv 0$ for $v \leq v_0$). In accordance with the above, the free-volume exchange does not require an appreciable activation energy in liquid-like clusters containing m cells at $m \geq 2$. The situation to some extent resembles a system of two types of sites (or bonds), black and white, which is considered (at $m \geq 1$) in the usual percolation theory [74]. Thus, there should be a critical value $f = f_c(m_M)$, the percolation threshold at $0 < f_c(m_M) < 1$, so that there exists an infinite liquid-like cluster percolating through the system and a series of finite-size liquid-like and solid-like clusters at $f_c(m_M) < f(<1)$; the situation is opposite for $f < f_c(m_M)$. In the free-volume model, the system under consideration is identified with a supercooled liquid at $f_c(m_M) < f(<1)$ and with a glass at $f < f_c(m_M)(<1)$. Then such "ideal" glass transition can be considered as a percolation transition at $f = f_c(m_M)$ between two different phases: liquid and glass. Actually, the quantity f is assumed to depend on temperature and it decreases monotonously as the system cools down. The thermal-equilibrium behaviour of $f(T)$ is found from the free energy minimum. An essential contribution to the free energy $F(f, T, p)$ is due to the communal entropy [77, 78], a kind of configurational entropy which is determined by the number of ways in which the free volume can be distributed among the cells in liquid-like clusters. The percolation transition occurs at a finite $T = T_c(p)$ at which $f(T_c) = f_c(m_M)$. The "ideal" glass transition is a thermodynamic first-order phase transition at which the communal entropy has a finite discontinuity from an average free volume $\overline{v}_f > v_c$ in the liquid to a smaller free volume $\overline{v}_f < v_c$. The macroscopic molecular transport vanishes for $T < T_c$. In other words, the "ideal" glass transition is considered as a classical percolation transition from delocalized states of "mobile" atoms to localized states of "immobile" atoms (cf. the Anderson-Mott quantum transition near the mobility edge for electrons in disordered systems; see, e.g., [1, 79, 80, 81]). Time-dependent relaxational properties of the supercooled liquid near T_g and kinetic aspects of the real glass transition due to the viscoelastic transition, including the deviation of T_g from T_c and the dependence of T_g on the cooling rate R_c, are also investigated in the free-volume model (see also, e.g., [82]). The free-volume model describes some basic properties of glass-formers near T_g and the glass transition (e.g., the Doolittle formula, the dependence of T_g on R_c) is in qualitative agreement with the experimental data. At the same time, this model meets with difficulties in interpreting other essential features of these phenomena. As noted in above discussions, the Vogel–Fulcher law can hardly be considered to result from the Doolittle formula. In this respect, the free-volume model may not lead in a straightforward way to the Vogel–Fulcher behaviour. Unlike the configurational-entropy model, the model under consideration does not give a correct pressure dependence of the viscosity η near the "ideal" glass transition, with $\eta = const.$ at $T = T_c(p)$ [4, 85]. In fact, the free-volume model modified to account for the pressure dependence of $\overline{v}_f$ gives rise to the relation $\eta(T,p) \propto \exp[v_c/\overline{v}_f(T,p)]$, so that $\overline{v}_f(T,p) = const.$ and thus, $\eta(T,p) = const.$ at $T = T_c(p)$ only if

$$\frac{d\overline{v}_f}{dp}(T_c) = \frac{\partial \overline{v}_f}{\partial p}(T_c) + \frac{\partial \overline{v}_f}{\partial T}(T_c)\frac{dT_c}{dp} = 0. \qquad (3.15)$$

Equation (3.15) is equivalent to the Ehrenfest relations (1.5), which actually are violated for the glass transition. An important problem arises when comparing the experimentally observed dynamical viscoelastic behaviour of the liquid with that in the free-volume model. In the latter, the behaviour of the complex shear modulus appears to be described by Eq. (1.7) while the experimental behaviour rather corresponds to a different relation close to Eq. (1.8). Although the thermodynamic free-volume model reflects an important aspect of the "ideal" glass transition, the model does not seem to take into account at least two other basic factors, the non-ergodicity of glass and the fairly extensive random network of directed bonds in the liquid near T_g.

In addition, a model of mobile "defects" has also been considered which seems to take into account phenomenologically some basic properties of the "ideal" glass transition and associated phenomena in three-dimensional glass-forming materials. The basic idea of the model is as follows. A "site" (an atom or a group of atoms) can relax to a state (a configuration) with lower energy only when a kind of mobile topological structure called a "defect" succeeds in diffusing to the "site". Although the "defect" should be associated with some extra volume, it is not a free-volume void (a statistical accumulation of extra volume which occurs with equal probability anywhere). On the contrary, the "defects" are assumed to have a nearly permanent structure and satisfy a certain conservation law (and to be frozen in the glass). The "defect" contributes to the relaxation of the "site" by inducing a distortion of its environment. The potential energy of the neighbouring "sites" increases at the distortion, resulting in an increase of the rate of overbarrier transition of the "sites" to states with low energies. The concentration of the mobile "defects" is assumed to be low enough, $c_d \ll 1$, so there exists a broad distribution of relaxation times $\tau(r_0)$ depending on the random spacing r_0 between the relaxation "site" and the initial position of a "defect", $\tau(r_0') \gg \tau(r_0'')$ when $r_0' > r_0''$. This relaxation is essentially inhomogeneous and in this respect, qualitatively differs from the Maxwell relaxation in viscous liquids. The viscoelastic relaxation due to the "defects" in a glass-forming liquid can then be described by Eq. (1.8), which fits the experimental data rather well, differing from the Maxwell formula in Eq. (1.7). Taking into account [33], the "defect" can be identified with an odd disclination-like line (loop), an intrinsically topologically linear structure characteristic of a glass-forming material. The average length of the odd line per unit volume can be estimated as follows, $\bar{\lambda} \equiv ma_1/(na_1)^3 \equiv ga_1^{-2}$ with $n \gg m \gg 1$ and $g \equiv m/n^3 \ll 1$, so that the average spacing between a "site" and the nearest "defect" is

$\bar{r}_0 \propto (ga_1^{-2})^{-1/2}$. It is supposed that the "defect" undergoes a Brownian diffusion due to random activationless jumps, so it can be concluded that the average time of "defect" diffusion to a "site" and thus the relaxation time of the "site" can be estimated as $\tau \propto r_0^{-2} \propto a_1^2 c_d^{-1}$. The Vogel–Fulcher law (3.2) for τ results when a kind of thermodynamic phase transition takes place for the concentration c_d of mobile "defects":

$$c_d = \exp[-\varepsilon_0/(T - T_0)], \quad \text{for } \varepsilon_0 > 0 \quad \text{and} \\ T > T_0 > 0; \quad c_d = 0, \quad \text{for } T < T_0. \tag{3.16}$$

The phase transition in the three-dimensional system in question essentially differs from a conventional first- or second-order phase transition and is sometimes called a topological transition since a topological feature related to the mobile odd lines disappears at $T = T_0$, i.e., $c_d = 0$. In fact, this phase transition is closely similar to the well-known topological phase transition in two-dimensional systems in which no long-range order of conventional type occurs [87, 88]. Then, the following relation can be obtained by applying the common condition that the free energy of the system has a minimum in a metastable equilibrium state. The expressions derived in this theory for the entropy S, internal energy E and free energy F are as follows:

$$F(c_d, T) = E - TS = (A_1 - A_2 T)c_d + (A_3 T - A_4)c_d \ln c_d, \tag{3.17}$$

where

$$E = (A_1 c_d - A_4 c_d \ln c_d) + E_0 \quad \text{for } T > T_0 = A_4/A_3 > 0, \tag{3.18}$$

while

$$F = F_0, \quad E = E_0, \quad \text{for } T < T_0, \\ \partial F_0/\partial c_d = \partial E_0/\partial c_d = 0, \tag{3.19}$$

where A_i is the material constant ($i = 1, 2, 3, 4$). In accordance with the model under discussion, two quantities characteristic of glass-forming systems satisfy conservation laws: the material density (as common) and the total number density of odd lines. It is worth noting that the peculiar term $\propto c_d \ln c_d$ in the entropy or the internal energy is found to be due to the contributions of the mixing entropy of odd loops with small size

($\ll a_1 c_d^{-1/3}$) or the elastic energy of self-screening large odd loops with typical size $a_1 c_d^{-1/3}$ [the average spacing between the loops, in a semi-dilute solution with $c_d \ll 1$ and $a_1 c_d^{-1/3} \ll L$ respectively (L is the sample size)]. Then the odd loops are their own antidefects in the sense that the related 2π disclinations are characterized by parity, rather than by intensity, and by the group Z_2 containing two elements. Therefore, the odd loops can screen each other's elastic energies, so that pairing of the loops gives rise to a decrease of the total elastic energy. On the other hand, dissociation of the pairs leads to entropy increase. The competition of the two factors lowering the free energy just results in the phase transition which can be essentially related to Vogel–Fulcher relaxation processes. That is meaningful if the parameter T_0 and ε_0 are positive, with the expressions

$$T_0 = A_4/A_3 > 0 \quad \text{and} \quad \varepsilon_0 = (A_1 A_3 - A_2 A_4) A_3^{-1} > 0 \qquad (3.20)$$

found in this model. The positive sign of the expression for T_0 is associated with the fact that only the bulk elastic energy of the odd lines can noticeably change under extrinsic forces, though the disclination-core energy exceeds the bulk energy; T_0 is, in fact, identified with the shear elastic energy, $T_0 \simeq \mu_0 a_1^3$. If this is true, an "ideal" glass structure at $T < T_0$ can really exist as an amorphous solid with a significant shear elasticity (of the same scale as in crystals), which differs from the supercooled liquid phase by the absence of mobile odd lines ($c_d = 0$). This random structure is presumably highly degenerate because of the large residual entropy S_0 which can also be revealed in the model in question. However, once established, macroscopic changes of the structure at $T < T_0$ are assumed to occur only by bond-breaking and bond-rearranging processes. Then, T_0 appears to be the highest temperature at which such "ideal" random structure is stable relative to generation of mobile odd lines (loops). Because of the conservation law for the total number density of odd lines, disappearance of the latter ($c_d = 0$) for glass at $T < T_0$ means that, generally speaking, these "defects" do not annihilate but form strongly coupled pairs which are frozen in the proper glass structure. It is worth noting in this model that macroscopic relaxation associated with large-scale rearrangements takes place for finite c_d at $T > T_0$ in Eq. (3.17) (for liquid) while local relaxation processes also occur at $T < T_0$, i.e., at $c_d = 0$ (for glass).

It is also argued in the theory under discussion that the residual entropy in a glass is related to the so-called topological entropy S_t, which is directly related to the difference in topological properties between a crystal and its melt and (together with the entropy due to volume change) essentially contributes to the total melting entropy. The entropy S_t is identified with the sum of the configurational and mixing entropies of linear topological structures like disclinations which can be spontaneously generated and move freely in a liquid, unlike in a crystal. It is shown in reasonable agreement with experimental data, that $S_t = \ln 2 \cdot \bar{z}(/2 - 1)$ for a liquid with average coordination number $\bar{z}$ and that S_0 is of the same scale as S_t.

3.3 Mode-coupling model: Dynamic liquid-glass transition

The models considered above, which can reflect some essential aspects of the glass transition as a dynamic transition (i.e., a crossover) from an ergodic simple liquid state at high temperatures to a non-ergodic glassy state at low ones, can hardly account explicitly for the non-ergodicity of glass and for the collective motion of interacting particles in a supercooled liquid (see, e.g., [64, 65]). However, both factors might be taken into account in an approach in which the basic variables are density fluctuations $\rho_{\mathbf{q}}$ of wavevectors $\mathbf{q}$ and the basic dynamic characteristic of an isotropic system is the density auto-correlation function $\phi_q(t) = \langle \rho_{\mathbf{q}}(t)\rho_{\mathbf{q}}(0)\rangle / S_q$, where $S_q = \langle |\rho_{\mathbf{q}}|^2 \rangle$ is the (static) structure factor at $q = |q|$. The approach is related to the so-called mode-coupling theory (MCT) described in a series of works (see, e.g., [11, 12], [89] and [90]; see also Sec. 12.1 below). Although rigorous results and an explicit small parameter for expanding an appropriate dynamic characteristic of the system like $\phi_q(t)$ around a "critical" point do not seem to be available, the MCT model appears to give rise to a theoretical model of an "ideal" glass transition as a dynamic (not thermodynamic) transition from ergodic liquid dynamics to non-ergodic glass dynamics, at least in systems in which the positions of interacting particles do not change with increasing time t (i.e., the particles are not hopping between neighbouring positions, so the phonon-assisted relaxation of particles is neglected). In fact, theoretical interpretations of neutron scattering data and dynamics simulations for the systems in question (see, e.g., [10, 92]) gave rise to MCT models related to the earlier mathematical approach [13]. In this approach, a closed set of nonlinear equations of motion for correlation functions of density fluctuations are approximately derived (at given static correlations and interparticle interaction potentials) for explaining properties of liquids with interacting density fluctuations. In other words, the system of closed nonlinear equations for supercooled liquid dynamics is the foundation of MCT results. A special version of the mode-coupling equations gives rise to an "ideal" dynamic transition on a hypersurface in an appropriate space of variables, described by a singularity at "critical" temperature $T = T_c$ or particle concentration (density) $n = n_c$. As the dynamic transition exhibited qualitative similarities to the glass transition under discussion, it was assumed in the MCT that the distance from the critical point could be used as a small parameter for an analytical solution and, moreover, the critical point was argued to be a bifurcation point of the cuspoid type on the mentioned hypersurface. However, the nonlinear equations, describing such a dynamic transition, are so complicated that so far, only a few attempts have been made to solve those numerically, with numerical solutions that agree with appropriate experimental data to an accuracy of $\leq 20\%$ (see, e.g., [11, 12]). In what follows, the basic concepts of MCT will be formulated, and the formulae will mainly be quoted, which can be directly connected with possible experiments (see also Sec. 12.1 below).

As mentioned, the basic dynamic characteristic used in the MCT is the density autocorrelation function $\phi_q(t) = \langle \rho_{\mathbf{q}}(t)\rho_{\mathbf{q}}(0)\rangle / S_q$, where $S_q = \langle |\rho_{\mathbf{q}}|^2 \rangle$ is the (static) structure factor of an isotropic liquid at a wave number $q = |q|$ and $\rho_{\mathbf{q}}(t)$ describes density fluctuations of wave vector q in collective motions of interacting particles. Expecting that the blocking of any particle's motion by surrounding particles is the microscopic origin of the slowing down of structural relaxation in liquid, one can reduce the calculations of quantities like longitudinal (l) and transverse (tr) sound moduli and shear viscosity to calculations of $\phi_q(t)$. It is taken into account in the calculations that the microscopic density $\rho_q(t)$ contains information on all particles positions and that the spectrum of ϕ_q is proportional to that of the dynamic structure factor $S(q, \nu)$, which can be directly measured in neutron scattering experiments.

In what follows, the basic results are briefly described by taking into account the leading order contributions to $\phi_q(t) = (\phi_q(z = t + i\eta, \eta \to +0))$ while all intermediate calculations of the MCT model are omitted. Hence, the dynamic compressibility χ_q of the system can be found, $\chi_q(z) = [1 + z\phi_q(z)]S_q/T$, which is reduced to the isothermal compressibility at low $q(\to 0)$ and $z(\to 0)$. The dynamic compressibility is shown to characterize the longitudinal (l-) phonon propagator for the system vibrational dynamics, with eigenfrequencies $\Omega_q = (q^2 V_T^2 S_q^{-1})^{1/2}$, as follows:

$$\chi_q(t) = [1 + z\phi_q(t)]\, S_q/T = -(q^2/\rho)\{t^2 - \Omega_q^2 + tM_q^l(t)\}^{-1}. \quad (3.21)$$

Here, $V_T = (T/m)^{1/2}$ is the thermal velocity of identical particles of mass m in the liquid, and the function $M_q^l(t)$ is an analogue of the polarization operator in quantum electrodynamics that describes the influence of surrounding harmonically vibrating particles in thermal-equilibrium liquid on the phonon with bare dispersion law Ω_q. The influence gives rise to nontrivial features of the liquid vibrational dynamics in the same way that the dielectric function leads to nontrivial features of field propagation in dielectric media; for example, the phonons exhibit a linear dispersion, $\Omega_q \simeq sq\ (\to 0)$ and the polarization operator $M_q^l \propto q^2(\to 0)$ at small $q \to 0$. In the generalized theory of classical liquid dynamics (see, e.g., [11, 12] and references therein), and the leading approximation, the operator $M_q^l(t)$ can be expressed in terms of two functions, $M_q(t)$ and $\delta_q(t)$, as follows:

$$M_q^l(t) = M_q(t)[1 - \delta_q(t)M_q(t)]^{-1} \quad \text{at} \quad M_q(t) = m_q^0(t) + \Omega_q^2 m_q(t). \quad (3.22)$$

The function $m_q^0(t)$ describes the effect of low frequencies in normal liquid dynamics (e.g., due to uncorrelated binary collisions of particles, e.g., at actual $h\nu_q < 10\,\text{meV}$) at $m_q^0(t) \approx const$, while $\delta_q(t)$ describes phonon assisted, hopping and relaxation of particles. However, for supercooled liquids, the most important part of the theory is shown to be a polynomial function $m_q(t)$ that can be in the simplest approximation described by products of functions $\phi_q(t)$ at different q,

$$m_q(t) = F_q[V, \phi(t)] \approx (1/2) \sum_{k,p} V(q;k,p)\phi_k(t)\phi_p(t). \qquad (3.23)$$

The function $V(q;k,p) \geq 0$ is the so-called mode coupling vertex for longitudinal sound modulus, independent of the particle mass in classical liquids, which is an equilibrium quantity expressed in terms of the structure factor S_q. The vertex characterizes three-particle correlations and is related to coefficients in the coupling of pairs of density fluctuations to fluctuating forces. Temperature and inter-particle interaction potentials can implicitly enter the coefficients via the structure factor S_q (and possibly other canonically averaged equilibrium quantities). The approximation corresponds to factorizing averages of products of density fluctuations into products of averages determining the function $V(q;k,p)$. For the transverse (tr) sound modulus, a similar approximation can be $M_q^{tr}(t) = M_q(t)\{1 - \delta_q^{tr}(z)M_q(t)\}^{-1}$ with $m_q^{tr}(t) \approx (1/2)\sum_{k,p} V^{tr}(q;k,p)\phi_k(t)\phi_p(t)$ and $M_q(t) = m_q^0(t) + \Omega_q^2 m_q^{tr}(t)$ while the characteristic function $V^{tr}(q;k,p) \geq 0$ is the respective coupling vertex. For the function $\delta_q(t)$, a similar mode coupling expression can be obtained. The equation for $m_q(t)$ is an example showing how correlation functions can be expressed in terms of $\phi_q(t)$, so that longitudinal dynamical fluctuations can be reduced to density fluctuations and low-frequency anomalies, e.g., for dynamical compressibility χ_q. The most important contributions to $m_q(t)$ and $m_q^{tr}(t)$ are found to come mainly from density fluctuations of microscopic scale, $k^{-1} \sim p^{-1} \sim 3\text{Å}^{-1}$.

The terms of Eqs. (3.21)–(3.23) depend regularly on control parameters like density n, temperature T and inter-particle interaction potentials (not containing singularities). With the given control parameters, one can estimate the structure factor S_q and vertices and then approximately express complicated correlators in terms of the basic dynamic characteristic $\phi_q(t)$. Then, the equations become a closed set and can determine the appropriate specified correlation functions for a supercooled classical liquid. It is important to note that no assumptions on the existence (or absence) of glass transition are made in the MCT. The main problem now is to understand how a relevant model for $m_q(t)$ gives rise to a singularity describing the above-mentioned dynamical transition in the relevant solution of the closed set of nonlinear equations in MCT.

It is known from the liquid dynamics [10, 92] that the function $m_q(t)$ actually describes the so-called cage effect. The effect means that in a densely packed liquid, a particle cannot move freely (unlike a particle in a gas); rather it rattles in traps provided by its neighbours and their fluctuating barriers. The cage effect becomes stronger as the density increases and can provide both retardation effects (density fluctuations at some time t' influence the particle positions and interparticle forces for all times $t > t'$) and nonlinear interaction effects for the particles. If the cages fluctuate slowly and interact strongly, both kinds of effects can give rise to a self-consistent dynamic transition which is the basic phenomenon (for $\phi_q(t)$) in the MCT model. It is important that the mathematical essence of the model is not contained in the detailed form of the vertices or wavevector sums, as shown for instance, in [11, 12] by introducing so-called schematical models, e.g., with $m_q(t) = V_1\phi_q(t) + V_2(\phi_q(t))^2$. It was derived in such models in detail that indeed, the MCT gives rise to a crossover from ergodic liquid dynamics to non-ergodic glassy dynamics described by the solution of the equations of motion for $\phi_q(t)$. Since no singularities exist in the equations of motion, there is an important difference of the MCT as compared with ordinary phase transition theories of the glass transition. In the latter, the anomalies are determined by singularities of thermodynamic functions like S_q, and the mode coupling equations describe how the singularities of vertices influence the spectra whereas in the dynamic MCT under discussion, the singularities of averaged spectra are available with thermodynamic functions being regular ones. The approximations giving rise to the MCT are not controllable in the sense that no calculations of the accuracy of the solutions to the basic equations are available.

If the space of wave vectors is assumed to be discrete with M values $(q_1, q_2, \ldots, q_M)$, there are $N = M(M+1)/2$ coupling constants $V(q; k, p)$ that can be combined into the control parameter vector $V = (V_1, \ldots, V_N)$ a point in the control parameter space K. Then the vertex $m_q(t)$ which describes the cage effect can be specified in connection with Eq. (3.23) by an N-dimensional vector V of the control parameters which summarizes the vertex constants. Moreover, the equation $m_q(t) = F_q[V, \phi_k(t)]$ can be transformed (by ignoring the δ-term describing phonon assisted relaxation of particles) for small enough times t to $\phi_q(t) = 1 - (1/2)(\Omega_q t)^2 + (1/6)(\Omega_q t)^2(\nu_q/\Omega_q) + \cdots$, in which the frequencies ν_q and Ω_q govern the initial decay in dynamics. The expression $\phi(t)$ can specify a well-defined mathematical problem for nonlinear dynamics and thus, a model for the dynamic transition from a classical liquid to a glass. In the simplest meaningful example, the space K is reduced to a $2D$ space, e.g., $m_q(t) = V_1\,\phi_q(t) + V_2\,(\phi_q(t))^2$, as introduced above for a schematic model. In the latter, as well as in more complicated schematic models of such kind, the equations of motion for $\phi_q(t)$ can yield a dynamic transition from an ergodic system (e.g., a classic liquid) to a non-ergodic (glassy) system. In accordance with the MCT approach, one can explain how the solution of the closed set of equations in MCT with coefficients being regular functions of control parameters (Eq. (3.23)), can possess singularites describing an "ideal" glass transition (neglecting phonon assisted relaxation) or a crossover from ergodic liquid dynamics to non-ergodic glassy dynamics. For the sake of simplicity, one can consider that there is a subset D_G of the N-dimensional space, specified by vectors V, where

$$\phi_q(t \to \infty) = f_q, \quad 0 < f_q < 1, \tag{3.24}$$

with f_q depending smoothly on V. Equation (3.24) means that the system behaves as a non-ergodic one [93]. The Laplace transform of the correlation functions for the non-ergodic states exhibits an exact simple pole at $z = 0$ with strength factor f_q, $-z\phi_q(z) = f_q$ at $z \to 0$. Then, the points in D_G can be identified as ideal glass states which would be true glass states if no phonon assisted (thermal activated), hopping relaxation of particles occurs, so the factor $f_q \neq 0$ can be considered as the non-ergodicity parameter. On the other hand, in the interior D_L of the complement to D_G the states can be characterized as ergodic liquid states with $\phi_q(t \to \infty) = 0$. Then boundary points of the ideal glass region in the n-dimensional space can be called glass transition singularities at critical values $V = V_c$. For $V \to V_c$ and $t \to \infty$, the solution of Eqs. (3.21)–(3.23) exhibits subtle singular behaviour. These subtleties are the mathematically interesting features of the MCT. The singularities imply anomalous dynamics for parameters $V \to V_c$ for large enough times or low enough frequencies. These anomalies can be related to the abovementioned crossover phenomena.

Moreover, it is found in the MCT that typical ideal glass transition singularities can be described as a bifurcation of the type of so-called cuspoids [94]. Since the dimensionality of D_G is smaller than that of the complete control parameter space, it is concluded that generically, the system is in liquid state. This means that realistic systems in which phonon assisted relaxation exists are ergodic (i.e., are not characterized by ideal glassy dynamics). However, if the activated relaxation is rare enough, the system behaviour becomes similar to that close to an ideal glass transition. In addition, it is argued in MCT, the same singular dynamics can appear in all liquids, irrespective of whether they are monatomic ones or complicated polymers. It is therefore useful to check whether or not experimental data for complicated systems indicates that the actual parameters are close to such singularities. There is also experimental evidence indicating that the MCT may be relevant for approximately describing the dynamic transition as a crossover at a critical temperature $T \approx T_c$, from ergodic supercooled liquid dynamics to non-ergodic glass dynamics.

On the other hand, the theory can give some predictions for experiments in supercooled simple liquids (but not necessarily in complex network liquids) [10, 11, 75, 89, 91, 92, 93]. An important result of the MCT model also seems to be a derivation of the stretched-exponential relaxation close to the glass transition (although the original Kohlrausch phenomenon, at $\upsilon = 1/2$, does not seem to be obtained as an analytical result, the calculated functions are so close to it that they could fit the representative data as well as the empirical law used earlier in the literature). At the same time, the MCT model seems to meet some problems. For instance, this model seems to be relevant for systems of dimensionality $d \geq 3$ but hardly for two-dimensional systems in which the critical behaviour of viscosity is described by $\eta \propto |T - T_c|^\mu$ at $\mu \approx 1$, rather than by the Vogel–Fulcher law (Eq. (3.2)) which probably fits the experimental data better.

4

KOHLRAUSCH–WILLIAMS–WATTS (KWW) RELAXATION

4.1 General features of slow relaxation processes

The relaxation processes in a supercooled liquid for small deviations of the macroscopic characteristics $P(t)$ (e.g., shear stresses, electrical fields, sudden temperature changes) from their thermal-equilibrium values P_{eq} near T_g differ essentially from the exponential relaxation typical of a Maxwell viscous liquid or a Debye insulator (see, e.g., Sec. 1.2 and [26, 27, 28]). These macroscopic processes are commonly called "linear relaxation" and are described by the time dependence of the parameters $X(t) \equiv [P(t) - P_{eq}]/[P(0) - P_{eq}]$, varying from $X(0) = 1$ to $X(\infty) = 0$. The linear relaxation appears to be well-described by the stretched-exponential Kohlrausch–William–Watts (KWW) law (1.6) in the experimentally important range of times (t), where $X(t)$ changes from $X(t) \simeq 1$ to very small $X(t)$ (say, from $X(t) = 0.99$ to $X(t) = 0.01$) see, e.g., [4, 29, 30]. The behaviour of the average relaxation time $\tau(T)$ in the KWW formula is observed to be of Vogel–Fulcher type (Eq. (3.2)). A similar type of linear relaxation is experimentally found in a large variety of materials and phenomena (see, e.g., [27, 28, 57]).

An important feature of linear relaxation in a supercooled liquid near T_g is the observed time–temperature ($t - T$) scaling: a variation of $X(t, T)$ with t at $T = const.$ can be reproduced by appropriately changing T at $t = const.$ In other words, $X(t, T)$ depends only on $t/\tau(T)$ [or on $T/\psi(T)$] and the form of $X(t, T)$ does not change, although the average relaxation time $\tau(T)$ changes with T by many orders of magnitude [29]. For the KWW formula (1.6), this means that the fractional exponent is independent of both T and t and that τ does not depend on t. In fact, $X(t)$ is a solution of the equation

$$dX(t)/dt = -X(t)/\tau_{\text{eff}}(t, T), \tag{4.1}$$

with an effective time τ_{eff} also depending on the observation time, $\beta\tau_{\text{eff}}(t) = t^{1-\beta}\tau^{\beta}$ at $0 < \beta = const. < 1$. It follows from Eq. (4.1) that $X(t)$ can be formally written as a superposition of exponential terms $\exp(-t/\xi)$ for different atomic degrees of freedom or modes (see, e.g., [4, 28]):

$$X(t) = \int_0^{\infty} d\xi\, g_X(\xi)\exp(-t/\xi),\ X(0) = 1. \tag{4.2}$$

From the point of view of the microscopic theory, the representation (4.2) contains an ambiguity in the choice of the relevant spectral density $g_X(\xi)$. The problem is to determine theoretically in an independent way the density $g_X(\xi)$ that corresponds to the KWW behaviour of $X(t)$ and its time–temperature scaling. It is argued in some works that a spectrum of relaxation times in an amorphous solid can be related to the random character of local atomic configurations, even when locally the relaxation is of the Debye type. However, the local relaxation can also be a nonexponential type, due to the cooperative nature of the atomic motion or of the correlated motions which involve increasing numbers of atoms in a supercooled liquid. Generally speaking, real linear relaxation involves both parallel elementary events composing the relaxation, which are locally of the Debye type and uncorrelated with each other, and correlated nonexponential elementary events. The problem is whether the relaxation times ξ corresponds to actual physical processes. Moreover, the problem is whether the elementary relaxation events are mainly of the Debye type and occur in a parallel way in different random configurations or whether they occur in series corresponding to cooperative relaxation involving the correlated motion of many atoms and correlations between events. It is worth noting that the time–temperature scaling property of the KWW formula can rule out some models of stretched-exponential relaxation.

4.2 Parallel-diffusion relaxation models

For parallel relaxation events, each essential degree of freedom $f_i(i = 1, 2, \dots)$ relaxes in an exponential way with a time $\xi_i \equiv \xi(f_i)$. In a random inhomogeneous continuum in each homogeneous local region (r), the relevant mode $f(r)$ relaxes in the Debye way with a time $\xi(f)$. Then,

$$g_X(\xi) = \int (dr)\rho[f(r)]\delta(\xi - \xi[f(r)]) \equiv g(\xi), \tag{4.3}$$

where $\rho[f(r)]$ stands for the density of the local configuration parameters $f(r)$. It is not easy to reconcile formulae (4.2) and (4.3) in general with the time–temperature scaling noted above; for instance, this is difficult to achieve in terms of a distribution of local temperature independent activation energies of the relaxation times (see, e.g., [4]). Nevertheless, relaxation associated with parallel events or related processes can still be of the KWW type. Two models can be mentioned in this respect: the defect diffusion model and the free-volume model.

The defect diffusion model of KWW relaxation can actually be considered as more or less associated with parallel relaxation events. This model can be related to the defect model of the Vogel–Fulcher law (Eq. (3.2)) considered in Sec. 3.4. As mentioned above, the model implies that randomly walking topological stable structures favour an effective lowering of the potential barriers in macroscopic relaxation and thus, play the role of "defects" assisting atomic rearrangements and relaxation. The model is largely cooperative as far as the "defect" is associated with collective degrees of freedom as noted in Sec. 3.4; these "defects" are essential constituents of the proper glass structure. It is worth noting that proper atoms with a "wrong" coordination number z' can also play the role of such a "defect" in a computer simulation of BeF_2 supercooled liquid (see, e.g., [100]). The problem of KWW relaxation in supercooled liquids has been considered in both continuum [85, 86] and lattice [97, 98, 99] defect diffusion models.

The situation in the continuum model is as follows. Let us consider, for the sake of definiteness, the relaxation of shear stresses $\sigma_s(t) = \sigma_s(0)X(t)$

in a liquid containing N "defects" in its volume V. The "relaxation function" $X(t)$ is identified with the probability that at time t, no defect has reached the relaxing region under consideration or that relaxation does not occur — $0 \leq X(t) \leq 1$, with $X(0) = 1$ and $X(\infty) = 0$. The probability $X(t)$ is expressed through the probability $p(t, r_0)$ that a "defect" localized at a distance r_0 from the centre of the relaxing region at $t = 0$ has reached this region at time t at least once (which is sufficient for occurrence of relaxation in this model). Assuming that the "defects" are statistically independent at any time, at low enough concentrations at least, the expression for $X(t)$ is obtained to be the average of the probability $1 - p(t, r_0)$ over the initial uniform distribution of "defects". In the macroscopic limit for $V \to \infty$ and $n = N/V = const.$,

$$X(t) = \prod_{i=1}^{N} \int_{(V)} \frac{(dr_0^{(i)})}{V} [1 - p(t, r_0)] = \left[1 - \int_{(V)} \frac{(dr_0)}{V} p(t, r_0)\right]^N$$
$$\simeq \exp[-N_a(t)], \tag{4.4}$$

with

$$N_a(t) = n \int_{(V)} (dr_0) p(t, r_0)$$

standing for the average number of "defects" which have reached the relaxing region at time t at least once. It follows that $X(t)$ finally depends only on $p(t, r_0)$ and does not depend on the particular property P considered. As usual, the probability $p(t, r_0)$ is found by solving the diffusion equation, with an absolutely adsorbing boundary of the relaxing region (see, e.g., [101]). The result obtained from Eq. (4.4) is

$$X(t) = \exp\left(-\frac{4}{3}\pi n R_0^3\right) \exp[-(t/\tau)^{1/2} - t/\tau_D],$$
$$\tau^{-1} = 64\pi (nR_0^3)^2 D/R_0^2, \quad \tau_D^{-1} = 4\pi (nR_0^3)^2 D/R_0^2, \tag{4.5}$$

where R_0 is a typical radius of the relaxing regions and D is the "defect" diffusion coefficient. It follows that the relaxation with $\beta = 1/2$ at the experimentally important times t noted above when the "defect" number density n is high enough, $nR_0^3 \gg 1$, while it is of the Debye type for $nR_0^3 \ll 1$.

However, some problems appear in this model of KWW relaxation, in which an expected competition of the "defects" to reach the relaxation region first has not been taken into account. In fact, a very short-time relaxation [$\sim \exp(-\frac{4}{3}\pi n R_0^3)$ for $t \to 0$] appears in $X(t)$ which is significant in the KWW relaxation range for $nR_0^3 \gg 1$ and can hardly be physically understood. Moreover, the time–temperature scaling mentioned above holds in formula (4.5) when τ/τ_D and thus nR_0^3 is independent of T. This means, however, that R_0^3 should increase just as n decreases with decreasing T, at $T \approx T_g$ (see Sec. 3.4), which is rather difficult to understand (e.g., [4]).

An essentially inhomogeneous lattice defect diffusion model which was based on some ideas of dispersive transport theory in semiconductors (see, e.g., [102]) has also been developed [97, 98, 99]. The defect diffusion is characterized by a distribution $g(\xi)$ of statistically independent times of a series of diffusion jumps between discrete lattice sites, the relaxing region being identified with a site. It was assumed that $g(\xi)$ exhibits a non-Gaussian tail at long times ξ, $g(\xi) \propto \xi^{-(1+\upsilon)}$ for $0 < \upsilon < 1$ (cf. $g(\xi) \propto \exp(-const.\xi^2)$ for usual diffusion), so that $N_a(t) \propto t^\beta$ in Eq. (4.4). Thus, KWW relaxation, $X(t) \propto \exp[-(t/\tau)^\upsilon]$, is realized for asymptotically long times which do not necessarily overlap the experimentally important range of t. Unlike Eq. (4.5), in this model, the KWW law appears for low n [98].

It seems difficult to understand the origin of the long-time, non-Gaussian tail of $g(\xi)$, since no trapping centres for the "defects" are involved and neither thermally activated hops with activation energy $\varepsilon(\rho)$ dependent on strongly fluctuating intersite separations (ρ) nor tunneling seem to determine the situation related to the Vogel–Fulcher law (3.2). It is worth noting that the behaviour (3.2) or $\tau(T)$ in KWW relaxation is neither obtained in the continuum nor in the lattice defect diffusion models.

An explicit example of KWW relaxation determined by parallel processes is the relaxation mechanism in the free-volume model of a supercooled liquid. In this model, it is assumed that the relaxation is related to uncorrelated activationless atomic hops through surfaces (S) of liquid-like (l) cell clusters with relaxation times $\xi = \xi_0 \exp(\upsilon_c/\overline{\upsilon}_f) \propto (V/S)_l$ (Eq. (3.12)). Based on the percolation theory used in this model [74], it is argued that (in a three-dimensional system)

$$(S/V)_l = const. \cdot \nu^{-x} \quad \text{with } 0 < x \leq 1/3,$$

while the probability of a finite cluster containing ν atoms is

$$W(\nu) \propto \exp(-\,const. \cdot \nu^{y}), \quad \text{with } 2/3 < y \leq 1.$$

Equations (4.2)–(4.5) can give rise to the following relationship:

$$X(t) = \int_0^{\infty} d\xi g(\xi) e^{-t/\xi} \sim \int_0^{\infty} d\nu W(\nu) \exp[-t/\xi(\nu)] \propto \exp[-(t/\tau)^{\upsilon}], \tag{4.6}$$

with $0 < \beta = y(x+y)^{-1} < 1$ and $\tau \propto \exp(\upsilon_c/\overline{\upsilon}_f)$, for asymptotically long times $t > \tau$. However, the temperature dependence of $\tau(T)$ can hardly be identified by Eq. (3.2) which is characteristic of Kohlrausch relaxation. It is also not clear whether the asymptotic region $t \gg \tau$ practically coincides with the empirical range of t mentioned above; experimentally, as a rule, $1/2 \leq \beta \leq 1$ while $1/2 < \upsilon < 4/3$ in Eq. (4.6).

4.3 Correlated, hierarchically constrained, relaxation models

An alternative type of KWW relaxation model is based on the idea that relaxation is a cooperative process involving a series of locally correlated events. Each event occurs for a number of configurational degrees of freedom (modes) and implies some atomic rearrangements when the preceding event is completed. The relaxation is described by the same formula (4.2) and the spectrum, $[g(\xi)]$, of relaxation times ξ is related to the correlations between the different modes rather than to local inhomogeneities, with ξ corresponding to real physical processes (see, e.g., [95, 96]).

An example of this type of relaxation is dipole relaxation in insulators which is associated with subsequent steps: slow degrees of freedom relax in an environment formed after relaxation of faster ones [95]. A similar model is considered for the relaxation in supercooled polymer liquids in which individual polymer chains move independently and the chain dynamics is determined by harmonic coupling between neighbours. The result is similar to the KWW law with $\upsilon = 1/2$ [96]. The general problem of correlated relaxation processes in glass-forming liquids, as well as in general systems of strongly interacting particles with a hierarchy of fast to slow degrees of freedom has recently been analysed for a particular class of models of hierarchically ordered structures of states, though in rather abstract terms [96]. It is argued that the relevant theory of the relaxation under discussion should satisfy the following three conditions:

(i) It should be based on dynamics rather than on statistical mechanics, since ergodicity in phase space for the systems considered is violated for actual, not too large observation times t. This is related to the change of free-energy surface with varying t: fast modes ($\xi \ll t$) contribute to the thermal-equilibrium free energy and slow modes ($\xi > t$) are frozen in while modes with $\xi \sim t$ are active in the relaxation processes.

(ii) The theory should contain dynamical constraints for the important models. The constraints approximate interactions between the essential modes since diagonalization of the nonlinear Hamiltonian of the strongly interacting system in question which gives rise to independent "normal" modes, is actually unrealizable. These constraints are important in a very large range of t, $\xi_{\min} \lesssim t \lesssim \xi_{\max}$, from the smallest $\xi_{\min}(\gtrsim 10^{-13}\,\mathrm{s})$ to the much larger $\xi_{\max}$, a macroscopic time of ergodicity (the system is ergodic for $t > \xi_{\max}$). In a glass-forming liquid near T_g, $\xi_{\max}$ is expected to be larger than any realistic t.

(iii) The theory should consider a hierarchy of modes, from very fast to very slow, the latter essentially varying only after the former has changed in an appropriate way. The hierarchy is assumed to generate a broad spectrum $g(\xi)$ in Eq. (4.2). (It is also possible that the slow modes can constrain the fast ones.)

It is argued in the above references that for real macroscopic systems, e.g., a glass-forming liquid, the spectrum $g(\xi)$ should be continuous (not discrete). However, a continuum model satisfying conditions (i)–(iii) was analyzed by considering relaxation in appropriate discrete models and using the continuum limit in the resulting relations, by replacing sums by integrals. In this sense (not in the sense that it corresponds to actual glass-formers), a class of simple discrete models is proposed in the work (see also the discussion in [30] and [96]).

Hierarchically constrained systems are explicitly considered, in which the degrees of freedom are Ising spins $S_i = \pm 1$ distributed over a discrete set of energy levels ($n = 0, 1, 2, \ldots$) with N_n spins at the nth level. A spin at the $(n+1)$th level can flip (change its state) only if a certain condition for the spins at the nth level is satisfied which just describes the constraint. The condition is that $\mu_n (\leq N_n)$ spins should form one particular configuration of the 2^{μ_n} possible configurations (e.g., with parallel spins in this particular configuration). This means that the modes at the nth level are fast compared to those at the $(n+1)$th level and the slow modes do not constrain the fast ones. The average relaxation times at different levels are related by the formulae

$$\xi_{n+1} = 2^{\mu_n} \xi_n, \quad \text{or } \xi_n = \xi_0 \exp\left(\sum_{k=0}^{n-1} \overline{\mu}_k \right), \quad \overline{\mu}_k = \mu_k \ln 2, \qquad (4.7)$$

if all $2\mu^n$ configurations are equivalent. Then, the relaxation function $X(t)$ is the spin correlation function:

$$X(t) = N^{-1} \sum_{i=1}^{N} \langle S_i(0) S_i(t) \rangle = \sum_{n=0}^{\infty} g_n \exp(-t/\xi_n), \qquad (4.8)$$

where $N = \sum_{n=0}^{\infty} N_n (< \infty)$, $g_n = N_n/N (< 1)$ and $\xi_n = 2/w_n$, with w_n the probability (per sec) of a spin flip at the nth level. As well as the above, the continuum limit of Eq. (4.2) is important in the analysis of $X(t)$. As explained in [96], the realistic physics of dynamically constrained atomic motion suggests that μ_k do not exceed values of about 10 and $\xi_{max} = \lim_{n\to\infty} \xi_n$ should be finite though macroscopically large in a glass-forming liquid at $T > T_g$. It is also concluded that $X(t) \propto \exp(-t/\xi_0)$ at weak constraints, e.g., with $\xi_0 \gtrsim 10^{-12}$ s and that nonexponential, Kohlrausch-like relaxation takes place at

$$\xi_0 \ll t \lesssim \xi_{max}, \qquad (4.9)$$

while $X(t) \propto \exp(-t/\xi_{max})$ at $t \gtrsim \xi_{max}$ [the whole range $t \gg \xi_0$ is expected to be important at $T < T_g$ or at $T < T_0 < T_g$ with no finite ξ_g in Eq. (3.2)]. The parameter ξ_0 may, in fact, be macroscopic; the lower limit for the shortest relevant relaxation time for a configurational degree of freedom is $\xi_{0m} \approx 10^{-12}$–$10^{-13}$ s (too short a time for ξ_0 would not give rise to the Kohlrausch law for finite-size systems and realistic observation times).

The character of the relaxation in the class of models under discussion depends on the particular dependence of the two basic functions g_n and μ_n on n. A variety of simple assumptions on the form of g_n and μ_n are considered in [130]. The most interesting assumption for glass-forming liquids is considered to be

$$\mu_n = \mu_0 n^{-p}, \ \ p \geq 1 \quad \text{and} \quad N_n = N_0 \lambda^{-n}, \ \ \lambda > 1, \tag{4.10}$$

ξ_{max} being finite for $p > 1$. In the continuum limit of the case (4.10), it is obtained in the models [30] and [96] that $X(t)$ is of the Kohlrausch type in the range of times given by (4.9),

$$X(t) \propto \exp[-(t/\tau)^{\upsilon}], \tag{4.11}$$

with υ and ξ/ξ_0 being functions of the system parameters (μ_0, g_0, λ) only, not of the property (X) considered and υ close to $1/(1+\overline{\mu}_0)$, $0 < \upsilon < 1$. It is argued that the obtained dependence of X on t fits the KWW relaxation data well in the important range of t mentioned above (for representative values $g_0 = 1/2$, $\lambda = 2$ and $\mu_0 = 2$), with actual $\upsilon \simeq 0.563$. It is also assumed in [96] that ρ is a smooth, monotonously increasing function of T, in accordance with the growth of the number of decaying constraints with increasing T for a given time scale, and that $p(T) \geq p(T_0) = 1$. The result obtained is

$$\xi_{max}^{-1} \simeq \xi_0^{-1} \exp[-\varepsilon_0/(T - T_0)], \quad \text{for } T > T_0; \ \ 0, \quad \text{for } T < T_0, \tag{4.12}$$

where $\varepsilon_0 = \overline{\mu}_0/(dp/dT)_{T=T_0}$, so that $\xi_{max} < \infty$ at $T > T_0$, while $\xi_{max} \to \infty$ when T approaches a certain T_0 characteristic of the system. Then Eq. (4.11) would correspond to the empirical Kohlrausch relaxation if ξ_{max} is substituted for the Kohlrausch relaxation time τ. However, in Eq. (4.11), τ does not practically depend on T and differs essentially from ξ_{max} of Eq. (4.12) in the hierarchical model under discussion. Another problem of the models is to find explicit connections between the dynamic constraints, their parameters (β, τ) and microscopic features of atomic dynamics in actual glass-formers. In spite of these and other problems in the models, it seems that hierarchically constrained models may be relevant for describing slow KWW relaxations in glass-forming liquids near glass transition.

4.4 Concluding remarks

In conclusion, it appears that slow KWW relaxation processes observed in supercooled liquids near the glass transition (see, e.g., [2, 28]), at long enough times or low enough frequencies, can be associated not only with parallel relaxation but also with a hierarchy of correlated processes and states in a random inhomogeneous medium. On the other hand, in glasses ($T < T_g$), macroscopic Vogel-Fulcher relaxation processes corresponding to theoretically infinite relaxation times and to non-ergodicity of the system appear to be suppressed, though local relaxation processes can occur at long finite relaxation times. Moreover, the origin of KWW relaxation around T_g, with a Vogel-Fulcher behaviour of the average relaxation time, might be expected to relate to the origin of glass transition. In this connection, the mode-coupling theory for slow relaxation processes in simple liquids has also been developed. One can also ask whether such a theory gives rise to a consistent description of slow relaxations in a supercooled liquid with a Vogel-Fulcher average relaxation time.

In a series of cases, at least in dielectrics, two distinct slow relaxation processes are revealed at long enough times t (or low enough frequencies ν), which are usually referred to as the primary or α-processes and the secondary or β-processes (see, e.g., the review [11] (1992)). The simplest model of the α-relaxation was first used by Maxwell in his viscoelastic theory [18] and later applied by Debye for describing the relaxation of electric dipoles in polar liquids, $X_\alpha(t) \propto \exp(-t/\tau_\alpha)$, corresponding to KWW relaxation in the case at $\upsilon = 1$ in the function of Eq. (1.6). However, the stretching exponential effect in Eq. (1.6) at $1/2 \leq \upsilon \leq 1$ is still a remarkable feature of the α-relaxation at long enough times t or low enough frequencies ν. On the other hand, in the glass-forming liquids, resonance peaks in dielectric loss spectra have also been detected, of which characteristic frequencies are higher than the α-peak ones but lower than the typical frequencies in the band of microscopic excitations. The excitation resonances may be related to reorientational or vibrational motions of characteristic microscopic clusters in the glass-formers. It is worth adding again that both α- and β-relaxation spectra do not seem to show any pronounced anomaly at the calorimetric glass-transition temperature T_g.

PART II

ANOMALOUS LOW-ENERGY DYNAMICS OF GLASSES

5

ORIGIN OF ANOMALOUS LOW-ENERGY PROPERTIES OF GLASSES

The primary goal of the second part of the present book is to discuss a consistent description of the theoretical *soft-mode model*, the nature of atomic-motion *soft modes* and the role of the modes in anomalous properties of glasses, far below the liquid–glass transition. As seen in what follows, the properties (at least of a series of glasses) can be consistently described within the framework of this model, and some predictions can be made for experimental tests. The properties include the experimentally observed and studied dynamic and thermal properties of glasses at low frequencies $\nu \ll \nu_D$ and/or low temperatures $T \ll T_D = h\nu_D/k_B \equiv h\nu_D$, where ν_D and T_D are the effective Debye frequency and temperature respectively (as well as in part A, in what follows the Boltzmann constant $k_B \equiv 1$). These properties are often referred to as anomalous properties in the sense that they are not characteristic of the crystalline counterparts. The most widely known anomalous properties of glasses appear to be the T- and T^2-temperature dependence of specific heat and thermal conductivity respectively at very low temperatures, $T \lesssim 1\,\mathrm{K}$, discovered by Zeller and Pohl in 1971 [104], and the so-called "boson peak", a broad asymmetric peak in photon or neutron inelastic scattering intensity at a moderately low frequency shift (below referred to as frequency) $\nu = \nu_{BP} \approx 1\,\mathrm{THz}$, which is not characteristic of crystals. Since that time, the anomalous properties of glasses have been investigated in numerous works (e.g., [105, 106, 107, 109, 110, 111, 112]) and are referred to as universal properties in the sense that they are also weakly sensitive to differences in chemical composition and local atomic structures of the materials. Let us emphasize that the above does not contradict the experimental fact that in a few crystals, e.g., in the cristobalite polymorph of SiO_2, a peak is also observed in the reduced inelastic neutron scattering intensity, i.e. in the reduced vibrational density of states [7], at a noticeably higher $\nu = \nu_p$, e.g., at $\nu_D > \nu_p \gtrsim 2\,\text{THz} > \nu_{BP}$, which can be sometimes related to the lowest van Hove singularity of the crystal [113]. This and other differences between low-energy properties of glasses and those of respective crystals show that the anomalous properties under discussion in what follows characterize glassy, amorphous systems, rather than crystals. In other words, the amorphous, glassy state of the system in general appears to be a necessary condition for the boson peak existence.

Taking into account the above-mentioned primary goal and, on the other hand, the very large number of investigations of the anomalous dynamic and thermal properties of glasses, the present review also describes, for comparison, other (not explicitly related to the soft-mode model) recent theoretical models of the glassy properties. The soft-mode model appears to be the only model that is able to describe, qualitatively and in scale at least, the glassy properties at both very low and moderately low frequencies and temperatures. In the section below, we concentrate on the soft-mode model and anomalous, universal properties of nonmetallic (insulating and semiconducting) glasses, for which the role of electronic excitations and their interaction with other excitations are not so important (for metallic glasses the properties are considered in detail in a number of reviews, in particular, in [8, 9]).

In accordance with a general concept of solid state theory [115] that explains the properties of a material as due to the existence of appropriate elementary excitations and their interactions, the deviations of dynamic and thermal properties of glasses from Debye ones of crystalline materials at low $\nu \ll \nu_D$ and $T \ll T_D$ can be attributed to the occurrence of excess excitations of low energy $\varepsilon \ll h\nu_D$, which are essentially different from the ubiquitous low frequency acoustic phonons. At the same time, the nature of excess low-energy nonacoustic excitations has been unclear for a long time. In the very-low-energy limit, at $T \lesssim T_m = h\nu_m \approx 1$ K and $\varepsilon \lesssim \varepsilon_m = h\nu_m$, a commonly accepted "standard tunneling model" (STM) was proposed by P. Anderson, Halperin and Varma [116] and W. Phillips [117] in 1972 (Sec. 9.1), in which so-called "two-level system" (TLS) excitations or "tunneling-state" excitations (related to atomic tunneling processes) are essential and give rise in particular to the observed linear temperature variation of the specific heat. In fact, the TLSs are considered as "pre-existing defects" in the glass structure (dynamics), between which universal weak elastic interactions are expected to occur, while the STM is relevant at such values of ε (or T) at which the contribution of interactions to glass properties can be negligible (see below). However, the microscopic nature of the TLS excitations is still unclear and even less is known on the excess vibrational excitations of moderately low frequencies $\nu \approx 1$ THz. The latter might be related to the boson peak and to the observed maximum ("bump") of the reduced specific heat $C(T)/T^3$ at moderately low $T \sim 30$ K, which have also been investigated in a lot of experiments [105, 107]. The results have shown that the "excess" vibrational excitations coexist and interact with the Debye acoustic phonons. Some detailed, but material dependent models of the excess low-energy excitations have been developed [105], e.g., with the coupled rotation of SiO_4 tetrahedra for vibrational excitations determining the boson peak in v-SiO_2 [118]. Since the anomalous properties due to the excess excitations were observed practically in all glasses studied, more general theoretical models of glass atomic dynamics at low excitation energies $\varepsilon \ll h\nu_D$ have been developed, including the soft-mode model [119], the "defect-interaction model" [120] and the "fractal model" [121].

In the latter, a fractal was defined as a self-similar structure (like, e.g., a percolating network) of different length scale $L_{fr} \gg a_1$. The fractal model suggested an explanation of the origin of excess vibrational excitations: a crossover was predicted at a critical frequency from standard Debye phonons, occurring at frequencies lower than a critical one, to "fractons", localized vibrational excitations of higher frequencies in fractals. In this sense, the vibrational DOS in the model was found to contain the fractons as excess, non-Debye, states. It was assumed in this model that the glass structure consisted of fractals, which eventually prevented the propagation of short-wave excitations of length $\lambda_{exc} < L_{fr}$. However, in this model, some questions remained unanswered, e.g., whether and how one could map a continuous glassy random network of particles with its medium-range order (MRO) structures of a typical length L_{MRO} onto the self-similar structure of fractals. Therefore, and because the basic experimental data for the specific heat and thermal conductivity at very low $T \lesssim 1\,\text{K}$ (Ch. 6) cannot be explained in this model, the fractal model is not discussed in more detail in what follows.

On the other hand, the defect-interaction model suggested that in a glass, universal dipolar elastic interactions between effective structure "defects" essentially contributes to the excess low-energy excitations, determining the experimentally found essential deviation of the behavior of specific heat and thermal conductivity in glasses from the Debye one at low $T \ll T_D$. In this connection, a recent advanced model of TLS excitations and related dynamic and thermal properties of glasses at very low $T \lesssim 1\,\text{K}$ [122], which takes into account the contribution of both the "pre-existing" TLS defects and their relatively weak universal elastic interactions with each other, is an essential extension of both the STM and the earlier defect-interaction model [120]. In the models relevant at very low $T \lesssim 1\,\text{K}$, as well as in the soft-mode model at very low excitation energy (Sec. 8.1), the viewpoint appears to be that the contribution of the TLS excitations in the limit case of vanishing interactions, which is described by STM, and is most essential for the glass dynamics at not too low ε, $T'_m < \varepsilon \lesssim \varepsilon_m \ll h\nu_D$, where the experimental value $T'_m \sim 0.01$ K rather is implied (Sec. 9.1). At lower $T < T'_m$, new types of excitations and related dynamic phenomena in glasses can be predicted which are due to weak elastic interactions (Sec. 9.2).

The theoretical concept of soft modes as nonacoustic atomic motion modes, generally characterized by anharmonic potential and their low-energy excitations, acts as the base of the original soft-mode model. It was proposed in 1979–1980 [119]. Then, it was developed and extended in 1980–1983 [123] and it was described in more detail and applied in some papers and reviews (see, e.g., [124, 125, 126, 127, 128]) for explaining universal dynamic and thermal properties of glasses at low $\nu \ll \nu_D \sim 5\,\text{THz}$ and/or low $T \ll T_D \sim 5 \cdot 10^2\,\text{K}$. Although the soft modes were expected by their origin to occur for a minority of atoms in any disordered solid, the low atomic concentration $c_{sm}(\ll 1)$ of soft modes was argued to reach its upper limit of a universal scale just in glasses. In particular, the soft-mode excitations at very low T and are shown (Sec. 8.1) to be the TLS or tunneling-state (TS) postulated earlier in the standard tunneling model [116, 117]. As can be concluded from what follows, the existence of numerous soft atomic modes in an amorphous solid can be considered as a universal feature of glass at temperatures considerably lower than the liquid–glass transition temperature. By definition, a soft mode in a glass is introduced as an atomic-motion mode characterized by a low effective frequency, $\nu \ll \nu_D$ and a small effective spring constant, $k \ll k_0$, which coexist and interact with low-frequency acoustic phonons. As well as in crystalline materials, the low-frequency phonons are identified as Debye phonons associated with atomic-motion modes for the vast majority of atoms with standard scale spring constants $k \approx k_0$ (e.g., for covalent or ionic bonding, $k_0 \approx 10\text{–}20\,\mathrm{eV/Å^2}$). Similar to the viewpoint of STM, the basic concept of the soft-mode model (and of the present survey) is that the contributions of soft-mode excitations and their weak interactions with acoustic phonons, as well as the resulting weak elastic interactions between the soft modes, generally speaking, can be essential in glassy dynamics at low $\{\varepsilon, T\} \ll h\nu_D$. Due to the coupling to elastic and possible, electric fields, the soft-mode excitations can change both dynamic and thermal properties of glasses. In particular, the energy of a soft-mode excitation can be modified by an external elastic field, giving rise to a transition of the excitation into another state, with a finite phase shift compared to the field phase [105]. The result is that in a glass the attenuation of acoustic or electromagnetic waves, as well as the macroscopic sound velocity s_0 and dielectric susceptibility ϵ_0, can be changed due to interactions between the excitations and waves. In particular, the "bare" sound velocity s_0 changes to a complex quantity $s = s' + is''$, where the real part s' corresponds to the reactive response while s'' to the dissipative behavior of the system [106]. In what follows the properties of glasses are discussed, in which a deviation from thermal equilibrium is small, in the usual sense that kinetic characteristics like a thermal current are linear in appropriate thermodynamic forces proportional to the deviation of thermodynamic parameters (e.g., temperature) from the equilibrium values. In other words, the low energy properties of (almost) thermal-equilibrium glasses are mainly discussed, in which most numerous thermal excitations of energy $\varepsilon \approx T$ are of basic importance.

At very low temperatures, $T \lesssim 1\,\mathrm{K}$, and/or appropriate frequencies, the universal dynamic properties of both thermal-equilibrium and non-equilibrium glasses have recently been discussed in an excellent review

[122], in which the concept of dipolar elastic interactions between structural "defects" is significantly developed (the soft mode concept is not explicitly applied). Earlier several comprehensive reviews of experimental works have been published [105, 106, 107, 110, 111], concerning anomalous very-low-energy properties of nonmetallic glasses and their theoretical interpretation. Taking this into account, we focus on recent experimental data, in particular for moderately low energies, and on their theoretical analysis within the framework of the soft-mode model. It can be concluded from what follows that the soft mode excitations can be expected to determine in general the dynamic and thermal properties at very low excitation energies (Ch. 8), corresponding to very low ν and T. The situation is more complex at moderately low frequencies and temperatures: the soft-mode excitations, interacting with Debye phonons, appear to be decisive for the origin of boson peak (BP) in many glasses, in particular SiO_2, in which the BP is followed by the so-called high-frequency sound [112], as well as for the origin of this sound (Ch. 10). However, the soft-mode excitations are hardly decisive for the origin of BP in glasses in which no high-frequency sound is observed above the BP (e.g., in B_2O_3).

The content of the second part of the book is as follows. The basic experimental data concerning dynamic and thermal properties of (mostly almost thermal-equilibrium) glasses at low ν and T are discussed in Ch. 6, focusing on recent experiments. The main purpose of this book is to consider and summarize the basic results of the soft-mode model of low-energy excitations, actually with Anderson non-localized states, and to show that the model is able, from a unified viewpoint, to describe qualitatively and scale-wise, the universal anomalous dynamic and thermal properties of (nonmetallic) glasses, at all low temperatures and frequencies far below the liquid–glass transition temperature. The soft-mode model (Chs. 7–11) is shown to be an important generalization to all low $T < T_D$ and $\nu < \nu_D$ of the standard tunneling model (Sec. 9.1) that characterizes the anomalous properties only at very low $T \lesssim T_m \approx 1\,\mathrm{K}$ and $\nu \ll \nu_m = T/h$. Other recent models of the low-energy dynamic and thermal properties, not explicitly related to low-energy soft-mode excitations, are considered in Chs. 12–13 and compared to the soft-mode model properties. On the other hand, anomalous electron phenomena have been experimentally discovered in semiconducting glasses (Ch. 14), which were also theoretically described by applying, and in the framework of, the soft-mode model (Ch. 15). Moreover, the soft-mode model also predicted (Ch. 16) some anomalous changes of semiconducting glass properties (Ch. 16) both under moderately high hydrostatic pressures and/or under light of frequency $\nu \approx \nu_g$ corresponding to the interband, mobility-gap width $E_g = h\nu_g$ of the scale of 1–3 eV. Finally, some conclusions are suggested in Ch. 17.

The basic issues of theoretical derivations are mainly expounded in the book and only references are given for mathematical details.

6

EXPERIMENTAL BACKGROUND FOR ANOMALOUS LOW-ENERGY ATOMIC DYNAMICS

One of the well-known basic results of atomic dynamics of a crystalline lattice is that its atomic vibrations can be described as superpositions of "normal vibrational modes", which are the same plane waves as in an appropriate elastic continuum for long wavelengths $\lambda(\nu) \gg a_1$ [113]. This model, called "Debye model", is successful at interpreting the experimental data for the specific heat $C(T) \propto T^3$ of crystalline solids at low $T \ll T_D$, which is due to contributions of low-frequency acoustic phonons with $\lambda(\nu) = \lambda_{ph}(\nu) = s_0/\nu_1$. However, numerous experimental data demonstrate that the specific heat and other dynamic and thermal properties of (almost) equilibrium glasses at low $T \ll T_D$ are dramatically different from those of respective crystalline materials. Moreover, dynamic and thermal properties of glasses at very low $T \lesssim T_m \approx 1$ K and frequencies of applied acoustic (electromagnetic) waves, $\nu \lesssim \nu_m$, are essentially different from those at moderately low T and ν, $10 \lesssim T \lesssim T_M \sim 10^2$ K and $0.1 \lesssim \nu \lesssim \nu_M \sim 1$ THz (as noted above, in what follows temperature units are used, in which the Boltzmann constant $k_B \equiv 1$). Moreover, dynamic and thermal properties of various types of glasses at low $T \ll T_D$ and/or $\nu \ll \nu_D$ can be characterized by similar temperature/frequency dependencies and scale of magnitude.

6.1 Very low temperatures and frequencies

One of the most well-known examples of the difference of dynamic and thermal properties of glasses from those of their crystalline counterparts

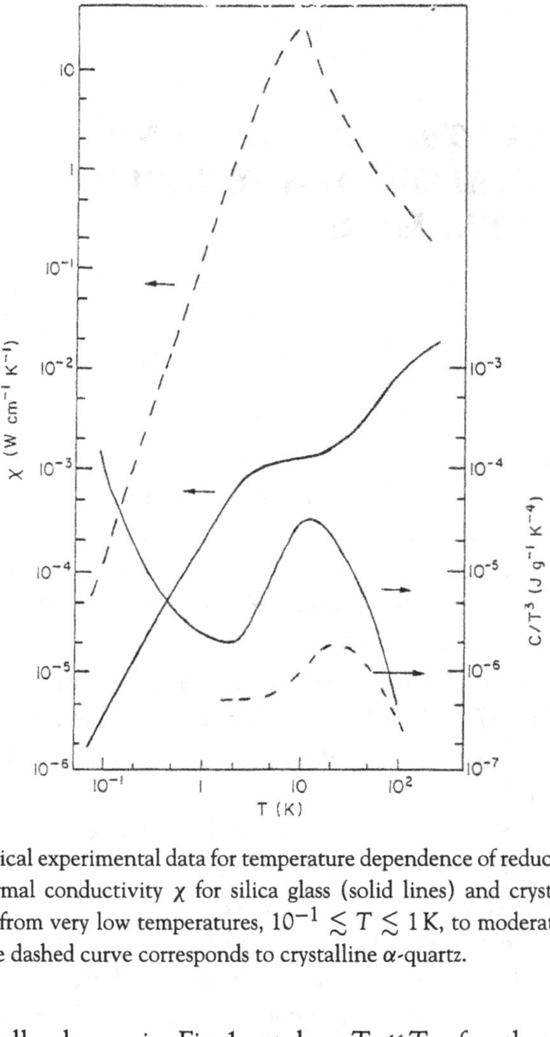

Figure 1: Typical experimental data for temperature dependence of reduced specific heat C/T^3 and thermal conductivity χ for silica glass (solid lines) and crystalline α-quartz (dashed lines), from very low temperatures, $10^{-1} \lesssim T \lesssim 1$ K, to moderately low $T, 1 < T \lesssim 10^2$ K. The dashed curve corresponds to crystalline α-quartz.

is schematically shown in Fig. 1, at low $T \ll T_D$ for the temperature dependence of $C(T)$ of a nonmetal, e.g., silica glass (v-SiO_2, solid curve), which is considerably larger than that of the crystalline counterpart (c-SiO_2, dashed curve). At very low (and not too low) $T, 0.1 < T \lesssim 1$ K, the specific heat of heat-carrying acoustic phonons in a glass is well approximated by the following expression [104, 107, 108]:

$$C(T) = a^{(1)}T^{1+\sigma} + a^{(3)}T^3 \simeq a^{(1)}T^{1+\sigma}, \quad 0.1 \lesssim \sigma \leq 0.3, \qquad (6.1)$$

where the coefficients $a^{(1)}$ and $a^{(3)} \approx T_D^{-3}$ are finite constants, whereas $a^{(1)} \equiv 0$ and the Debye term $C_D = a^{(3)}T^3$ accounting for the contribution of low-frequency acoustic phonons is only important for (non-metallic) crystalline materials. With such typical $\sigma \ll 1$ (e.g., $\sigma = 0.22$ for Suprasil and $\sigma = 0.30$ for Suprasil W), the temperature dependence of the specific heat is close to a linear one. This dependence was first explained by the standard tunneling model (STM, in what follows) [116, 117], assuming that a glass contains, in addition to the ubiquitous Debye phonons (harmonic excitations), also essentially anharmonic excitations (Chs. 8, 9). The latter are characterized by very low energies $\varepsilon \lesssim \varepsilon_m \equiv h\nu_m$ and might be approximated as randomly distributed local two-level systems (TLS, in what follows), each being described by only two energy levels, E_0 (ground state) and E_1 (excited state) at excitation energy $\varepsilon = \Delta E \equiv E_1 - E_0 > 0$. As far as the excitations were independent (not interacting with each other), their total density of energy levels per unit volume and per unit energy interval, i.e., the density of states (DOS) $G^{(TLS)}(\varepsilon)$ was assumed to be finite and uniform, $G^{(TLS)}(\varepsilon) \approx G^{(TLS)}(0) > 0$, determining the specific heat $C(T) \approx a^{(1)}T$ for very low T. However, unlike the Debye parameter $a^{(3)}$, the coefficient $a^{(1)}$ was predicted and then found in experiment, to be a material constant $a^{(1)} \approx G^{(TLS)}(0) \approx 3 \cdot 10^{20}$ eV^{-1} cm^{-3} if thermal TLS excitations (at $\varepsilon \approx T$) were characterized by an internal relaxation time $\tau_1(T)$ that is small compared to the measurement time t_m. At larger $\tau_1(T)/t_m > 1$, the specific heat was observed to logarithmically depend on t_m, $a^{(1)} \propto f(\ln t_m)$, while this regular function becomes a constant at $t_m \gg \tau_1$, in qualitative accordance with the STM prediction (Sec. 9.1).

Another property of essential interest at very low T is the thermal conductivity $\kappa(T)$ of nonmetallic glasses schematically shown also in Fig. 1 (v-SiO_2, solid curve), which is commonly accepted to be due to acoustic phonons as heat carriers. It is found to be considerably smaller than that of a crystalline counterpart (c-SiO_2, dashed curve), at both very low (and not too low) $T, 0.1 < T \lesssim 1$ K, and moderately low T, $5 < T \lesssim 10^2$ K. As usual, the property can be described approximately as $\kappa(T) = (1/3)C_D(T)s_0 l_{ac}(T) \propto T^3 l_{ac}(T)$, in terms of the phonon mean-free-path $l_{ac}(T)$ of which the typical behavior in glasses (and crystals) is schematically shown in Fig. 2 [130]. In particular, at $0.1 < T \lesssim 1$ K, the

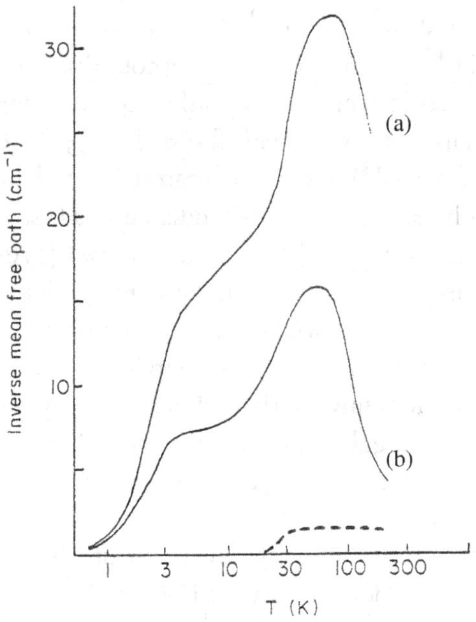

Figure 2: Typical experimental data for temperature dependence of inverse mean-free-path (absorption) of longitudinal acoustic phonons in SiO_2 glass at low $T \ll T_D$ (solid curves): (a) $2\pi\nu = 0.93$ GHz; (b) $2\pi\nu = 1$ GHz.

measured $\kappa(T)$ can be approximated as follows [104, 111]:

$$\kappa(T) = a^{(2)} T^{2-\rho}, \quad \text{at } 0.1 \lesssim \rho \lesssim 0.3, \tag{6.2}$$

increasing monotonically with T and corresponds to $l_{ac}^{-1}(T) \propto T^2$. Let us add that, unlike the specific heat, the kinetic and dissipative properties (e.g., thermal conductivity and wave absorption) of glasses are not observed to considerably depend on the measurement time t_m at very low T and ν.

It is worthy to note that the temperature variation and the scale of both specific heat and thermal conductivity depend relatively weakly on chemical nature and specific short- and medium-range order of glasses in the sense that they may differ by a factor in the range of $\approx 1-2$.

It has been shown in [131] that the thermal conductivity and mean-free-path of phonons in different glasses, when scaled by a single material parameter, have quite similar temperature dependence and magnitude from the very low temperatures, $0.1 \lesssim T \lesssim 1$ K, yet measured at that time, to around the melt temperature. The scaling parameter for temperature was an effective Debye temperature $T_D = h\nu_D/k_B \equiv h\nu_D$, with the highest phonon frequency ν_D that in principle might be estimated from the known atomic (molecular) composition of a glass, while the scaling parameter for thermal conductivity was chosen to be $\kappa^0 = 4\pi T_D^2/h^2 s_0$. Then the kinetic expression for $\kappa(T)$ was transformed to the scaled, dimensionless form:

$$\kappa/\kappa^0 = (T/T_D)^3 \int_0^{T_D/T} dx \cdot [T_D l_{ac}(xT)/s_0 h] x^4 \exp(x) \cdot [\exp(x) - 1]^{-2}$$

$$\propto \int_0^{\nu_D} d\nu g(\nu) D_0(\nu) \cdot x^2 \exp(x)[\exp(x) - 1]^{-2}, \qquad (6.3)$$

where $x = h\nu/T$. The related "energy diffusivity" $D_0(\nu) \propto l_{ac}(\nu)$ can be considered as a classical property as far as anharmonic interactions of phonons are unimportant at not too low temperatures. Strictly speaking, the applicability of this expression to glasses is debatable because there is no well-defined procedure for computing the effective Debye temperature. Therefore, the latter can be considered as an adjustable scaling parameter by varying the number N of effective atomic cells in the standard expression $T_D = hs_0(3N/4\pi)^{1/3}$ for crystalline materials, for bringing the data of different glasses into a close register on a plot of κ/κ^0 versus T/T_D. For calculating the thermal conductivity due to (almost) thermal equilibrium acoustic phonons from Eq. (6.3), it is useful to take into account [113] that in general, the inverse mean-free-path $l_{ac}^{-1}(\nu) = (l_{ac}^{(el)}(\nu))^{-1} + (l_{ac}^{(in)}(\nu))^{-1}$ at $h\nu = h\nu_T \equiv T$, where $(l_{ac}^{(el)})^{-1}$ and $(l_{ac}^{(in)})^{-1}$ are contributions of elastic (*el*) and inelastic (*in*) scattering of the phonons. For example, in non-perfect crystalline materials containing static defects, $(l_{ac}^{(el)}(\nu))^{-1} = l_b^{-1} + l_R^{-1} = C_b + C_R\nu^4$ due to scattering from the sample boundary, $C_b = const.$, and the Rayleigh scattering with long wavelengths $\lambda_{ac} = s_0/\nu \gg L_d(> a_1)$ from the defects of a typical size L_d. Mechanisms of inelastic scattering phonons are also involved here which are due to interaction of phonons with some excitations responsible for the "excess", non-Debye, specific heat in glasses at low temperatures. The "excess" excitations are either anharmonic non-vibrational ones related to the TLS introduced in STM, at very low energies or temperatures (Chs. 8 and 9), or harmonic vibrational excitations at moderately low ones (Ch. 10). Generally speaking, the contribution $(l_{ac}^{(in)}(\nu))^{-1}$of inelastic scattering can be approximated by a superposition of power-law functions, $(l_{ac}^{(in)}(\nu))^{-1} = \sum_\alpha C_\alpha \nu^\alpha$, where C_α are material dependent constants, at $1 \leq \alpha \leq 4$. In particular, at very low $\nu \lesssim \nu_m$ and $T \lesssim T_m \approx 1\,\text{K}$, $(l_{ac}^{(in)}(\nu))^{-1} \approx const. \cdot \nu \gg (l_{ac}^{(el)})^{-1} \approx C_R \nu^4$ and $\kappa(T) \propto C_D(T) l_{ac}(\nu_T) \propto T^2$ [116], in approximate agreement with Eqs. (6.2) and (6.3), at $C_D(T) \propto T^3$ (for a quite different behavior of $l_{ac}^{(in)}(\nu, T)$ and $\kappa(T)$ at moderately low ν and T, see Ch 10). On the other hand, in [131], the measured temperature dependence of thermal conductivity of six different glasses, two oxide (SiO_2, B_2O_3) and four polymeric (PB, PET, PS, PMM) were transformed into scaled data for κ/κ_0 versus T/T_D. As revealed in this investigation, three distinct temperature regimes are seen from the experimental data with boundaries around $T/T_D \approx 10^{-2}$ and 10^{-1}. In what follows, only the regions $T/T_D \lesssim 10^{-2}$ and $10^{-2} < T/T_D \lesssim 10^{-1}$ are under discussion (at very high $T/T_D \gtrsim 1{-}10$, the type of excitations carrying heat does not seem to be well-defined, though there are some experimental and theoretical indications that the excitations may still be like acoustic ones in the sense that they are related to sound waves, which can persist even the melt).

At $T/\Theta \lesssim 10^{-2}$, i.e., at very low $T \lesssim T_m \approx 1\,\text{K}$, the phonons are certainly and are well-defined heat carriers, corresponding to large mean-free-path l_{ac}, e.g., $l_{ac} = 150\lambda_{ac}$ and are supposed to be determined by weak inelastic scattering due to "resonant" interaction of a phonon of energy $h\nu$ with anharmonic TLS excitations. Furthermore, it is found in [132] that in all measurements published (at least before 2002) on the thermal conductivity and acoustic attenuation at very-low temperature or frequency ($\nu = \nu_T$) on a total of over 60 different compositions of glasses (amorphous solids), the ratio of the acoustic phonon wavelength to mean free path can be characterized by a universal scale, $l_{ac}^{-1}\lambda_{ac} \approx 10^{-3}{-}10^{-2}$. By applying the standard deformation-potential interaction [133] $V_{DP}(x) \approx \beta x e_{ac}$, with β the coupling constant, x the TLS coordinate and $e_{ac}(\ll 1)$ the weak acoustic strain field, the acoustic phonon mean-free-path $l_{ac} = l_{ac}^{(res)}$ for "resonant" scattering, with an absorbed phonon and emitted TLS excitation of close energy $\varepsilon \simeq h\nu$ or vice versa, was estimated in STM to be $l_{ac}^{(res)} \approx l_0 \cdot (\nu_D/\nu) \coth(h\nu/2T) \gg l_0$ at $\nu \lesssim \nu_m \sim 10^{-3}\nu_D$, where $l_0 \approx M(2\pi\nu_D)^2/G^{(TLS)}(0)\beta^2$ is a material constant and M is an average atomic mass. The mean-free-path determines at appropriate very low ν and T both the thermal conductivity $\kappa(T)$ due to acoustic phonons and the ultrasonic attenuation $\alpha_{ac}(\nu)$. Comparing an estimate of l_{ac} which can be obtained from the experimental data for κ at very low $T \lesssim 1$ K, with the theoretical estimate of $l_{ac} = l_{ac}^{(res)} (\gg l_0)$ and the empirical value $G^{(TLS)}(0) \approx 3 \cdot 10^{20}\,\mathrm{eV}^{-1}\,\mathrm{cm}^{-3}$, one can assess the coupling constant scale, $|\beta| \sim 1$ eV [106, 116, 122]. Thus $|\beta|$ is found to be much larger than the Debye energy $h\nu_D \sim 0.03$ eV that determines the energy scale for acoustic phonons, in particular their interaction with standard localized vibrations.

The universality of glass dynamic properties is also observed in numerous experiments at very low T and ν, which characterize the behavior and scale of changes of the properties in an external acoustic or electromagnetic wave field. The macroscopic response of implied excess, non-Debye, excitations to such a field can be described by a complex, elastic or electric, parameter $z = z' + iz''$ and generalized equations of motion (e.g., [106, 122]), of which solutions can be applied for analyzing experimental data. In acoustic experiments, usually the change δs of sound velocity and the related dissipative characteristic are measured, which are linked correspondingly to the real part of the complex elastic parameter, $\delta s(\nu, T) = s'(\nu, T) - s_0 = -\rho s_0^3 z'_{el}/2$, and to its imaginary part, the so-called internal friction $Q^{-1} = \rho s_0^2 z''_{el}$, where $s_0 \equiv s(\nu = 0)$ and ρ is mass density. In electromagnetic experiments for very-low-ν photons, the change of dielectric susceptibility, $\delta\epsilon'(\nu, T) = \epsilon'(\nu, T) - \epsilon_0$ at $\epsilon_0 \equiv \epsilon(\nu = 0)$, and the change of dielectric losses and loss tangent change $(\Delta \tan(\delta))\ldots$ are generally measured. Experimental data show a similarity, within the sign, in temperature dependence of the relative change in the real part of sound velocity, $\delta s/s_0$, and dielectric susceptibility, $\delta\epsilon'/\epsilon_0$, for v-$SiO_2$, for example, in typical curves for $\delta s(\nu, T)/s_0$ and $\delta\epsilon'(\nu, T)/\epsilon_0$ at realistic $\nu \approx 1 \div 10$ GHz. In fact, the relative change of sound velocity is a nonmonotonic function of T and weakly depends on ν with increasing T, rising first logarithmically, exhibiting a maximum at a temperature dependent frequency $\nu_{max}(T)$ and finally also decreasing logarithmically, whereas the relative change of dielectric susceptibility, $\delta\epsilon'(\nu, T)/\epsilon_0$, exhibits a minimum instead of a maximum around the frequency $\nu_{min}(T)$. In particular, the temperature dependence of $\delta s(T)/s_0$ at a given low enough $\nu = const. \ll 1$ THz can be

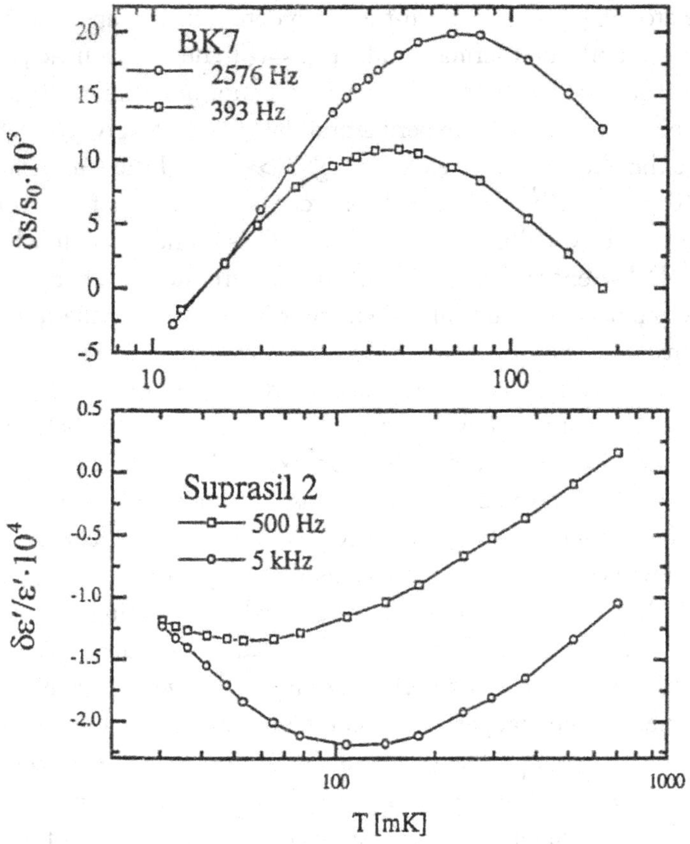

Figure 3: Typical experimental data for temperature dependence of relative change for sound velocity, $\delta s/s_0$ (for Suprasil 1 glass v-SiO_2), at very low $T \lesssim 1$ K and (given) frequency $\nu \simeq 0.1$ GHz $\lesssim \nu_m$.

described approximately as (Fig. 3):

$$(s - s_0)/s_0 \equiv \delta s(\nu = const., T)/s_0 = B(T) \ln (T/T_0), \tag{6.4}$$

with $B(T) \approx B = const. > 0$ at sufficiently low $T \ll h\nu$ while $B(T) \approx -B/2$ at sufficiently high $T(\gtrsim h\nu$, if still $T \lesssim 1$ K). Here, T_0 is an arbitrary reference temperature in the experimental range of interest and the parameter B seems to depend weakly on chemical nature and specific short- and medium-range order of glasses. A similar expression, with $-B$

substituted for B, appears to be able to approximate similarly the variation of $\delta\epsilon'(\nu, T)/\epsilon_0$ with increasing T, at a given ν.

At first glance, Eq. (6.4) seems to be relevant for describing the experimental results. However, a more precise observation reveals that the ratio of the slopes in the experimental semi-logarithmic plot for $\delta s(T)/s$ vs $\ln(T/T_0)$ at least at ultra-low $T < 0.1$ K is not 2:(– 1) but rather 1:(– 1) (see in Fig. 1 and especially have also been in Fig. 2 of [134]).

Dissipative effects in glasses measured in numerous works at low $T \ll T_D$ and $\nu \ll \nu_D$, particularly at very low $\nu \ll 1$ THz. The typical temperature dependence of l_{ac}^{-1} for low-intensity longitudinal acoustic waves at both moderately low and very low T and a given low $\nu \ll \nu_D$ is presented in Fig. 2. An example of important dissipative effects is the absorption coefficient $\alpha_{ac} = l_{ac}^{-1}$ of acoustic phonons or of electromagnetic waves (photons) at $\alpha \equiv \alpha_{em} = l_{em}^{-1}$. For phonons of very low ν, the useful equivalent dimensionless parameter is the "internal friction", $Q^{-1} = \alpha_{ac}\lambda_{ac}$, with the wavelength much shorter than the mean-free-path at low $\nu \ll 1$ THz. Since the behavior of dissipative properties and reactive response ones (like the relative change $\delta s(\nu, T)/s_0$) are related to each other by the well-known Kramers-Kronig relations [135, 136], and both types of properties are similar for acoustic and electromagnetic waves, there is a close analogy between the acoustic and dielectric properties [105, 106, 107]. In fact, the identity of excitations giving rise to acoustic or electromagnetic absorption was explicitly established by cross experiments in which both types of waves act in turn upon a glass sample [106]. In what follows, the acoustic properties are considered in detail. In the range of very low T and ν, typical experimental data are shown in Fig. 4 for the internal friction as a function of ν, T and of the wave intensity I, both at low $I \ll I_c$ and at higher $I \gtrsim I_c$, with a "critical" wave intensity I_c being a material constant. For the very low $T \lesssim 1$ K under discussion, the experimental results can be approximately described as follows [106, 122, 134]. Two limiting cases can be distinguished, one at sufficiently high $T/h\nu \gtrsim 1$ and another at sufficiently low $T/h\nu \ll 1$. In the former case, $Q^{-1}(\nu, T; I)$ is observed to be practically independent of I and to exhibit a "plateau" as a function of both ν and T:

$$Q^{-1}(\nu, T; I) \simeq Q^{-1}(\nu, T) \equiv Q^{-1}(\nu, T; I = 0) \approx const. \qquad (6.5)$$

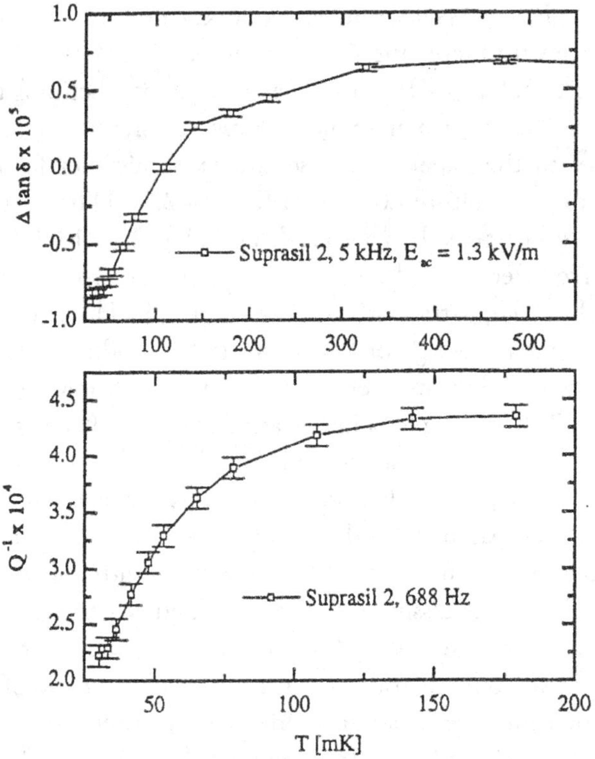

Figure 4: Typical experimental data for temperature dependence of internal friction Q^{-1} of sound waves and of dielectric absorption α of electromagnetic waves, at very low $T \lesssim 1\,\mathrm{K}$ and (given) very low frequency $\nu = 10\,\mathrm{GHz} \lesssim \nu_m$.

In the latter case, in contrast,

$$Q^{-1}(\nu, T; I \ll I_c) \simeq Q^{-1}(\nu, T) \propto \nu^{-1} T^{\alpha}, \tag{6.6}$$

for low $I \ll I_c$, with $1 \leq \alpha \leq 3$ when increasing temperature from ultra low $T < 0.1$ K to $T \lesssim 1$ K. In this case for high $I \gg I_c$, $Q^{-1}(\nu, T; I \gtrsim I_c)$, it is found in a rather good approximation to decrease with growing I as $I^{-1/2}$. The effect of a considerable decrease of $Q^{-1}(\nu, T; I \gtrsim I_c)$ with growing I is called "saturation" and gives rise to a reduction of acoustic loss, corresponding to an effective "hole" of finite width $\Delta\nu$ burnt into the attenuation spectral distribution around the phonon energy $h\nu$ of the applied acoustic wave. The typical scale of I_c can be found from experiments, for example, $I_c \approx 10^{-6}$ w/cm^2 for v-SiO$_2$ at $\nu = 550$ MHz, $T = 25$ K [137]. It is worth adding that the nonlinear behavior of Q with increasing I at very low ν and T cannot be characteristic of absorption of phonons (photons) by some harmonic excitations and thus, indicates that extra anharmonic excitation of very low energy $\varepsilon(\lesssim h\nu_m)$, e.g., two-level system ones, exist. For simplifying the explanation of the behavior of the dynamic properties at very low T and ν, the relatively high T corresponds to $\nu\tau_1(T) \ll 1$ while the relatively low T to $\nu\tau_1(T) \gg 1$, where $\tau_1(T)$ is the above-mentioned characteristic internal relaxation time of thermal anharmonic excitations related to the two-level systems [116, 117] (Secs. 8.1 and 9.1). In particular, if one supposes that $\tau_1^{-1}(T)$ is proportional to the density of most numerous thermal Debye phonons interacting with independent anharmonic excitations or, equivalently, to the Debye specific heat $C_D(T) \propto T^3$, then, $\alpha = 3$ could be expected in Eq. (6.6). However, as can be seen from recent experimental data, the increase of the internal friction appears to be less steep than T^3, $\alpha < 3$, at ultra-low temperatures (Sec. 9.2).

Many other experiments [106, 134, 137] at very low temperatures show a saturation effect in acoustic absorption similar to that mentioned above, when a high intensity pulse of time t and phonon energy $\varepsilon_p = h\nu$ is applied to a glass. The saturation effect in absorption of such an intense acoustic pulse actually gave the first evidence of the important role of interactions between the anharmonic excitations, two-level systems similar to 1/2 spins, in the absorption. In the original "hole burning" experiment performed on a common optical BK 7 by Arnold and Hunklinger [106] (1975) and illustrated in Fig. 5, the intense acoustic pulse was applied for equalizing the occupations of anharmonic two-level system excitations of energy close to that of the pulse, so that the excitations could no longer effectively absorb the radiation of almost resonant frequency. This nonlinear effect can be interpreted as a generation of a "hole" of width $h\Delta\nu$ created in the occupation of energy levels of the two-level systems. If a second, weak, probing pulse of time t and slightly different energy $\varepsilon_p' \approx \varepsilon_p = h\nu$ is applied immediately afterwards, measurements of its attenuation as a function of frequency difference for both pulses can reveal the spectral character of the hole created by the intense pulse. Then, the hole width $h\Delta\nu$ is observed to increase linearly with growing T and with increasing weak pulse time t_p,

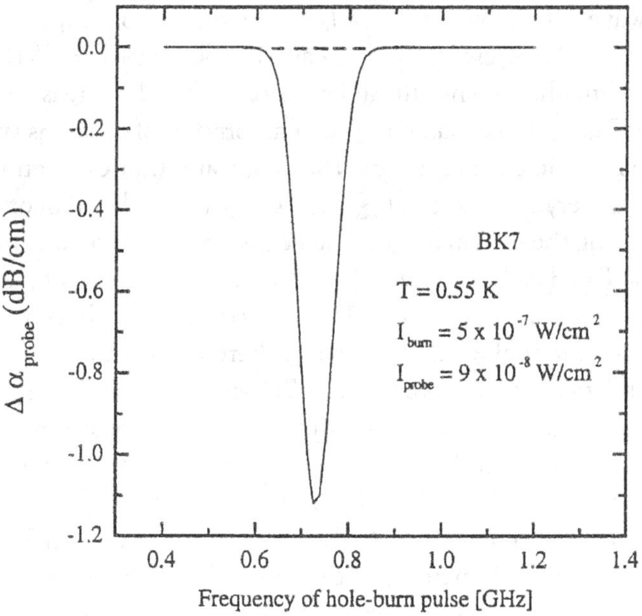

Figure 5: Typical behaviour of high-intensity acoustic-absorption hole burning at GHz scale wave frequencies and given very low $T = const. \lesssim 1$ K: variation of absorption of extra, low-intensity and acoustic pulse.

so that finally, a broad hole occurs that is much wider than that expected from the pulse spectrum [122, 137]. The hole broadening mechanism was called "spectral diffusion", by analogy with the mechanism known from magnetic resonance experiments and was theoretically analyzed to be in agreement with experimental data [138].

Another spectral diffusion effect was observed in the investigations of spontaneous and stimulated coherent effects of acoustic echoes generated in glasses (e.g., fused silica glass) at ultra-low $T \ll T_m \approx 1$ K by a sequence of short pulses, typically of time interval 100 ns and of GHz scale frequency below $T \approx 0.1$ K (see, e.g., [110, 137] and references therein). The dependence of two-pulse (spontaneous) and three-pulse (stimulated) echo amplitude on the amplitude, frequency, width and separation time of the generating pulses have also been studied in a number of glasses for which the typical value and dependence of echo decay time t_d on T were measured; e.g., $t_d(T) \propto T^{-2}$ at $\nu \approx 0.7-1.5$ GHz and $T \approx 0.02-0.08$ K at a typical $t_d \approx 20\,\mu$s at $T \approx 0.02$ K, were found for a two-pulse echo in v-SiO_2. In the cited works, important experimental data are available which concern temperature dependence of two characteristic relaxation times of the anharmonic, 1/2 spin-like, excitations similar to relaxation times of the Bloch dynamics of many-spin systems: "longitudinal" relaxation time τ_1 for independent excitations and "transverse" relaxation time τ_2 for interacting excitations (see, e.g., [105, 106, 122]). It is worth adding that, as well as for the specific heat, thermal conductivity and changes in sound (light) velocity, the scale and dependencies (on T, ν and I) of the dissipative and the coherent dynamic properties, including the "hole" in a wave absorption spectrum, of the glasses at very low ν and T appear to weakly depend on their chemical nature and specific short-range and medium-range order, i.e., they are also universal in this sense.

Surprising memory effects have also been discovered in dielectric experiments at ultra-low $T < 0.1$ K [122, 139]: the dielectric constant ϵ', determined at audio frequencies, jumps rapidly upward after suddenly applying a DC electric field F and then decays with a logarithmic rate depending on the time elapsed after the field was applied (the decrease was monitored during a whole day). As observed explicitly, ϵ' had a minimum at $F = 0$ at which the sample was cooled down from high temperatures. Moreover, the response of a thin film of an amorphous photoresist (cooled down in the absence of an electric field) to the electric field F exhibits its minimum at $F = 0$. However, when F was finite, e.g., at 5 MV/m, for several minutes after applying the field, a new local minimum appeared with its depth increasing logarithmically with time. It was argued that the jump of ϵ' corresponded to a rise of a non-equilibrium state while the decay was associated with the formation of a new hole at the applied electric field. In addition, a weak dependence of the dielectric constant on the measurement frequency was found at low enough temperatures while the dependence disappeared at higher ones at which the minimum of ϵ' appeared. In the present book, in which we only characterize the universal properties of almost equilibrium glasses, we will not discuss the origin and theory of these very interesting effects, which are pronounced non-equilibrium effects.

6.2 Moderately low temperatures and frequencies

Let us now consider moderately low temperatures (significantly lower than the calorimetric liquid-to-glass transition temperature T_g) and frequencies, $3T_m < T \lesssim T_M \sim 10^2$ K and $3\nu_m < \nu \lesssim \nu_M \sim 3$ THz, at which the observed behavior of dynamic and thermal properties of a glass entirely differs from the one at very low $T \lesssim T_m \simeq 1$ K and $\nu \lesssim \nu_m \approx 3 \cdot 10^{-2}$ THz. In particular for the thermal conductivity, this region at $10^{-2} < T/T_D \lesssim 10^{-1}$ is referred to as the "plateau" since the property weakly depends on temperature, while acoustic phonons can still be essential heat carriers that are characterized rather by much smaller $l_{ac}/\lambda_{ac} \approx 1$ characterizing strong scattering of the phonons. Both the plateau for thermal conductivity, $\kappa(T) \approx const$, and the observed broad maximum, "bump" of the reduced specific heat $C(T)/T^3$ around $T = T_{BP} \equiv h\nu_{BP}/2 \approx 10–20$ K, correlated with the boson peak frequency $\nu_{BP} \approx 1$ THz , appears to be universal in the same sense that the properties relatively weakly depend on chemical nature and specific short- and medium-range order of glasses (Fig. 1, solid lines). In contrast, in non-perfect crystalline materials at the same moderately low temperatures (Fig. 1, dashed curves) no bump of specific heat typically occurs while the thermal conductivity exhibits a maximum at $T = T_{MAX}$, marking a transition from acoustic phonon scattering by the sample boundary at $T < T_{MAX} \ll T_D$ to Rayleigh scattering at $T > T_{MAX}$.

According to some experimental data for glasses, e.g., for SiO_2 glass in an acoustic wave of a given moderately low frequency, like $\nu = 930$ MHz (Fig. 2(a)) and $\nu = 1$ THz (Fig. 2(b)), the temperature dependence of acoustic (electromagnetic) wave attenuation $\alpha(\nu, T)$ exhibits a broad maximum around $T \sim 10$ K [130]. The broad peak of acoustic (or electromagnetic) attenuation can be estimated by a well-known general expression for classical Debye relaxation induced absorption [106] (1986):

$$\alpha(\nu, T) \sim \int d\tau g(\tau) f_D(\nu, T), \quad \text{at } f_D(\nu, T) = (2\pi\nu)^2 \tau [1 + (2\pi\nu)^2 \tau^2]^{-1}. \tag{6.7}$$

The relaxation time $\tau = \tau_0 \exp(V_B/T)$ characterizes a classical thermal activated hopping over a barrier of a random height V_B between potential wells (e.g., in double-well potentials), with a characteristic vibration frequency τ_0^{-1}, while $g(\tau)$ is an appropriate distribution density of the barrier heights. In other words, the attenuation can be described by a superposition of temperature dependent Debye functions $f_D(\nu, T)$ exhibiting a maximum at $\nu = \tau^{-1}(T)$ while $g(\tau)$ is acting as a weight.

The most well-known universal dynamic property of a glass at such frequencies (Ch. 5) is a broad asymmetric peak around a characteristic, almost temperature independent frequency $\nu_{BP} \approx 1$ THz observed in measured inelastic photon (light, x-rays) and neutron scattering spectra [140]. Such experiments are relevant for studying processes in which photons (neutrons) of initial energy $h\nu_i$, momentum $\hbar\mathbf{Q}_i$ and polarization e_i are inelastically scattered by moderately-low-energy excitations of the glass and transformed into those of final energy $h\nu_f$, momentum $\hbar\mathbf{Q}_f$, and polarization e_f. The spectrum of scattered photons (neutrons) characterizes the spectrum of excitations, which is described by the energy and momentum shifts $h\nu = h\nu_i - h\nu_f$ and $\hbar\mathbf{Q} = \hbar\mathbf{Q}_i - \hbar\mathbf{Q}_f$. In what follows, it is useful to restrict ourselves to the simplest realistic case in which the contribution of Umklapp processes [113] to scattering events and momentum conservation can be neglected. Then, the conservation law for the scattering momentum shifts for acoustic phonons, $\mathbf{q}$, and photons (neutrons), $\mathbf{Q}$, and reads $\mathbf{q} = \mathbf{Q}$. The scattering process is usually described by the standard linear response theory in terms of differential scattering cross-section (per unit frequency shift and angle ranges) or related total scattering intensity (integrated over scattering frequencies and angles), which can be calculated by applying the perturbation theory for a relatively weak photon-electron (neutron-nucleus) interaction. The total scattering intensity $I(\nu, \mathbf{Q} = \mathbf{q}; T)$ of the peak in a glass is observed to depend on T as the so-called boson factor, $1+N(\nu, T)$ or $N(\nu, T) = [\exp(h\nu/kT) - 1]^{-1}$, for vibrational excitations of energy $\varepsilon \equiv h\nu$ that scatter photons (neutrons) in a glass. In this sense, the moderately low-energy excitations in question are suggested to be harmonic vibrational (*HV*) ones. Therefore the broad and asymmetric peak in a glass is generally referred to as the universal *boson peak* (BP, in what follows). Several features of the observed BP can be noted.

(I) The BP frequency ν_{BP} is considerably lower than the effective Debye frequency ν_D of the material, e.g., the frequency shift in the inelastic scattering is smaller than about $\nu_D/3$. In this sense, the peak is determined by excess, non-Debye, excitations and their DOS that considerably (e.g., by about an order of magnitude in silica) exceeds the Debye DOS at such frequencies.

(II) The BP is determined by harmonic vibrational excitations, also explicitly follows from its normal mode analysis for a silica glass model [141].

(III) The BP is broad and asymmetric: a lower threshold exists at a finite harmonic excitation energy $\varepsilon_{\min}^{(h)} \equiv h\nu_{\min}^{(h)} \approx 5T_m$ and vibrational excitations occur only at $\nu_{\min}^{(h)} \lesssim \nu \lesssim \nu_M \approx 3\,\text{THz}\ (\ll \nu_D)$, so that the high frequency wing of the peak is broader than the low frequency one, $\nu_{BP} - \nu_{\min}^{(h)} < \nu_M - \nu_{BP}$.

(IV) Unlike the well-known Brillouin spectrum for inelastic scattering of acoustic phonons with $\nu \approx s_0 q/2\pi = s_0/\lambda$, the BP frequency and intensity do not exhibit a noticeable dependence on the wave number $Q = q$, where $Q \equiv |\mathbf{Q}|$ and $q \equiv |\mathbf{q}|$.

It follows from above that the total scattering intensity in a glass, a symmetric tensor $I_{ij}^{(HV)}(\nu, q; T) = I_{ji}^{(HV)}(\nu, q; T)$, can be defined as the Fourier transform of a space $(\mathbf{r})$ and time (t) correlation function of fluctuations $\delta\psi(\mathbf{r},t)$ of appropriate random parameters $\psi(\mathbf{r}, t)$ in a glass,

$$I_{ij}^{(HV)}(\nu, q, T) = \Omega^{-1} \int dt \exp(2\pi i\nu t) \iint d^3r_1 d^3r_2 \exp[i\mathbf{q}(\mathbf{r}_1 - \mathbf{r}_2)] \cdot \langle \delta\psi_i(\mathbf{r}_1, t)\delta\psi(\mathbf{r}_2, 0)\rangle. \tag{6.8a}$$

Here Ω is the system volume, i, j are tensorial indices, $\mathbf{q}$ is the photon (neutron) wave vector shift and $\langle \cdots \rangle$ is the average over both the canonical thermal-equilibrium ensemble of systems and static disorder. The experimental observation that the total scattering intensity around the BP is almost independent of q, $I_{ij}^{(HV)}(\nu, q; T) \simeq I_{ij}^{(HV)}(\nu, T) = c_{ij}(\nu)g(\nu)\nu^{-1}[1 + N(\nu, T)] \simeq I_{ij}^{(HV)}(\nu; T)$, was theoretically interpreted [142] by suggesting that in the frequency range around the BP, a glass behaves as an amorphous system for which the wave-vector shift $Q = q$ in a scattering event is not a well-defined quantum number in the usual sense that its quantum-mechanical uncertainty due to mechanical and electric disorder induced scattering is large, $|\delta\mathbf{q}| \approx q$. This rather means that the correlation length $\xi(\nu)$ over which a harmonic vibrational excitation can be approximated by a plane wave, is of the same scale as its wavelength $\lambda(\nu)$, i.e.,

$\xi(\nu) \approx \lambda(\nu)$, in contrast to the standard weak scattering of excitations for which $\lambda(\nu) \ll \xi(\nu) \approx l(\nu)$, the excitation mean-free-path.

Taking into account the above, one can present the harmonic vibrational contribution to $I_{ij}^{(HV)}(\nu, T)$ in the following form:

$$I_{ij}^{(HV)}(\nu, T) = c_{ij}(\nu) g(\nu) \nu^{-1} [1 + N(\nu, T)], \qquad (6.8b)$$

where $g(\nu) \equiv h\, \mathrm{G}_{\mathrm{exc}}^{(HV)}(\varepsilon = h\nu)$ by definition is the DOS for frequencies of (non-acoustic) harmonic vibrational excitations and $c_{ij}(\nu)$ is the coupling coefficient of the harmonic vibrational excitation of frequency ν to a photon (neutron). For the sake of simplicity in what follows, the total scattering intensity around the BP and the coupling coefficient are approximated by scalar quantities, $I_{ij}^{(HV)}(\nu, \mathbf{q}, T) \simeq I^{(HV)}(\nu, T)$ and $c_{ij}(\nu) \simeq c(\nu)$, where the coefficient $c(\nu) \equiv c_R(\nu)$ depends on ν for photon (Raman, x-ray) scattering, whereas for neutron scattering, the coefficient $c(\nu) \equiv c_n = const$ does not (for x-ray scattering, $c(\nu)$ might be a weaker function than for Raman scattering). Most of the experimental data on inelastic scattering spectra, largely on neutron scattering, were interpreted in terms of the reduced DOS, $g(\nu)/\nu^2$, of the harmonic excitations. However, at least for light (Raman) scattering and a collection of glasses, the characteristic of the peak measured in experiment was the so-called Raman dynamic susceptibility $\chi_R''(\nu)$, defined below and related to the total vibrational DOS by $J(\nu^2) = g(\nu)/2\nu$ for (non-acoustic) harmonic vibrational excitations scattering photons (neutrons). In this connection, the BP was defined by the relationships

$$g(\nu)/\nu^2 = \max \text{ or, at least in a collection of glasses, } J(\nu^2) = \max. \qquad (6.9)$$

For neutron scattering with $c(\nu) \equiv c_n = const.$, the experimental data for the BP in scattering intensity is described as a peak of the reduced density of harmonic vibrational states. Since the Raman coupling coefficient $c_R(\nu)$ depends considerably on ν, the BP in Raman scattering cannot necessarily be defined as a peak of only reduced DOS. In this connection, we concentrate on the Raman scattering data, of which comparison with different theoretical models seems to be most helpful for choosing relevant models of vibrational dynamics in glasses. Indeed, in numerous Raman scattering experiments, the total scattering intensity $I_R(\nu, T)$, as well as the reduced intensity $I_R^{(r)}(\nu)$ and Raman dynamic susceptibility $\chi''_R(\nu)$ (by definition, both independent of T), are measured, particularly for ν around the BP frequency ν_{BP}. Then, the phenomenological relationships for $I_R(\nu, T)$, $I_R^{(r)}(\nu)$ and $\chi''_R(\nu)$ are as follows:

$$\begin{aligned} I_R(\nu, T) &= c_R(\nu)g(\nu)[1 + N(\nu, T)]/\nu \equiv \chi''_R(\nu)[1 + N(\nu, T)] \\ &\equiv I_R^{(r)}(\nu)\nu[1 + N(\nu, T)], \end{aligned} \tag{6.10}$$

$$\text{e.g., } I_R(\nu, T) \propto c_R(\nu)g(\nu)\nu^{-\beta}, \tag{6.11}$$

where $\beta \simeq 2$ for high $T > h\nu_{BP}/k_B \equiv T_{BP}$ or $\beta \simeq 1$ for low $T \lesssim T_{BP}$. According to a recent analysis of $c_R(\nu)$ from scattering intensity data for different glasses in the range between 10 and 50 cm^{-1} around the BP, two groups of glasses can be distinguished [143]. For the first group (e.g., SiO_2, B_2O_3, CKN, PC),

$$c_R(\nu) \propto [(\nu/\nu_{BP}) + B], \quad \text{at } B \approx 0.5, \tag{6.12}$$

with its high frequency limit varying from about $2\nu_{BP}$ for SiO_2 and Se glasses up to about $4\nu_{BP}$ for PC glass. However, for the second group (e.g., As_2S_3, $GeSe_2$, GeO_2),

$$c_R(\nu) \propto \nu, \quad \text{with } B \approx 0. \tag{6.13}$$

It is worth concluding that the BP in $I_R(\nu, T)$ at high T is the same as the one in $I_R^{(r)}(\nu)$, i.e., in the reduced DOS $g(\nu)\nu^{-2}$, for the first group of glasses with $c(\nu) \approx const.$, or in $g(\nu)\nu^{-1}$, for the second group with $c_2(\nu) \propto \nu$. On the other hand, the BP in $I_R(\nu, T)$ at low T is the same as the one in $\chi''_R(\nu)$, i.e., the one in $g(\nu)\nu^{-1}$ for the first group of glasses or in $g(\nu)$ for the second group. Moreover, in addition to numerous BP experimental data in scattering spectra of $I_R^{(r)}(\nu)$ (see, e.g., [118, 140, 144] and references therein), the BP also appears to be observed in the spectra of $\chi''_R(\nu)$, in a collection of different glasses like diglycidyl bisphenol A (DGEBA), polymethyl methacrylate (PMMA) and calcium potassium nitrate (CKN), polysterene and polycarbonate [145] with the characteristic frequency $\nu_{BP} \sim 1$ THz.

In general, the characteristic BP and related collective vibrational excitations in glasses can be studied by measuring not only the DOS of excitations, $g(\nu)/\nu^2 = 2J(\nu^2)/\nu$ or $J(\nu^2)$ (Eqs. (6.8)–(6.10)), but also the dynamic structure factor $S(Q, \nu)$ that depends on the incident particle wave number $Q = q$, as glasses are isotropic disordered materials [112, 128, 147, 148]. The dynamic structure factor $S(q, \nu)$ is the space (r) and time (t) Fourier transform of the self-correlation function $S(r, t) = \langle n(r, t)n(0, 0)\rangle$ of particle density $n(r, t)$, which roughly describes the spectrum of single excitations produced by an incident plane wave with a given wave number q in the applied approximation. This function of (q, ν) has been studied in glasses earlier on by applying the Brillouin light scattering or inelastic neutron scattering. For the Brillouin light scattering, the accessible momentum shifts are limited to the range $Q \ll 10^7$ cm^{-1} due to small momenta of light phonons and the related excitations are low-frequency acoustic phonons with the dispersion relation $\nu_B^2(q) \approx const. + (s_0 q)^2$ at $const. = 0$. On the other hand, standard neutron scattering experiments, because of the kinetic limitations related to the neutron mass, have been applied only at large Q. Thus, the techniques left an unexplored gap at $10^7 \lesssim Q \lesssim 10^8$ cm^{-1} for collective excitations which has been investigated by recent high-resolution inelastic x-ray and neutron scattering techniques [112, 147, 148]. Since one can find necessary details in the cited references we briefly summarize only some general features of $G_{exc}(\nu)$ and $S(q, \nu)$ found in recent experiments at moderately low frequencies.

(A) It has been discovered in high resolution x-ray scattering experiments [149] in a variety of glasses that the collective excitations are propagating acoustic-like waves ("high frequency sound", or HFS in what follows) with an acoustic-like dispersion relation, $\nu_B^2(q) = const. + (s_0 q)^2$ at $const. \approx \nu_{BP}^2$, which occur in the terahertz frequency range, e.g., at 1 THz $\approx \nu_{BP} < \nu \lesssim \nu_M \approx 2\text{–}3$ THz. In fact, "primary", Brillouin-like peaks of frequencies $\nu_B(q) \approx [const. + (s_0 q)^2]^{1/2}$ $(const. \geq 0)$, corresponding to sound-like excitations, are observed in $S(q, \nu)$ not only at small $q \ll 10^7$cm for standard sound waves but also at relatively large $q < q_D/2 = \nu_D/2s_0$ in the range $10^7 \lesssim q \lesssim 10^8$ cm^{-1}, for frequencies higher than the BP one, e.g., in v-SiO_2, glycerol, LiCl:H_2O [149, 150]. Such a Brillouin peak may characterize a single sound-like excitation of the HFS. In addition, the effective sound velocity s_0 is found to increase slightly with ν from $s_0 = s_1$ for $q \ll 10^7$ cm^{-1}to $s_0 = s_2$ for $10^7 \lesssim q < 10^8$ cm^{-1}, at $0 < s_2 - s_1 \ll s_2$. Each Brillouin peak has a finite width that is proportional to a finite excess density of states $\Delta G_{exc}(\nu = s_0 q)$ of non-Debye, harmonic vibrational excitations, of which the maximum is usually identified as the BP [112, 140]. However, the HFS peaks do not seem to be detected in some other glasses, e.g., in v-B_2O_3 glass [151].

(B) Another "secondary" peak was also detected in $S(q, \nu)$ at relatively large q and at frequencies substantially lower than the corresponding $\nu_B(q)$, which is found at a large enough $q \approx 0.5 \cdot q_D$ to exhibit practically all essential properties of the BP at $\nu = \nu_{BP}$ in the total vibrational DOS $J(\nu) = g(\nu)/2\nu$ or in the reduced DOS $g(\nu)/\nu^2$ (see, e.g., [112]). Moreover, the typical frequency of the BP is located in a range in which the dispersion relation for the Brillouin peak is still an acoustic-like one.

(C) The acoustic phonon attenuation parameter $\alpha_{ac} = l_{ac}^{-1} = s_0/2\gamma_{ac}(\nu)$ [112, 152, 153], or the width $2\gamma_{ac}(q)$ (full width at half maximum) of related Brillouin line around a vibrational frequency $\nu = qs_0/2\pi$, measured in glasses, was often found to be temperature independent and characterized by a small width, $2\gamma_{ac}(\nu) \ll \nu$, and a power law frequency dependence, $2\gamma_{ac}(\nu) \propto \nu^\kappa$ with $\kappa \approx 2$, for low but not too low ν, e.g., at $0.45\,\text{THz} \lesssim \nu < \nu_{BP} \sim 1\,\text{THz}$ (Fig. 6). Moreover, some evidence was found experimentally in SiO_2 glass for a crossover in the frequency dependence by applying a new inelastic ultraviolet light scattering technique, for wavelengths λ in the range $50 \leq \lambda \leq 80\,\text{nm}$ [153]. The crossover occurred at a certain value $q = q_{co}$, e.g. $q_{co} \simeq 0.15\,\text{nm}^{-1}$, in the q range under discussion between a temperature dependent attenuation mechanism at lower frequencies and a temperature independent mechanism at higher ones. The temperature dependent attenuation at low enough frequencies has been attributed to Akhiezer-like sound attenuation mechanisms [113] due to inelastic scattering of phonons determined by weak anharmonic phonon–phonon interactions, with the width $2\gamma_{ac}^{(A)}(\nu, T) \to 0$ at $T \to 0$. On the other hand, the temperature independent mechanism with $2\gamma_{ac}(\nu) \propto \nu^\kappa$ at $\kappa \approx 2$ at higher ν could not be ascribed to Rayleigh elastic scattering from static defects, for which the temperature independent phonon width is $2\gamma_{ac}^{(R)}(\nu) \propto \nu^4$. Instead, because of the absence of any signature in the static scattering factor at $q = q_{co}$, it has been suggested that the attenuation at $q > q_{co}$ was associated with some spatial fluctuations of local elastic constants, rather than with dynamic effects.

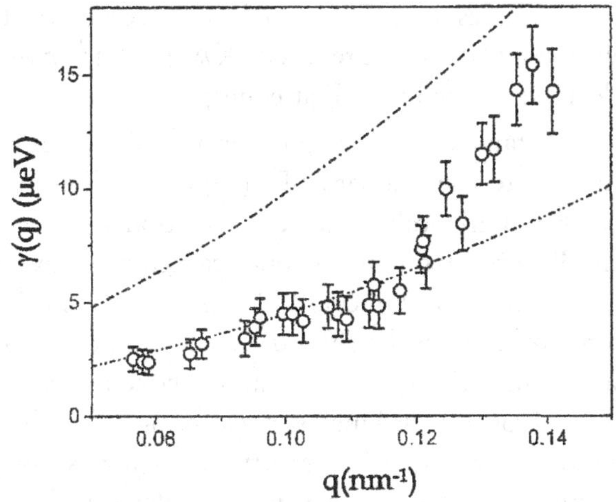

Figure 6: Experimental data for Brillouin peak widths at $T = const \ll T_D$ (e.g., at room temperature in v-SiO_2) as a function of the acoustic phonon momentum transfer q, $\gamma_{ac}(q) \propto q^{\kappa}$ (i.e., also of the frequency ν at $\nu \propto q$, $\gamma_{ac}(\nu) \propto \nu^{\kappa}$), with $\kappa \approx 2$ for $\nu < \nu_{BP}$. The dot-dashed and dot-dot-dashed lines correspond to an extrapolation of the $\kappa = 2$ behaviour measured at low $q \ll q_D$ and at higher q by applying Brillouin inelastic UV light spectroscopy and IXS technique, respectively.

(D) Some recent experimental data [154], which were able to reveal contributions of collective (non-acoustic) harmonic vibrational excitations to the BP, appeared to be important for discriminating relevant theoretical approaches describing correlated BP and HFS. In the cited experimental investigation, the glass dynamics were studied by using probe molecules embedded in a glass matrix which had to follow collective modes with a correlation length $L_c > L_p$. The probe molecule was chosen to exhibit inelastic resonant scattering of X rays due to low-energy nuclear transitions for a resonant nucleus in the centre of mass of a ferrocene molecule with a neutral resonant nucleus, ^{57}Fe, or an ionic one, $^{57}Fe^{2+}$. The vibrational spectrum of the probe was found from the data to be independent of the matrix local modes as the probe was not chemically bound. For detecting collective modes of a glass, the probe molecule had to be affected by correlated forces at least from the nearest glass molecules. Four glasses (toluene, ethylbenzene, dibutylphtalate, glycerol) have been studied. In all the glasses, no HFS seemed to be seen up to the inelastic scattering energy 6 meV, and the reduced DOS $g_c(\nu)/\nu^2$ of the collective excitations clearly exhibited a peak at energy $h\nu_{BP}^{(c)}$ with basic properties similar to those of standard BP. The position $\nu_{BP}^{(c)}$ of the peak was consistent with and close to the standard BP energy ν_{BP}. Then, the data was suggested to show that a significant part of vibrational excitations related to the standard BP correspond just to non-localized collective modes with the correlation length L_c of the scale and larger than ~ 20 Å. Moreover, the temperature evolution of the BP showed the same features (including its disappearance with increasing temperature above the glass transition) like those observed for standard inelastic scattering data. The basic question was whether the standard BP essentially originates from collective harmonic vibrational modes or if it is mainly related to local (or quasi-local) ones. At $\nu > \nu_{BP}$, it was observed experimentally [154] that the reduced DOS $g_c(\nu)/\nu^2$ of the collective harmonic vibrational excitations at $\nu > \nu_{BP}^{(c)} \approx \nu_{BP}$ and temperatures far lower than the glass transition one exhibited very precisely over 2–3 decades of values which showed temperature independent exponential decrease with increasing frequency:

$$g_c(\nu)/\nu^2 \propto \exp\left(-\nu/\nu_t\right), \qquad (6.14)$$

where the decrease scale ν_t is correlated with the BP frequency, $\nu_t \approx 0.5 \cdot \nu_{BP}$. It was also noted in this study that similar features seemed to be found for the standard BP in the total reduced DOS $g(\nu)/\nu^2$ of harmonic vibrational excitations in neutron data for the glasses under discussion, though with a less steep slope. The difference between $g(\nu)$ and $g_c(\nu)$ was suggested to give rise to two different interpretations: either attributed to quasi-local vibrations in $g(\nu)$ at higher frequencies $\nu > \nu_{BP}$ (Sec. 10.2) or to collective vibrations, assuming a lower sensitivity of relatively large molecular probes to collective vibrations with shorter wavelengths. While the four glasses studied do not seem to exhibit HFS excitations above the BP, at least at $0 < h\nu - h\nu_{BP}^{(c)} < 6$ meV, it seems of interest to experimentally verify whether Eq. (6.14) holds true for glasses that exhibit HFS excitations above the BP as can be predicted (Eq. (10.24)) in the theory discussed in Sec. 10.6, in which the characteristic correlation length L_c of non-localized collective modes may plausibly be about 20 Å.

In conclusion, it is particularly important to emphasize that both qualitative behavior and scale of the properties of glasses at moderately low ν and T are universal in the same sense that they weakly depend on chemical nature and specific short-range and medium-range order of the materials. A number of experiments were made to understand whether the unusual universal dynamic and thermal properties were due to impurities or were intrinsic. Introduction of high concentrations of impurities actually gave rise to changes in the specific heat, thermal conductivity and other properties but the changes were found to be essentially related to the corresponding change ΔT_g of the glass transition temperature T_g (see [106, 155] (1986)). These and other experimental data appear to indicate that the universal dynamic and thermal properties of glasses are largely of intrinsic origin, i.e., are weakly related to extrinsic defects, including those associated with impurity induced ones. Therefore, in the present book effects of extrinsic defects are not considered.

7

SOFT-MODE MODEL OF LOW-ENERGY ATOMIC DYNAMICS

At present, no consistent quantitative microscopic theory of the anomalous low-energy dynamics and related excitations of glasses appears to be available. However, several theoretical mean-field models of the dynamics and excitations have recently been developed, of which the interrelations have been discussed but are not yet fully understood. In the present book, one of the mean-field theories, the soft-mode model (SMM in what follows) of low-energy excitations is presented in more detail in and Chs. 7, 8 and 10, while other models are briefly discussed and compared in some aspects to the SMM in Chs. 5 and 11–13. The SMM and its basic concepts were proposed in [119], worked out in [123] and the basic results were presented in several reviews, e.g., [124, 125, 126] and [127, 128]. In the last two references, as well as in some other ones published after 1983, the model was referred to as "soft-potential model"; in the present review the original title "soft-mode model" is applied, in particular because the physical meaning of the term "soft potential" does not seem to be clear enough.

The basic concept of the original SMM is that non-acoustic, soft-mode excitations of low energy $\varepsilon_{sm} \equiv h\nu_{sm} \ll h\nu_D$ coexist and interact with standard low-energy acoustic phonons of energy $\varepsilon_{\mathrm{ph}} \equiv h\nu_{\mathrm{ph}} \ll h\nu_D$ in a glass (but not in a crystal) and that both types of excitations and their interactions determine the anomalous and mostly universal, low-energy glass dynamics. The non-acoustic excitations describe atomic motions in "soft (motion) modes" that are by definition characterized by small effective spring constants k, $0 < k \ll k_0 = M\omega_{\mathrm{D}}^2$, with a typical average atomic mass

M in the range at 10^{-22}–10^{-21} g and k_0, the standard scale of k for the vast majority of atoms involved in acoustic-motion modes. The potential energies of the vast majority of atoms in most amorphous solids, including glasses, are characterized by spring constants k of the same scale k_0 as those in most crystalline materials, as follows from the well-known similarity of elastic properties of the amorphous solids and crystals. The soft modes can be characterized by a spectrum of small random k, $0 < k \ll k_0$, in the corresponding tail of the probability distribution density (PDD), $P(k)$, which contains the minority of atoms of low concentration $c_{sm} \ll 1$. As far as the atomic structure of a glass is approximated as a system of randomly located atoms coupled with each other by harmonic interactions (Ch.1), it is characterized by both short-range order local structures of a length scale $L_{SRO} \approx a_1 \approx 3$ Å and medium-range order local structures of a noticeably larger, nanometric length scale $L_{LRO} \approx 3a_1 \sim 10$ Å, with their microscopic parameters (bond length, bond angles, dihedral angles, etc.) spatially fluctuating around the average values (the fluctuations are experimentally detected to be very small, e.g., for bond lengths, or moderately small, e.g, for dihedral angles in medium-range order structures. It also seems plausible (see below) that the soft-mode concentration in amorphous solids reaches its highest value in glasses (g), i.e., $c_{sm}(g) = (c_{sm})_{max} \ll 1$.

7.1 Atomic soft modes and related potentials

A "soft mode" is actually suggested in what follows to be associated with a rare and large anisotropic low-density, void-like fluctuation of atomic configuration parameters, e.g., bond angles and dihedral angles, in the glass structure. In fact, a group of n_{sm} atoms referred to as an effective "atom" in a MRO scale atomic configuration around such a structural fluctuation, can be considered as a subsystem which moves in a "soft mode" x_{sm}, or dimensionless $x \equiv x_{sm}/a_0$, and in an effective local potential $V_{sm}(x)$ with a small random spring constant $k = k(x_{min})$ in a potential well at $x = x_{min}$ ($a_0 = 1$ Å is the atomic length scale; x_{min} or x_{max} is the coordinate of a potential minimum or maximum). Such a "soft configuration" is mainly an intrinsic atomic configuration (although it can also be a point defect related configuration) close to its harmonic instability at small values of the spring parameter, $|k(x)| \equiv |d^2V_{sm}(x)/dx^2| \ll k_0 \equiv 4\pi^2 M\nu_D^2$. In fact, this may mean that the "atom" in its soft-mode motion is weakly coupled with the surrounding atoms in a void-like structure generated at the liquid-glass transition. One can readily see that either the spring parameter $k(x) > 0$ for x around a potential minimum in a single-well or a double-well potential, or $k(x) < 0$ for an unstable configuration around a potential maximum of a double-well potential. Thus, a soft mode, particularly in a double-well potential, can be considered as a topological defect rather than a point defect and the characteristic size L_{sm} can roughly be estimated in the range $a_1 < L_{sm} \lesssim L_{MRO}$. In what follows, a soft mode is characterized by two basic dimensionless random parameters, $\eta(\gtrless 0)$ and $\xi(\lessgtr 0)$, with small $\{|\eta|, \xi^2\} \lesssim \eta_{sm} = const. \ll 1$, e.g., at $\eta_{sm} \approx 0.1$, for a low spring constant $k = k(\eta, \xi) \ll k_0$. Thus, the soft modes characterize states of collective atomic motion of the length scale L_{sm}, involving n_{sm} atoms typically at $5 \lesssim n_{sm} \lesssim (L_{MRO}/a_1)^3 \lesssim \sim 10$.

In general, the potential $V_{sm}(x)$ characterizes the change of the total potential energy $V_{sm}(x, X; Y)$ of an atomic soft configuration in the multi-dimensional "phase" space $(x \equiv x_0, X; Y)$ upon a displacement x in a soft mode at fixed values of other coordinates of the "atom", $X = \{x_1, x_2, \ldots, x_n\} (n > 1)$ and its environment, $Y = \{y_1, y_2, \ldots, y_m\}$ $(m > 1)$. By physical sense, a soft configuration can easily be rearranged with a significant atomic displacement x up to one comparable to 1, since the "atom" is weakly coupled to the environment. In other words, small variations of the parameters of a harmonically nearly unstable configuration in a soft mode may give rise to significant changes of the potential shape for the weakly coupled "atom". This indicates that the soft-mode potential energy $V_{sm}(x)$, generally speaking, is an anharmonic potential like $V_{sm}(x) \simeq A_1x + \frac{1}{2}A_2x^2 + \frac{1}{3}A_3x^3 + \frac{1}{4}A_4x^4 + \cdots$, in contrast to conventional harmonic potentials in which each atom of mass m_0 is characterized by a single harmonic potential well, $V(x) = k_0x^2/2$, and anharmonic corrections are small as the vibrational amplitude $x \approx (h\nu_h/k_0a_0^2)^{1/2} \ll 1$ at a typical harmonic vibrational frequency $\nu_h \approx (k_0/m_0)^{1/2}$. Most important in what follows is that anharmonic soft-mode potential energies generate unusual vibrational excitations and non-vibrational anharmonic excitations, in particular, those similar to above mentioned two-level system excitations, and a soft mode is characterized with respect to an applied force by a large generalized susceptibility, $\chi = k^{-1} \gg \chi_0 = k_0^{-1}$, providing a strong interaction of a soft-mode excitation with acoustic phonons.

The basic result of SMM can be described by two assertions which are justified below (see, e.g., [125]):

(I) a typical, most probable, random atomic soft configuration contains a single soft mode x, and the corresponding typical potential energy is that of a single soft-mode;

(II) substituting x' for x, one can well approximate the expression of a typical soft-mode potential energy operator at realistic, not too large $|x'| \lesssim 1$ by the general form of a quartic anharmonic potential energy (below often referred to as "potential"):

$$\widehat{V}_{sm}(x') \simeq A(h'x' + \eta' x'^2 + \xi' x'^3 + x'^4). \tag{7.1}$$

The random parameters $\{h', \eta', \xi'\}$ are small in magnitude compared to unity, and the reference point for x' can be located at an arbitrary point of the potential curve. A different equivalent (with accuracy to a constant term) form of the potential, in which the linear term disappears, can be obtained by displacing the reference point for the mode to an extremum of the function, by applying the relation $x' - x = (\xi' - \xi)/4$. Then,

$$\widehat{V}_{sm}(x) \equiv V_{sm}(x; \eta, \xi) \simeq A(\eta x^2 + \xi x^3 + x^4), \tag{7.2}$$

or

$$\widehat{V}_{sm}(x) \equiv \widetilde{V}_{sm}(\widetilde{x}; \widetilde{\eta}, \widetilde{\xi}) \simeq w(\widetilde{\eta}\,\widetilde{x}^2 + \widetilde{\xi}\,\widetilde{x}^3 + \widetilde{x}^4), \tag{7.3}$$

where $\{|\eta|, \xi^2\} \ll 1, \widetilde{x}/x = \eta_L^{-1/2}, \widetilde{\eta}/\eta = \eta_L^{-1}$ and $\widetilde{\xi}/\xi = \eta_L^{-1/2}$. In what follows, generally speaking, the parameters A, w and effective mass M_{sm} of a soft-mode "atom" are random parameters but their distribution densities (in accordance with qualitative arguments and simulations, see Sec. 7.3) is assumed sufficiently narrow so that in estimations, the parameters can be replaced by their average values. In Eq. (7.2), $A \approx M(2\pi\nu_D)^2 a_0^2$ is a typical atomic elastic energy ($A \sim 10\,\text{eV}$), $w = A\eta_L^2$ is a characteristic soft-mode energy (Sec. 7.4) related to the Debye energy as $w \approx 3 \cdot 10^{-4}$ eV $\approx 10^{-2} h\nu_D$ and $\eta_L = (\hbar^2/2M_{sm}a_0^2 A)^{1/3} \sim 10^{-2}$, where typical soft-mode effective mass and average atomic mass are respectively $M_{sm} \approx (5\text{–}10) \cdot M$ and $M \approx (10^{-23} - 10^{-22})\,\text{g} \sim 3 \cdot 10^{-23}\,\text{g}$ (Sec. 7.3). The basic random soft-mode parameters of atomic soft configurations are "softness" parameter $\eta(\lesseqgtr 0)$ at $|\eta| \ll 1$ and potential asymmetry parameter $\xi(\lesseqgtr 0)$ at small values $\xi^2 \ll 1$ corresponding to the mechanical stability condition. On the other hand, it can be concluded from Eq. (7.3) that characteristic soft-mode displacements, though noticeably larger than standard harmonic displacements, are still relatively small, $x \sim \eta_L^{1/2} \approx 0.1$. That is why the approximate Eqs. (7.1)–(7.3) are included in the expansion in x, besides small harmonic and cubic anharmonic terms, only the quartic anharmonic term Ax^4 in which a finite (not small) coefficient A is applied (generally speaking, Eq. (7.2) describes the potential energy of a quantum-mechanical anharmonic oscillator of which the energy spectrum problem does not yet have an analytical solution). In appropriate cases (see Eq. (7.7) below), the finite quartic anharmonic term provides the oscillator stabilization with a resulting positive finite spring constant at vanishing values of the random parameter $\eta \to 0$. By definition, the parameters vary in a continuous way due to the realization of any kind of random atomic configurations at $-\infty < \eta, \xi < \infty$, including $\{|\eta|, \xi^2\} \ll 1$ corresponding to $k \ll k_0$, and are characterized by a normalized (to the total atomic concentration $c_{tot} = 1$) continuous PDD function $F(\eta, \xi)$ (in principle, the parameters η and ξ might be explicitly found by averaging the corresponding potential energy derivatives of the second and third order over their probability distribution). The normalized PDD $P(k)$, including its soft-mode part, can also be derived by using the expression for $F(\eta, \xi)$ (Sec. 7.2) and the one for $k\ (\eta, \xi^2)$ following from $\widehat{V}_{sm}(x; \eta, \xi)$. In fact, physical properties Q of soft modes, including their small spring constants $k \ll k_0$, are even functions of the potential asymmetry ξ, $Q[\eta, \xi] = Q[\eta, -\xi] \equiv Q(\eta, \xi^2)$, in a glass that is an isotropic amorphous system in which no preferred directions are available. One can also suggest that the soft-mode motion of an "atom" is considerably slower than motions in standard modes with $k \approx k_0$; thereby, the contribution of the standard modes to physical characteristics, averaged over the "large" soft-mode motion time, can give rise to a qualitatively unimportant renormalization of the soft-mode potential parameters.

The above remarks agree with the suggestion that SMM is a mean-field (i.e., not a true microscopic) theory in which the expression for the single-mode potential energy $V_{sm}(x)$ is mathematically similar to the well-known "single-mode" expression, describing both the thermodynamic potential in the Landau phase transition theory around critical points [45] and some general characteristics of the mathematical "catastrophe theory" [94] for the behavior of a complex system with many degrees of freedom ("modes") near its critical states. In this connection, the occurrence of a soft-mode can be associated with an instability of a random "critical" local potential $V(x, X; Y)$ in an atomic-motion harmonic mode due to spatial fluctuations of microscopic parameters $Y = \{y_1, y_2, \ldots, y_s, \ldots\}$ of the environment of the "atom" with variables (x, X), in the disordered system. Furthermore, the single-mode character of a typical soft configuration and the related potential energy can be justified by applying some theorems and lemmas of the "catastrophe theory", in which the ensemble of subsystems under discussion can be described by a PDD $P(Y)$, which is regular in the sense that it does not exhibit singularities (or zeros) and is normalized to the total concentration of subsystems in the phase space Y. In this space, generally speaking, there is a hypersurface S separating the region of such values of Y at which potentials $V(x, X; Y)$ are single-well potentials from the alternative region with non-single-well potentials. Far from S, the potential near its extrema can be described as usual by a bilinear form $\sum A_{ij}x_ix_j = \sum_i \overline{A}_i\overline{x}_i^2$, whereas on the hypersurface, the matrix A_{ij} is degenerate with at least one eigenvalue $\overline{A}_i = 0$ or $\overline{A}_i$ (small in magnitude) close to S, so that the bilinear approximation is insufficient. By applying the catastrophe theory with the matrix A_{ij} close to its degeneracy in some modes $(\overline{x}_0, \overline{x}_1, \ldots, \overline{x}_r)$, the local description of a function near its extremum can be given by the splitting lemma of the catastrophe theory: $V(x, X) = \overline{V}(\overline{x}_0 \equiv x, \overline{x}_1, \ldots, \overline{x}_r) + \sum_{j>r} \overline{A}_j\overline{x}_j^2$, with $\overline{V}$ as a function of only modes $\overline{x}_i$ at $i = 1, \ldots, r$ corresponding to $\overline{A}_i$ such that $|\overline{A}_i| \ll \overline{A}_j$ at $i \leq r, j > r$; the modes $\overline{x}_j'$ are linked with $\overline{x}_i$, generally speaking, by a nonlinear transformation. Both $\overline{A}_j$ and parameters in $\overline{V}(x, \overline{x}_1, \ldots, \overline{x}_r)$ are functions of spatially fluctuating parameters Y.

Let us describe the smallness of $\overline{A}_i$ by a small parameter $\varsigma \ll 1$ and consider that the respective small phase volume $v_1 \propto \varsigma$ in a narrow layer around the hypersurface S, i.e., the probability of occurrence of a small $\overline{A}_i$ corresponds to the proximity of $V(x, X; Y)$ to degeneracy in one mode, e.g., $\overline{x}_0 \equiv x$. It is important that in the "catastrophe theory" the dimensions d_p of multitudes of points on the hypersurface S, at which $p \gtrsim 1$

linearly independent eigenvalues of matrices A_{ij} are zero, are related as $d_p - d_{p+1} = p+1$, the difference of numbers of independent terms of bilinear forms for p and $p+1$ variables. The consequence is that the ratio $v_p/v_{p+1} \sim \varsigma^{p+1} \ll 1$ of volumes v_p and v_{p+1}, in which correspondingly p and $p+1$ values of A'_i are small, are rapidly decreasing with increasing p (much more rapidly than typical $v_p/v_{p+1} \sim \varsigma$ in standard expansions). Since a probability W_p to realize a soft configuration with a given number p of soft modes in the phase space is proportional to the phase volume, similar relations $W_p/W_{p+1} \sim \varsigma^{p+1} \ll 1$ hold true also for the probabilities. One may conclude from above that, in particular, for a "critical" potential $V(x, X; Y)$ degenerate in one mode ($\overline{x}_0 \equiv x$), the following expansion is actually equivalent to Eq. (7.2) and can be applied close to the hypersurface:

$$V(x, Y) = \frac{1}{2}A_2(Y)x^2 + \frac{1}{3}A_3(Y)x^3 + \frac{1}{4}A_4(Y)x^4 + \cdots, \tag{7.4}$$

where the reference point for x is an extremum of the function. One can see in the space of coefficients A_i ($i = 2, 3, 4$) that a small realization probability, or phase volume $v' \propto \varsigma$ (around the plane $A_2(Y) = 0$), corresponds to a small A_2, whereas a considerably smaller probability $v'' \sim \varsigma^2$ corresponds to simultaneously small $A_2(Y)$ and $A_3(Y)$. On the other hand, one can see in a similar way that the occurrence of simultaneously small coefficients $A_2(Y)$, $A_3(Y)$ and quartic anharmonicity coefficient $A_4(Y)$ is characterized by a still smaller probability $v''' \sim \varsigma^3$ which is of the same scale as the above-mentioned probability for two nearly degenerate modes at $p + 1 = 3$. This result means that the most probable soft-mode configuration contains a single soft mode with a finite (not small) typical quartic anharmonicity $A_4 = 4A > 0$. The ultimate reason is that the concentration c_{nsw} of atoms in non-single-well (e.g., double-well) potentials is very small, $c_{nsw} \ll c_{sm} \ll 1$, as estimated in Ch. 8. The existence of the small parameter ς appears to be linked, first of all, with the smallness of c_{nsw}, provided that the contribution of local atomic configurations with two and more soft modes to physical properties of the systems under discussion is quite negligible. The above result concerning the isotropic amorphous systems under discussion may also be interpreted as follows: the region of soft configurations, each with two or more soft modes and respective non-single-well potentials, correspond to an isotropic tail of the

PDD $P(Y)$, far from its principal maximum around the average values $\overline{Y}$ for the vast majority of atoms (with single-well potentials at $k \approx k_0$, similar to standard ones in crystalline systems). Taking into account the above result, one can indeed justify Eq. (7.2), i.e., also Eq. (7.3). In fact, Eq. (7.4) coincides with Eq. (7.2) when defining $A_2 \equiv 2A\eta$, $A_3 \equiv 3A\xi$ and $A_4 \equiv 4A$ in which, generally speaking, the basic random parameters η and ξ are functions of the soft-mode environment coordinates, $\eta = \eta(Y)$ and $\xi = \xi(Y)$. As usual, one can suggest that the average values $\overline{Y}$ of Y characterize the vast majority of atoms (at $k \approx k_0$) in a macroscopic volume V and thus are mainly due to contributions of typical large atomic number densities n/V. In this connection, following the strong theorem of large numbers (see, e.g., [156]), the principal maximum of $P(Y)$ around the average values $\overline{Y}$ can be approximated by a Gaussian maximum, $P(Y) \sim \exp[-(Y-\overline{Y})^2/2(\Delta Y)^2]$, with an effective width ΔY.

As can be readily found from Eq. (7.2), depending on the value of $\eta/\eta_0 \equiv 32\eta/9\xi^2$, a typical soft-mode potential can be a single-well one around $x_{\min}^{(0)} \equiv 0$ for $\eta/\eta_0 > 1$, or a double-well potential with two minima at $x_{\min}^{(1,2)}$ and an interwell barrier of a considerable height at $x_{\max}$, $x_{\min}^{(1)} < x_{\max} < x_{\min}^{(2)}$, for $\eta/\eta_0 < 1$, as well as intermediate quartic anharmonic potential at $\eta \approx 0 \approx \eta_0$ [157] (see also [123]). Thus, a typical non-single-well soft-mode potential is a double-well potential (not a three- or more-well potential of which the realization probability W_p, at $p \geq 2$, is much smaller than the probability of a single-mode soft configuration, see above). Moreover, the spring constants $k(\eta, \xi^2) = k(x_{\min}) \equiv \{(d^2/dx^2)\widehat{V}_{sm}(x; \eta, \xi)\}_{x=x_{\min}}$ can be described by the following expressions: for a single-well potential with one minimum at $x = 0$ and $1 \gg \eta > \eta_0$,

$$0 < k \simeq k_0\eta \ll k_0. \tag{7.5}$$

and for a double-well potential with minima at $x = x_{\min}^{(1,2)}(\eta, \xi^2) = (3\xi/8)[-1 \pm \Lambda^{1/2}]$, $\eta < 0$ and $|\eta| \ll 1$,

$$0 < k = k_{1,2}(\eta, \xi^2) \simeq 2\eta_0 k_0 \Lambda^{1/2}(\Lambda^{1/2} \mp 1) \ll k_0, \tag{7.6}$$

while with minima at $x = 0$ and $x = x_{\min}^{(2)} > 0$ at $0 < \eta < \eta_0 \ll 1$,

$$0 < k \simeq k_0\eta \ll k_0 \text{ and } 0 < k = k_2(\eta, \xi^2) \ll k_0, \tag{7.7}$$

where $\Lambda \equiv \Lambda(\eta) \equiv 1 - \eta/\eta_0 > 0$. In other words, a single-well harmonic potential can become unstable to its transformation to a double-well potential with changing η, e.g., from $\eta > 0$ to $\eta < 0$, of which the stabilization at $k_{1,2}(\eta, \xi^2) > 0$ is provided by the finite quartic anharmonicity Ax^4 in Eq. (7.2).

Another important characteristic of a soft mode, its generalized susceptibility χ_{sm} with respect to a weak strain field f applied to the mode, can be obtained by minimizing the total potential energy $\Psi_{sm}(x) = V_{sm}(x) - fx$ and is straightforwardly related to the corresponding low spring constant, being anomalously high:

$$\chi^{(sm)} \equiv \chi^{(sm)}(\eta,\ \xi^2) = dx/df|_{x=x_{\min},\ f_x=0} = k^{-1}(\eta, \xi^2) \gg \chi_0 \equiv k_0^{-1}. \tag{7.8}$$

Actually, continuous spectra of values for spring constants and susceptibilities of soft modes are available, which correspond to the continuous spectrum of the soft-mode parameters η and ξ. Then, in the (η, ξ) plane, two lines $(\gamma = 1, 2)$ of critical points formally occur at $k(\eta, \xi^2) = 0$, i.e., $\chi_x(\eta, \xi^2) = \infty$ at $\eta = 0$ and $\xi^2 \ll 1(\gamma = 1)$ or at $\eta = 8\eta_0/9 \ll 1(\gamma = 2)$. In this sense, three classes of soft-mode potentials can formally be distinguished (Fig. 7): (i) single-well potentials, at $\eta_0 < \eta \ll 1$; (ii) two types $(\gamma = 1, 2)$ of pronounced double-well potentials with an interwell barrier of a considerable height, both at $\eta < 0$ and at $0 < \eta < 8\eta_0/9$; (iii) "transition" potentials in the close vicinity of the lines of critical points, which can be considered as double-well potentials with a vanishing barrier (i.e., with the maximum and one of the minima coinciding with each other) or as a

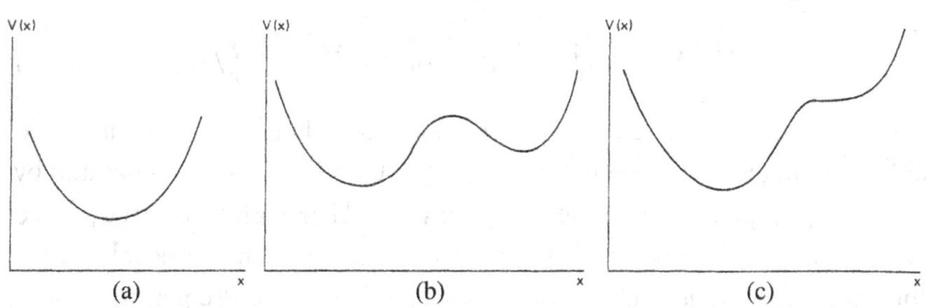

Figure 7: Three main types of soft-mode potentials: quasi-harmonic single-well, anharmonic double-well and intermediate ones.

single-well potential with an inflection point. Straightforward calculations show that the expression of $\widehat{V}_{sm}(x)$ in Eq. (7.2) describes a symmetric double-well potential with two identical wells, $\widehat{V}(x - x_{\min}^{(1,2)}) \equiv \widehat{V}_{1,2}(x)$, in two cases with

$$\rho_1 \equiv \eta/\eta_0 - 8/9 = 0, \quad \text{at } \eta_0 > \eta > 0, \quad \text{or}$$
$$\rho_2 \equiv (\eta_0/\eta_L)^{1/2} = 0, \quad \text{at } \eta < 0. \tag{7.9}$$

In particular, for practically important weakly asymmetric double-well potentials, at $0 < \eta/\eta_0 - 8/9 \ll 1$ or $(\eta_0/\eta_L)^{1/2} \ll 1$, one can readily find for the individual potential wells $\widehat{V}_{1,2}(x)$ of Eq. (7.2) the following approximate expressions:

$$\widehat{V}_1(x) \simeq A\{\eta x^2 + \xi x^3 + x^4\},$$
$$V_2(x) \simeq A\{5(\eta_0 - \eta)x^2 - 3|\xi|x^3 + x^4 - \eta_0(\eta_0 - \eta)\} \tag{7.10}$$

or

$$\widehat{V}_{1,2}(x) \simeq A\{2\eta_0(\Lambda \pm \Lambda^{1/2})x^2 - (|\xi|/2)(1 \pm 3\Lambda^{1/2})x^3 + x^4 + \eta_0^2\Lambda^{3/2}\delta_{i,2}\} \tag{7.11}$$

where a displacement x is referred to a minimum of the corresponding individual well. Moreover, the shape of different individual potential wells ($l = 1, 2$) in a double-well potential approach a harmonic one with increasing $|\eta|$ or η_0; this trend can readily be seen, in particular, for symmetric potentials at $\xi = 0$ or $\Lambda = 0$, for which the interwell barrier $V_B^{(\gamma)}(x)$ and barrier height $V_B^{(\gamma)}$ ($\gamma = 1, 2$) can be described as follows:

$$V_B^{(\gamma)}(x) \approx A(x^4 - q_\gamma x^2) \quad \text{or} \quad V_B^{(\gamma)} \simeq Aq_\gamma^2/16, \tag{7.12}$$

where $q_1 \simeq 2|\eta|$ or $q_2 \simeq \eta_0$. Thus, the double-well potentials occur in two different ranges of parameters η and η_0, $\eta < 0$ and $0 < \eta < \eta_0$, separated by a some gap at $\eta \approx 0 \approx \eta_0$ at which the interwell barrier is too low ($V_B^{(\gamma)} = 0$ at $\eta = 0$ or $\eta_0 = 0$). Therefore, no smooth transition seems to exist between the noted two ranges due to a continuous variation of the parameters; in this sense, two formally different types of double-well potentials and soft-mode excitations occur (cf. [138]).

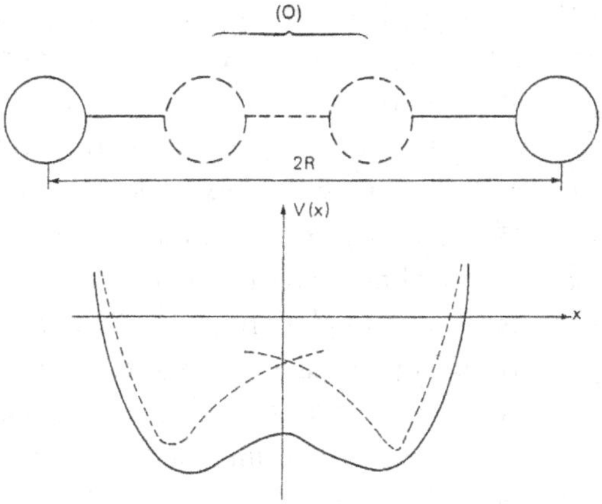

Figure 8: Simplest structural model of soft-mode potential $V_{sm}(x)$.

The explicit functional form of the parameters $\{\eta(Y), \xi(Y)\}$ and $F(\eta, \xi)$ can be found in specific microscopic models of the soft-mode structure in a glass [123, 124, 125]. At present, there are a number of microscopic structural models of soft modes in glasses, analytical and simulated, which seem to be rather "oversimplified" relative to realistic soft-mode structures. The simplest microscopic model of a soft-mode is a linear three-atomic fragment (Fig. 8) of a fixed length $R \equiv y_1 - y_{-1}$ (though fluctuating from one fragment to another around an average value $\overline{R}$), with the intermediate atom (0) of coordinate x moving between two boundary atoms $(-1, 1)$ with fixed but random locations $(Y \equiv \{y_{-1}, y_1\})$, in an atomic chain of a continuous random network (e.g., in a covalent glass) at an average coordination number $\overline{z}$ (e.g., $\overline{z} = 2$ in a-Se) [125, 157]. The mode "softness" parameter can be explicitly defined as $\eta \equiv (R_c - R)a_0^{-1} \gtreqless 0$; R_c is a critical value of R, which is determined by interatomic force constants in pair potentials $V_{-1,0}[r_{-1,0}(x)]$ and $V_{0,1}[r_{0,1}(x)]$ of the atomic pairs $(-1, 0)$ and $(0, 1)$. The asymmetry parameter ξ is determined by the difference ΔV of the pair potentials, and the explicit expression is defined as $\xi = \xi(R, R_c) \propto \Delta V = V_{-1,0}[r_{-1,0}(x = y_{-1} + R/2)] - V_{0,1}[r_{0,1}(x = y_1 - R/2)]$. In the limit of small R and large $\eta \approx 1$, the pair potentials strongly overlap and give rise to a standard, "non-soft", single-well potential with a large $k \approx k_0$, whereas in the limit of large R and $|\eta| = -\eta \approx 1$, the pair potentials do not overlap and rather lead to a double-well potential with $k \approx k_0$

in each well for the "intermediate" atom. However, at R close to R_c and $|\eta| \ll 1$, the "intermediate" atom moves in a soft-mode potential that can be either a single-well or double-well one, as well as a potential of intermediate shape. The model under discussion, formally with the bond angle $\theta = \pi$ at the "intermediate" atom, can be extended to a non-linear three-atomic fragment at a fixed, though random, bond angle θ in the range at $0 < \theta < \pi$, with the same qualitative results [157]. Other structural models of a similar type soft mode in a random network glass have also been proposed, e.g., in [160, 161]. In particular, an investigation of the potential energy of an oxygen atom in a linear three-atomic fragment Si –O – Si and some dynamic characteristics of a simplified structural model of SiO_2 glass has been performed, in which the O-Si interaction was described by a well-known anharmonic Morse potential and both soft ($k \ll k_0$) and "non-soft" ($k \approx k_0$) motion modes were described and analyzed. Some models of double-well potentials in a glass have also been developed (e.g., in [162]) which were related to void-like or "free volume" and structural defects (a correlation between the occurrence of atomic non-single-well potentials and violation of long-range order in a crystal has first been noted in a discussion of the origin of melting in solids [163]). It has to be noted that in the above mentioned models, a soft mode is an "expansion–compression" one. When a soft mode is associated with a shear or a rotation, the parameter η has a different meaning and may not be related to free volume, void-like defects or density fluctuations. In particular, the occurrence of soft-mode potentials, including double-well ones, is interpreted in a structural model of SiO_2 glass [118] in terms of coupled rotations of quasimolecular atomic groups of SiO_4 tetrahedra. A soft mode related to a rotational angle is characteristic of this model, in which η is actually determined by parameters of bond bending forces and elastic energies of dipolar (and/or possible quadrupolar) forces. Some essential features of soft-mode dynamics described in SMM can be confirmed in more sophisticated microscopic structural models of soft modes which follow from computer simulations on a model glass, first of all on a soft sphere glass. In fact, in this model glass, soft modes were revealed, with a large effective mass $M_{sm} \approx n_{sm}M$ at $n_{sm} \sim 10$ [164]. Similar scales of soft-mode effective mass have also been found from simulations on other models of non-metallic glasses, e.g., v-SiO_2 and v-Se [165]. Moreover, a soft mode has been established to be centred at a core, a structural irregularity with a large local strain. In [165], simulations were performed on large systems, so that the effects of interactions of localized soft modes with extended phonons and with each other could be studied in more detail. The vibrational frequency DOS $g(\nu)$ and related specific heat $C(T)$ of the one-atom glass, with mass M and well-known interatomic potential (e.g., the well-known Morse two-body interaction) in the soft sphere model, were directly calculated. The investigation of the results of calculations has shown that the occurrence of low frequency vibrational modes in glasses can be understood as due to interacting localized (soft) and extended (acoustic) modes, just as obtained in SMM (see Sec. 6). A method was developed in the cited study, which allowed one to deconvolute the low frequency vibrational eigenstates into their propagating and quasi-local or "resonant" components (see Sec. 10.2) and for the latter again, the effective mass $M_{sm} \sim 10\,M$ was found (the cores of these modes were low dimensional ones, e.g., chains with side branches). The anharmonicities of the soft-mode potential $V_{sm}(x; \eta, \xi)$ have also been calculated [164], the main result being in agreement with the results of SMM.

It seems of interest that molecular dynamic simulations are recently used to model two-level system excitations in silica glass [166]. The excitations' parameters are estimated to be broadly distributed in such a way that the tunneling amplitude varies the least in the 0.01–1 K range. The effective mass M_{TLS} of the excitations is found to correspond to approximately one to five coupled SiO_4 tetrahedra in a TLS that is rather like a $3D$ cluster. Simultaneous jumps of coupled SiO_4 tetrahedra are also found in different TLS at very low T, which are considered as evidence for strong interaction between the excitations below the mentioned temperature range (Sec. 9.2).

7.2 Probability distribution densities

Since physical soft-mode properties Q depend on random parameters (η, ξ) and $Q = Q(\eta, \xi^2)$ (see below Eq. (7.3)), the PDD $G(Q)$ are also important characteristics of the soft modes, which can be described by the following conventional probability theoretical expression in terms of the above introduced basic PDD $F(\eta, \xi)$:

$$G(Q) = \iint_{\Omega_{sm}} d\eta d\xi F(\eta, \xi)\delta[Q - Q(\eta, \xi^2)], \tag{7.13}$$

where the soft-mode range of $|\eta| \ll 1$ and $\xi^2 \ll 1$ is denoted as $\Omega \equiv \Omega_{sm}$. Since in the model under discussion, the microscopic variables Y (bond lengths, bond angles, dihedral angles, etc.) of the soft-mode configuration environment are reduced to two parameters $\{\eta(Y), \xi(Y)\}$, the function $F(\eta, \xi)$ can be considered to present a "two dimensional" description of the random local glass structure, more general than the "one dimensional" description given by the standard radial distribution densities $\rho(R)$ concerning a few (e.g., first and second) nearest neighbors [2, 7], and can be expressed in terms of the PDD $P(Y)$ as follows:

$$F(\eta, \xi) = \int \cdots \int dYP(Y)\delta[\eta - \eta(Y)]\delta[\xi - \xi(Y)], \tag{7.14}$$

at $\int_{-\infty}^{\infty}\int_{-\infty}^{\infty} d\eta d\xi F(\eta, \xi) = 1$. Specific expressions for $\{\eta(Y), \xi(Y)\}$ and $F(\eta, \xi)$ can be found in specific microscopic models of local glass structures (see some example in Sec. 7.4). In accordance with Sec. 7.1, the PDD should be an even function of the soft-mode potential asymmetry ξ as there is no preferred sign of asymmetry (as well as of direction) in glasses being isotropic amorphous systems, i.e., $F(\eta, \xi) = F(\eta, -\xi) = \widetilde{F}(\eta, \xi^2) \equiv F[\eta, \sqrt{(-\xi)^2}]$. Since in glasses, mechanically stable solids atomic configurations of small asymmetries appear to be most probable, the function $F(\eta, \xi) = F(\eta, \sqrt{\xi^2})$ may be expected to exhibit one maximum at $\xi = \overline{\xi} = 0$ or two symmetric maxima at finite $\xi = \pm\overline{\xi} \neq 0$ and $\overline{\xi}^2 \ll 1$, as schematically presented in Fig. 9, curves (b). For the sake of simplicity, we apply in what follows the PDD $F(\eta, \xi)$ at $\overline{\xi} = 0$ but this suggestion does not seem to restrict the generality of results. As noted in the introduction to the present Section, the vast majority of atoms and atomic potentials may be characterized by standard ("non-soft") single-well potentials with $k \approx k_0$, so that the PDD $F(\eta, \xi)$ must exhibit its highest maximum at $\eta \approx \overline{\eta} = 1 \approx \eta_0$ and $\xi = \pm\overline{\xi}, \overline{\xi} \geq 0$, corresponding to the highest maximum of $P(Y)$ at the average values $\overline{Y}$. In this connection, the soft modes should correspond to a tail of the PDD at small $|\eta| \ll 1$ and $\xi^2 \ll 1$, while their low atomic concentration, $c_{sm} \ll 1$, means that the tail

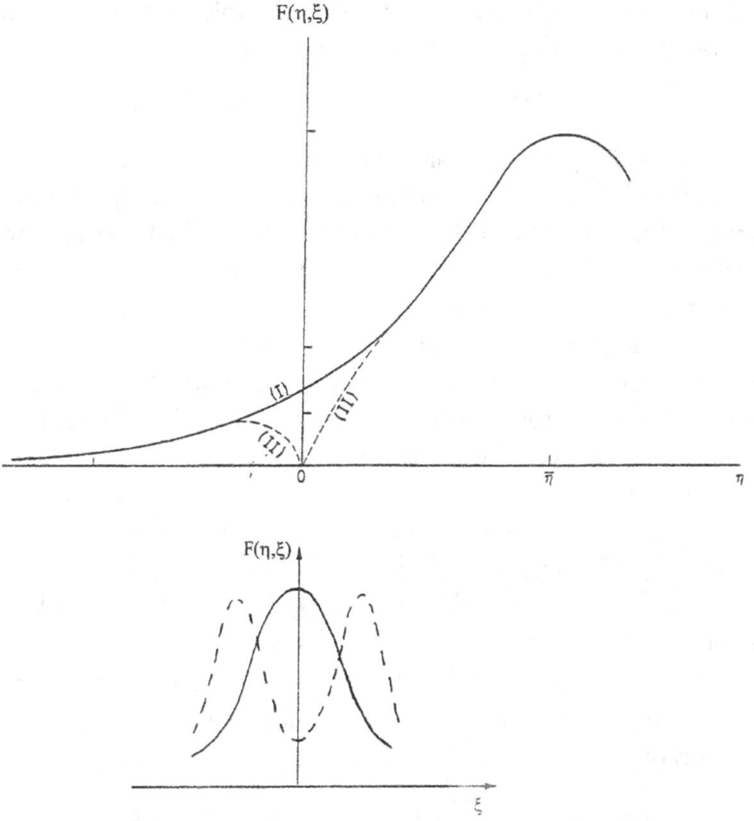

Figure 9: Two limit types of distribution density $F(\eta,\xi)$ for random soft-mode parameters (η,ξ) at small $|\eta|(\to 0)$, at $\bar{\xi}=\bar{\xi}(\xi=0$ or $0\neq\bar{\xi}\ll 1)$.

of $F(\eta,\xi)$ has to decrease sufficiently rapidly at $\eta\to-\infty$. The decreasing (not necessarily monotonously) tail of $F(\eta,\xi=const.)$ at $\eta\ll 1$ includes both soft modes at $|\eta|\ll 1$ and $\xi^2\ll 1$, with $k\ll k_0$, and "non-soft" modes with $|\eta|=-\eta\approx 1$ and $k\approx k_0$ in each well of non-single-well, first of all double-well, potentials of which the atomic concentration should be much lower than that of all soft modes.

In the original SMM [119, 123], it has been assumed for the sake of simplicity that the PDD $F(\eta,\xi)$ was a regular function of both η and ξ, i.e., $F(\eta,\xi)=F_1(\eta,\xi)\approx const.\neq 0$ at their small values. A general property of $F(\eta,\xi)$ for soft modes at small $|\eta|\ll 1$ and $\xi^2\ll 1$ has at first been established in [158] by an analysis, in which the typical soft-mode potential (7.2) was applied in the equivalent form of Eq. (7.4):

$$\widehat{V}_{sm}(x) = (1/2)A_2x^2 + (1/3)A_3x^3 + (1/4)A_4x^4, \tag{7.15}$$

where $A_n \equiv (d^nV/dx^n)_{x=0}$ are random parameters, e.g., $A_4 \equiv 4A$, $A_2 \equiv 2A\eta$ and $A_3 \equiv 3A\xi$. A simplified glass model was applied for calculations of the PDD $P(A_2, A_3)$ for sufficiently small $|A_2| \to 0$, in which a weakly coupled "atom" moves in a soft mode and the environment PDD $P(Y) \equiv P(\{\mathrm{y}_i\})$ is suggested to be regular, in the sense that it does not exhibit singularities and zeros (see above Eq. (7.4)). The potential V is approximated by a sum of energies $\phi(x - y_i)$ of central interaction forces of the "atom" with its environment ($\{y_i\}$). The same result was argued in the cited work to be obtained in a three-dimensional glass with interaction energies $\phi_i(|\mathbf{r} - \mathbf{y}_i|)$ related to central interaction forces. Taking this into account, one can find from the equation $V'(x_0) \equiv (dV/dx)_{x=x_0} = 0$ that at a minimum point $x = x_0$, the partial derivative $\partial x_0/\partial y_i = A_2^{-1}\phi_i'' \equiv A_2^{-1}\partial^2\phi(x_0 - y_i)/\partial x_0^2$ can be large at small A_2. Due to the strong dependence of x_0 on x_i, $A_2 = V''(x_0)$ and $2A_3 = V'''(x_0)$ strongly change with x_i: $\partial A_2/\partial x_i = 2A_3A_2^{-1}\phi_i'' - \phi_i'''$, $\partial A_3/\partial x_i = 3A_4V_2^{-1}\phi_i'' - \phi_i'/2$. Therefore, for the calculated resultant expression of

$$\begin{aligned} P(A_2, A_3) &= \int P(\{y_i\})\delta\{A_2 - A_2[x_0(\{y_i\}), Y \\ &\equiv \{y_i\}]\}\delta\{A_3 - A_3[x_0(\{y_i\}], y_i]\}\Pi_i dy_i, \end{aligned} \tag{7.16}$$

two limiting cases are available, corresponding to two different types of the behavior of $V = \sum_i \phi(x - y_i)$ in glasses of different structures. In one limiting case, all quantities $|\phi_i''| \lesssim |A_2| = |\sum_{j\neq i}\phi_j''|$ for the "atom" (e.g., in a relatively large void) weakly coupled with its environment, so that $|\nabla A_2|$ and $|\nabla A_3|$ are finite at $A_2 \to 0$, just as originally assumed in the SMM, $P(A_2, A_3) \to$ *const.* at $A_2 \to 0$. In another limit case, ϕ_i'' exhibit alternative signs so that $|A_2| \ll |\phi_i''|$ for some values of i at $A_2 \to 0$. Then, $|\nabla A_3| \sim |A_2|^{-1}$on a surface $A_3(x_0, x_i) = A_3 \equiv$ *const.*, whereas another surface $A_2(\{y_i\}) = A_2 \equiv$ *const.* the vector ∇A_2 is parallel to ∇A_3 (by definition perpendicular to A_3) and its projection onto a sub-surface defined by equations $V_2(x_0, x_i) = A_2$ and $V_3(x_0, x_i) = A_3$ is finite at $A_2 \to 0$. In this case, the PDD is readily seen to be noticeably different, $P(A_2, A_3) \to const.|A_2|$ at the small A_2 under consideration. Taking into account that $F(\eta,\xi)d\eta d\xi = P(A_2, A_3)dA_2dA_3$, i.e., $F(\eta,\xi) = 6A^2P(A_2, A_3)$, one can conclude that two different limit types of $F(\eta,\xi)$ in the soft-mode range can be found at $|\eta| \to 0$ and $\xi^2 \ll 1$ under discussion (Fig. 9, curves (a)): either curve (I)

$$F(\eta,\xi) = F_1(\eta,\xi) \approx F_1^{(0)} \equiv const. \neq 0, \tag{7.17}$$

or curve (II)

$$F(\eta,\xi) = F_2(\eta,\xi) \simeq |\eta|F_0^{(2)} \to 0 \text{ at } F_0^{(2)} \equiv const. \neq 0. \tag{7.18}$$

That is the essential feature of the soft-mode PDD, depending on the character of the "atom" interactions with its environment and the structure of environment in a glass, which is expected to hold true not only for central inter-atomic forces $\phi_i(|\mathbf{r} - \mathbf{y}_i|)$ assumed in the analysis for the sake of simplicity but also for non-central (e.g., covalent) forces. The case of Eq. (7.17) appeared to occur when the contributions of the environment atomic structure to the spring constant of the soft-mode "atom" were of the same sign and comparable magnitude, whereas the case of Eq. (7.18) was expected rather when the contributions are of alternating signs. Since no essential reasons seems to be available for the original PDD $P(Y)$ to have singularities (or zeros) in configuration space (Sec. 7.1), two kinds of the PDD, $F(\eta,\xi) = F_{1,2}(\eta,\xi)$, can be relevant in two limit types of inter-atomic forces. Therefore, generally speaking, both types of the "atom" interactions with different local structures in its environment can coexist in complex glasses, for which the PDD

$$F(\eta,\xi) = c_1F_1(\eta,\xi) + c_2F_2(\eta,\xi), \tag{7.19}$$

at $c_1 + c_2 = 1$ and $0 < c_{1,2} < 1$, can be relevant. Although the PDD $F(\eta,\xi)$ are different in two limit cases, the qualitative and scale difference in the properties of low-energy excitations, as well as associated low-temperature and low-frequency properties of glasses, do not appear to be drastic, as can be seen from direct calculations for very-low energy properties of glasses (Sec. 8) and for some essential moderately low energy properties (Ch. 10). In other words, either Eq. (7.17) or Eq. (7.18), or their superposition of

Eq. (7.19), can be used for describing the basic distribution density of random soft-mode parameters.

A quite different approach to the investigation of the features of $F(\eta, \xi)$ was later developed in [159]. In this approach one takes into account that the expansion in Eq. (7.1) is transformed to that in Eq. (7.2) by the following displacement of the reference point: $x' - x = (\xi' - \xi)/4$. The invariants of the transformation are as follows:

$$\begin{aligned} I_1 &\equiv \eta - 3\xi^2/8 = \eta' - 3\xi'^2/8, \\ I_2 &\equiv (1/2)(\eta - \xi^2/4) = -h' + (1/2)(\eta' - \xi'^2/4), \end{aligned} \tag{7.20}$$

with the Jacobian $|\eta|/2$. Then, one can express the normalized PDD $P(\eta, \xi, \xi')$ containing three "mixed" parameters (η, ξ, ξ') in terms of the PDD $P'(h', \eta', \xi')$, as follows:

$$P(\eta, \xi, \xi') = (|\eta|/2)\; P'[h'(\eta, \xi, \xi'), \eta'(\eta, \xi, \xi'), \xi'], \tag{7.21}$$

noting that the soft-mode potential in Eq. (7.2) and the related energy spectrum do not depend on ξ'. The result is that the PDD for the soft-mode energy spectrum and associated physical properties $Q(\eta, \xi^2)$ can be expressed in terms of this PDD averaged over the parameter ξ':

$$\begin{aligned} F(\eta, \xi) &= \int_{-\infty}^{\infty} d\xi' P(\eta, \xi, \xi') \\ &= (|\eta|/2) \int_{-\infty}^{\infty} d\xi' P'[h'(\eta, \xi, \xi'), \eta'(\eta, \xi, \xi'), \xi'], \end{aligned} \tag{7.22}$$

in which the last integral was not explicitly calculated. An additional assumption has been made here that the last integral is in general finite at $\eta \to 0$, from some physical arguments associated with specific experimental data concerning specific heat and thermal conductivity in glasses at moderately low $T \sim 10\,\mathrm{K}$.

If this assumption holds, then indeed, $F(\eta, \xi) = F_2(\eta, \xi) \simeq |\eta| F_2^{(0)}$ with $F_2^{(0)} = const. \neq 0$ as $\eta \to 0$. However, the assumption might be invalid for some glasses for which the integral in Eq. (7.22) could diverge at $|\eta| \to 0$ and in this case, in general, neither $F(\eta, \xi) = F_1(\eta, \xi)$ nor $F(\eta, \xi) = F_2(\eta, \xi)$ would be relevant. This feature of the soft-mode PDD, in particular, does not contradict the existence of universal dynamic properties determined by very low energy soft-mode excitations (Sec. 8.1), at least in the applied approximation in which independent soft modes (not interacting with each other) are only considered. Thus, so far, the conclusion is that there is no strict proof in favour of the universality of only one of the two types of the basic PDD, $F(\eta,\xi)=F_1(\eta,\xi)$ or $F(\eta,\xi)=F_2(\eta,\xi)$, at least for the independent soft modes under discussion.

In accordance with Sec. 7.1 (see below both Eq. (7.4) and Eq. (7.12)), the probability of a void-like fluctuation related to a soft mode in the structural model of Fig. 8 is expected to strongly decrease with increasing fluctuation length $[R-\overline{R})^2]^{1/2}$, plausibly by the Gaussian law $\exp[-(R-\overline{R})^2/(\Delta R)^2]$, so the concentration of single-well soft-mode potentials has to be significantly higher than that of double-well potentials, just as estimated above. In this model, $F(\eta,\xi)$ can be related to the radial distribution density $\rho^{(2)}(R)$ for second nearest neighbors (e.g., with $R=R_c-\eta a_0>0$ at $R_c>a_0$), and the PDD $F(\eta,\xi)$ can be described as $F(\eta,\xi)=F_I(\eta,\xi)\approx f_I(\xi)$ at $|\eta|\to 0$ (Eq. (7.17)), where $f_I(\xi)$ is a regular function (Fig. 9, curve (a I)). Moreover, it was found in simulations [164, 165] on soft sphere model glass that $F(\eta,\xi)=F_2(\eta,\xi)\approx|\eta|\cdot f_2(\xi)$ at $|\eta|\to 0$ (Eq. (7.18)) with $f_2(\xi)$ another regular function (Fig. 9, curve (a II)). It was also established in the simulations that the distribution function of other, generally speaking, random parameters of the soft-mode potential, like curvature (A) or quartic anharmonicity (w) were practically independent of other parameters and sufficiently narrow to replace estimates by their average values (e.g., $w\approx const.$). All the findings appear to support the main arguments, suggestions and results of SMM. In particular, one can conclude from the above results of simplified models and simulations that both limiting types (Eqs. (7.17)–(7.19)) of the PDD $F(\eta,\xi)$ can occur for soft modes in glasses.

Since both types of the function $F_\alpha(\eta,\xi)$ ($\alpha=1,2$) have to be normalized to the total atomic concentration $c_{tot}=1$, its soft-mode tail can approximately be normalized to the total soft-mode concentration c_{sm} that is lower by orders of magnitude, $c_{sm}\ll 1$. In other words, since by definition $\iint d\eta d\xi F(\eta,\xi)=c_{tot}=1$ for any type of $F(\eta,\xi)$, approximately

$$\iint_{\Omega_{sm}} d\eta d\xi F(\eta,\xi)\approx c_{sm}, \tag{7.23}$$

with $\Omega_{sm} \approx \{|\eta| \lesssim \eta_{sm}; \xi^2 \lesssim \eta_{sm}\}$ and η_{sm} a typical upper limit of $\{|\eta|, \xi^2\}$ for soft modes. In accordance with a comment in Sec. 7.1 about a Gaussian approximation of the PDD $P(Y)$ around its major maximum, one may estimate $F(\eta, \xi)$ in its soft-mode tail at $|\eta|, \xi^2 \ll 1$ (e.g., [125]):

$$F_\alpha(\eta, \xi) \approx \pi^{-1}(\Delta\eta)^{-1}(\Delta\xi^2)^{-1}F_\alpha^{(0)} \cdot |\eta|^{\alpha-1} \cdot \exp[-(\eta - \overline{\eta})^2/(\Delta\eta)^2]\exp[-(\xi - \overline{\xi})/(\Delta\xi)^2], \quad (7.24)$$

where for both limit types of PDD $F_\alpha^{(0)} \propto c_{sm}$ and in what follows $\overline{\xi} = 0$. A typical soft-mode tail of the PDD, at $|\eta| \lesssim \eta_{sm} \ll \overline{\eta} = 1$, may be approximated by the following function:

$$H_\alpha(\eta) \equiv \int -\infty^\infty d\xi F_\alpha(\eta, \xi) \approx (\Delta\eta)^{-1}F_\alpha^{(0)}|\eta|^{\alpha-1}\exp(-1/\eta_{sm}) \cdot \exp(2\eta\eta_{sm}^{-1}), \quad (7.25)$$

Here, the width $\Delta\eta$ of the distribution density is small in scale in the sense that the parameter $\eta_{sm} \equiv (\Delta\eta)^2 \ll 1$, and corresponds to a low concentration $c_{sm} \ll 1$, e.g., at typical $\eta_{sm} \sim 0.1$ and $c_{sm} \sim 0.01$. In fact, these and related estimations can be tested in concrete structural models of soft modes (see, e.g., Sec. 7.1) but much more such estimations still have to be made.

7.3 Low-energy excitations: Density of states and concentration

New types of elementary excitations of low energy, $\varepsilon \ll h\nu_D$, are characteristic of atomic soft-mode dynamics, which are quite different from acoustic phonons and can contribute to the universal low-temperature and low-frequency properties of glasses. In what follows, for the low energies implied, the single excitation approximation of the excitation spectrum is assumed to be relevant for calculations of the DOS. In the present Chapter, for the sake of simplicity, the soft-mode excitations, in general corresponding to low-energy excited states (Ψ_j, E_j) of a set of individual anharmonic soft-mode oscillators, are assumed to be independent of (i.e., not interacting with) each other. The effective Hamiltonian of a single excitation is suggested to be $\widehat{H}_{sm} = \widehat{T}_{sm} + \widehat{V}_{sm}$, in which the potential energy is $\widehat{V}_{sm}(x)$

from Eq. (7.2) while the kinetic energy operator is approximated in a standard way as $\widehat{T}_{sm} \approx -(\hbar^2/2M_{sm}a_0^2)(d^2/dx^2)$. The effective soft-mode mass M_{sm} can be estimated as $M_{sm} \sim 10M$, according to some simulations of soft modes [164] (Sec. 7.1) and a comparison of some formulae of SMM for dynamic and thermal properties of glasses with to respective experimental data. Then, the spectrum of single low-energy soft-mode excitations is identified with the one for lowest excited states $\Psi_j(x;\eta,\xi) \equiv |j\rangle$ and associated eigenvalues $E_j(\eta,\xi^2)$ of the Hamiltonian $\widehat{H}_{sm}$ at given values of the soft-mode parameters η and ξ in the Schrödinger equation

$$\begin{aligned}\widehat{H}_{sm}\Psi_j(x;\eta,\xi) &= E_j(\eta,\xi^2)\Psi_j(x;\eta,\xi), \quad \text{at } E_j(\eta,\xi^2)\\ &= E_0(\eta,\xi^2)+\varepsilon_j(\eta,\xi^2).\end{aligned} \tag{7.26}$$

Here $0<\varepsilon_j \ll h\nu_D$, with a discrete quantum number j, $1 \le j \le j_{\max}$. In practice $j_{\max} \ll 10$. In other words, the soft-mode excitations correspond to sufficiently low excited states of a soft-mode anharmonic oscillator,

$$\widehat{H}_{sm} \approx -T_0 d^2/dx^2 + A(\eta x^2 + \xi x^3 + x^4), \quad \text{at } |\eta| \ll 1 \quad \text{and} \quad \xi^2 \ll 1, \tag{7.27}$$

where $T_0 \equiv \hbar^2/2M_{sm}a_0^2 \approx 10^{-4}eV$ at a typical soft-mode effective $M_{sm} \equiv n_{sm}M$ at $n_{sm} \sim 10$. While the discrete quantum number j characterizes the excited state Ψ_j with eigenenergy level E_j above the ground state wave function Ψ_0 with energy level E_0, the function $\varepsilon_j(\eta,\xi^2)$ of continuously varying variables η and ξ describes an energy band of, generally speaking, anharmonic excitations, of which the energy spectrum is invariant to the above-mentioned transformations of $\{x;\eta,\xi\}$, e.g., $\varepsilon(\eta,\xi^2) \equiv \widetilde{\varepsilon}(\widetilde{\eta},\widetilde{\xi}^2)$. In this connection, the excitation energy spectrum under discussion can practically be considered as a continuous one (e.g., with overlapping energy bands), so that the basic spectral characteristic is a continuous function, the excitation DOS $G_{exc}(\varepsilon)$.

Although no general analytical solution of the excitation spectrum problem for the anharmonic quantum-mechanical oscillator is available at present, necessary essential information on the behavior of excitation spectrum $\varepsilon(\eta,\xi^2) \equiv \widetilde{\varepsilon}(\widetilde{\eta},\widetilde{\xi}^2)$ and related DOS $G_{exc}(\varepsilon)$ can still be obtained by applying appropriate perturbation theoretical approaches and the following formula:

$$\varepsilon_j^{(AV)}(\tilde{\eta}=0, \tilde{\xi}^2=0) = 3 \cdot 2^{-4/3} w[(j+1/2)^{4/3} - (1/2)^{4/3}], \quad (7.28a)$$

exactly describing the discrete excitation energy spectrum of soft-mode anharmonic vibrations in a single-well quartic anharmonic potential $\widetilde{V}_{sm}(\tilde{x}) = w\tilde{x}^4$ in Eq. (7.3) (see, e.g., [125] and references therein). The inter-level gap widths increase, $\varepsilon_{j,j-1}^{(AV)}(0,0) = 3 \cdot 2^{-4/3} w[(j+1/2)^{4/3} - (j-1/2)^{4/3}]$, with increasing j, and noticeably exceed at $j \geq 2$ the smallest gap width $\varepsilon_1^{(AV)}(0,0) = 3 \cdot 2^{-4/3} w[(3/2)^{4/3} - (1/2)^{4/3}] \simeq 1.5w$ for the first excited state in this spectrum, in contrast to the discrete harmonic spectrum of frequency ν with all gap widths being $h\nu$. Equation (7.28a) can be used for estimating the smallest finite vibrational single excitation energy $\varepsilon_{min}^{(V)} = \varepsilon_1^{(AV)}(0,0)$ in the continuous band spectrum produced by the perturbation operator $w(\tilde{\eta}\tilde{x}^2 + \tilde{\xi}\tilde{x}^3)$ in Eq. (7.3) at relevant values of the parameters, $0 < \tilde{\eta}, \tilde{\eta}_0 \equiv 9\tilde{\xi}^2/32 \lesssim 1$. Since excitations in a single-well potential can only be vibrational ones, the energy spectrum of vibrational excitations exhibits a finite gap between $\varepsilon = 0$ and $\varepsilon = \varepsilon_{min}^{(AV)} \approx w$, so w determines the lowest, anharmonic, vibrational excitation energy. In other words, the soft-mode excitation theory is characterized in energy variations by a new energy scale $w \approx 3 \cdot 10^{-4}$ eV, much smaller than the standard Debye energy $h\nu_D \approx 3 \cdot 10^{-2}$eV, and by a very small dynamic parameter $\eta_L = (\hbar^2/2M_{sm}a_0^2 A)^{1/3} \approx 10^{-2} \ll \eta_{sm} \approx 0.1$ for typical $|\tilde{\eta}| \approx \eta_{sm}/\eta_L \approx 10$. Then, the vibrations in each potential well become almost harmonic and as well-known in this case, the Schrödinger equation can be reduced to a standard wave equation, with the single excitation eigenvalues of harmonic vibrations (HV) in their first band, $j=1$, (Eq. (7.5)):

$$\varepsilon = (\nu_1^{(HV)}(\tilde{\eta}))^2 \sim (30w/h)^2 \eta, \quad (7.28b)$$

for typical average atomic mass $M \approx 3 \cdot 10^{-23}$ g. The reason for the occurrence of harmonic vibrations with the typical frequency of Eq. (7.28b) is that in Eq. (3.3) at $\eta \sim \eta_{sm} \gg \eta_L$, the operator $w\tilde{\eta}\tilde{x}^2 \approx kx^2/2$ becomes most essential, predominating over the quartic anharmonic operator, while the equal inter-level gaps are $h\nu^{(HV)}(\tilde{\eta}) \gg h\nu_{min}^{(AV)} \approx w$. On the other hand, non-vibrational excitations of lowest energy $\varepsilon_1 \equiv \varepsilon_1^{(NV)} \equiv E_1 - E_0 \ll w$ (Eq. (7.26)) occur in almost symmetric double-well potentials with typical small tunneling amplitudes $\Delta_0 \ll w$. In this case, the lowest excitation energy level is separated from the next excitation one by an interlevel gap of much larger width $\varepsilon_2 - \varepsilon_1 \equiv E_2 - E_1 > w \gg \varepsilon_1 = E_1 - E_0 \equiv \varepsilon_1^{(NV)}$, which is by definition characteristic of two-level systems. Let us add that in general, the soft-mode excitations can be well-defined in the standard sense that their interactions with each other and with acoustic phonons can be so weak such that their excitation life time τ_{exc} is large, $\tau_{exc} \gg \hbar/\varepsilon_{exc}$; for the harmonic vibrational excitations, this property holds for frequencies $\nu^{(HV)}$ in the range at $(3w/h)^2 \lesssim (\nu_1^{(HV)})^2 \lesssim \nu_0^2\ (\ll \nu_D^2)$, at

$$\nu_0 = \nu_1^{(HV)}(\eta = \eta_{sm}) \approx 10w/h \approx 0.1\nu_D \approx 1\,\text{THz}, \tag{7.29}$$

for typical values $\eta = \eta_{sm} \approx 0.1$ and $\nu_D \approx 10\,\text{THz}$. Taking all these into account, one can distinguish three energy ranges in the excitation spectrum:

$$(i)\ \varepsilon = h\nu \ll w, \quad \text{e.g., } \varepsilon \lesssim h\nu_m \approx 0.1w; \quad (ii)\ \varepsilon \sim w;$$
$$(iii)\ \kappa w \lesssim \varepsilon \lesssim h\nu_0(\ll h\nu_D). \tag{7.30}$$

at plausible $\varepsilon_m \equiv h\nu_m \sim 10^{-4}\,\text{eV}$, $w/\varepsilon_m \sim 3$ at $(0<)\ \kappa - 1 \ll 1$, and $h\nu_0/\varepsilon_m \sim 5h\nu_0/w \sim 30$.

Applying standard general expressions, the total DOS of soft-mode excitation spectrum can formally be described as follows:

$$G_{exc}(\varepsilon) = \iint_{\Omega_{exc}} d\eta d\xi F(\eta, \xi)\delta[\varepsilon - \varepsilon_1(\eta, \xi^2)], \tag{7.31}$$

where Ω_{exc} denotes the appropriate range of eigenvalues $\varepsilon = \varepsilon_1(\eta, \xi^2)$ for soft-mode excitations in their lowest band. Taking into account Eq. (7.30), the total excitation DOS can be approximated as

$$\begin{aligned} G_{exc}(\varepsilon) \approx\ & J^{(HV)}(\varepsilon = \nu^2)\Theta(h\nu - \kappa w)\Theta(\lambda h\nu_0 - h\nu) \\ & + G^{(AV)}(\varepsilon = h\nu)\Theta(\varepsilon - w)\Theta(\kappa w - \varepsilon) \\ & + G^{(NV)}(\varepsilon = h\nu)\Theta(\varepsilon_m - \varepsilon), \end{aligned} \tag{7.32}$$

where $\lambda = 1 + \upsilon$ and $0 < \upsilon \ll 1$, and the unit step function $\Theta(x) = \{1$, at $x > 0; 0$, at $x < 0\}$.

As can be seen more in detail in Chs. 8 and 10, the soft-mode model establishes that the typical soft-mode excitations at very low energies $\varepsilon = h\nu \lesssim \varepsilon_m \ll w$ are non-vibrational two-level systems (TLS), which are equivalent to the so-called tunneling states (TS) due to atomic tunneling in double-well potentials while at moderately low energies $\kappa w < h\nu \lesssim h\nu_0$ they are harmonic vibrational (*HV*) ones and at "intermediate" energies $\varepsilon \sim w$, they are anharmonic vibrational ones or partly related to overbarrier thermal activated hopping in double-well potentials. Thus, the soft-mode excitations display very different character in the three energy ranges. Correspondingly, the typical soft-mode excitations are described at moderately low energies by the total vibrational DOS, $G_{\text{exc}}(\varepsilon) \simeq J^{(HV)}(\varepsilon = \nu^2)$, while at very low energies by the DOS, $G_{\text{exc}}(\varepsilon) \simeq G^{(TLS)}(\varepsilon)$ (Sec. 8.1). A crossover can be expected to occur with increasing ε from the non-vibrational excitations to anharmonic vibrational excitations with the DOS $G^{(AV)}(\varepsilon)$ at $\varepsilon \sim w$, while another crossover to overbarrier thermal-activated hopping processes can be expected to occur with increasing T in the range $T_m < T \lesssim w$ (see Chs. 8 and 10). In principle, in the soft-mode model, the scale of partial concentrations $c_{sm}^{(1)}$ and $c_{sm}^{(2)}$ of soft modes in single-well (1) and double-well potentials (2), as well as $c_{sm}^{(\text{int})}$ for soft modes in "intermediate" anharmonic potentials [157], can be estimated from appropriate basic experimental data for glasses (Sec. 6.1), by applying equations like (7.23)–(7.25) and similar ones, in different ways for universal tunneling dynamics at very low energies and for another universal property (boson peak) at moderately low energies. However, it seems reasonable to assume that the following scale estimations of soft-mode concentrations are generally acceptable:

$$c_{sm} = c_{sm}^{(1)} + c_{sm}^{(\text{int})} + c_{sm}^{(2)} \simeq c_{sm}^{(1)} \gg c_{sm}^{(\text{int})} \gg c_{sm}^{(2)}, \qquad (7.33)$$

e.g.,
$c_{sm}/c_{sm}^{(\text{int})} \sim 10 \sim c_{sm}^{(\text{int})}/\ c_{sm}^{(2)}$, at $c_{TS} \approx 10^{-6}$, at $F_1^{(0)} \approx 10^{-2}$ and $\eta_{sm}^{-1} \approx 10$. In particular, the total soft-mode concentration can be estimated as $c_{sm} \approx \int_{-\infty}^{\eta_{sm}} d\eta H(\eta) \sim c_{sm}^{(2)} \exp(+1/\eta_{sm}) \sim 10^4 c_{TS} \sim 10^{-2}$ (the estimates appear to be compatible with Eq. (8.15).

7.4 Interaction of soft-mode excitations with acoustic phonons

In SMM, low-energy soft-mode excitations coexist and interact with essentially different acoustic phonons (phonons, in what follows). The interaction is expected to exist in general since at least, translational symmetry does not occur in glasses and gives rise to processes of resonant and relaxation associated scattering and attenuation (absorption) processes of propagating acoustic (phonons) and electromagnetic (photons) waves by low-energy soft mode. Generally speaking, two different types of soft-mode acoustic interaction mechanisms and related operators, $\widehat{V}_{\text{int}}$, can be distinguished by its origin.

One of the most general mechanism is the common "deformation-potential" mechanism [133] that is align associated with variations of the soft-mode potential shape in a wave field [123, 125, 127, 167]. In fact, an external field Φ changes the soft-mode parameters, e.g., $\eta \to \widehat{\eta}(\eta, \xi; \Phi) = \eta + \delta\eta(\eta, \xi; \Phi)$, and thus the excitation energy spectrum, $\varepsilon \to \varepsilon(\eta, \xi; \Phi)$, and the related DOS $G_{\text{exc}}(\varepsilon; \Phi)$. The resulting soft-mode-acoustic interaction can be described by the Hamiltonian $\widehat{V}_{\text{int}}(x; \Phi) = \widehat{V}_{DP}(x; \Phi)$ which is linear in a "small" acoustic strain tensor, $\widehat{e}_{\text{ac}} \equiv \widehat{e} = \Phi$, which by applying the most general form of the potential energy in Eq. (7.1), reads as follows:

$$\begin{aligned}
\widehat{V}_{\text{int}}(x) &\equiv \widehat{V}_{\text{int}}(x, \mathbf{r} = 0) \approx \widehat{V}_{DP}(x) \\
&= \widehat{V}_{sm}[x; \widehat{h}(\widehat{e}; h', \eta', \xi'), \widehat{\eta}(\widehat{e}; h', \eta', \xi'), \widehat{\xi}(\widehat{e}; h', \eta', \xi')] \\
&\quad - \widehat{V}_{sm}(h', \eta', \xi') \simeq \widehat{ef}(x; h', \eta', \xi'). \qquad (7.34)
\end{aligned}$$

Here, $|\widehat{e}| \ll 1$ and the soft-mode location is the reference point $r = 0$, while $\widehat{f}(x; \eta, \xi) = \widehat{\kappa}^{(1)}x + (1/2!)\widehat{\kappa}^{(2)}x^2 + (1/3!)\widehat{\kappa}^{(3)}x^3$ and $\widehat{\kappa}^{(\sigma)} = \partial q^{(\sigma)}/\partial \widehat{e}_{\text{ac}}$ are the respective tensorial functions with $q^{(1)} \equiv h', q^{(2)} \equiv \eta', q^{(3)} \equiv \xi'$. In other words, the interaction energy operator $V_{DP}(x)$ of acoustic phonons with a soft mode is determined by fluctuations $\delta\eta'(h', \eta', \xi'; \widehat{e}) \propto \widehat{e}$ of the parameters h', η', ξ' due to the small elastic strain $\widehat{e}$. Since for soft modes, typically $|x| \lesssim 1$ and there are no reasons for $|\widehat{\kappa}^{(\sigma)}|$ to increase with increasing σ, the simplest approximation for the interaction energy operator largely applied for estimations in the soft-mode model is the linear one:

$$\widehat{V}_{\rm int} = \widehat{V}_{DP}(x) \approx \widehat{\beta} x \widehat{e}. \tag{7.35}$$

Here, the tensorial coupling constant $\widehat{\beta} \equiv \widehat{\kappa}^{(1)}$ in soft-mode–acoustic interaction is a random parameter, generally speaking, with random magnitude and sign. The interaction of a soft mode with elastic media can be approximated by a "resonant" operator in the sense that acoustic phonons participating in the interaction are of energies close to the soft-mode excitation energy. The resonant interaction is characterized by absorption or emission of a phonon of energy $h\nu$ and corresponding emission or absorption of a soft-mode excitation of energy $\varepsilon \simeq h\nu$. The deformation-potential interaction is found from direct estimations to be most important at least for the above mentioned two-level system excitations in double-well potentials which predominate at very low enough energies $\varepsilon \ll w \ll h\nu_D$ (Ch. 9).

Another mechanism [127, 168] is similar to the "modulation" interaction of defect-related vibrations with acoustic phonons in crystals [169]. The mechanism is associated with displacements of soft-mode oscillators centres and resultant modulation of the excitation energy. Since by definition, a soft-mode "defect" of typical size L_{sm} is weakly coupled with neighboring atoms moving in standard, "non-soft" harmonic modes at $k \approx k_0$, these atoms move in a low-frequency acoustic wave of a typical length $\lambda_{\rm ac} \gg L_{sm}$ as a rigid solid so that for harmonic soft-mode excitations, the system can be considered as an oscillator with a vibrating centre. Since the phonons are not eigenmodes of the system of interacting acoustic and soft modes, the respective interaction Hamiltonian approximated to first order in a small vectorial displacement u can be described as

$$\widehat{V}_{\rm int}(x) \approx \widehat{V}_{\rm M}(x) = \widehat{V}_{sm}(x+\Delta x, \mathbf{r}) - \widehat{V}_{sm}(x, \mathbf{r}) \simeq \Delta x d\widehat{V}_{sm}(x, \mathbf{r})/dx, \tag{7.36}$$

where $\Delta x = \sum_{\alpha=1,2,3} u_\alpha(t)\zeta_\alpha$ is the oscillator centre displacement due to acoustic vibrations, $u_\alpha(t)$, in a direction defined by a vector with components ζ_α. The transition matrix elements are $\langle i|\, d\widehat{V}_{sm}/dx\, |i'\rangle = (\varepsilon(i) - \varepsilon(i'))\cdot\langle i|d/dx|i'\rangle = (m_{sm}/\hbar^2)(\varepsilon(i)-\varepsilon(i'))^2\cdot\langle i|x|i'\rangle$, where $\langle i|$ and $|i'\rangle$ are the initial and final phonon states, and the "modulation" mechanism can be most important for harmonic vibrational excitations mainly in single-well potentials for not too low excitation energies at $h\nu_D \gg \varepsilon \gg w$.

In fact, the interaction operators in Eqs. (7.34)–(7.36) give rise to similar relationships for the transition probabilities at both $\varepsilon \ll w \ll h\nu_D$ and $h\nu_D \gg \varepsilon \gg w$, taking into account that $<j|dV_{sm}/dx|j'\rangle = M_{sm}(\varepsilon_j - \varepsilon_{j'})^2 \hbar^{-2}\langle j|x|j'\rangle$ in the representation of the soft-mode excitations eigenstates $\Psi_j(x;\eta,\xi) \equiv |j\rangle$ and eigenvalues $\varepsilon_j(\eta,\xi^2)$. Then one can use the deformation-potential interaction operator, for estimations of the transition probabilities and related properties at the low $\varepsilon \ll h\nu_D$ under consideration, in particular, the simplest version of Eq. (7.34) in which for the sake of simplicity the scalar approximation is applied, which does not distinguish longitudinal and transverse acoustic phonons and substitutes the dimensionless scalar strain, e.g., $e_{\text{ac}} \equiv e = \partial u_z/\partial z$, for the tensorial acoustic strain $\widehat{e}$, where u_z is a microscopic acoustic displacement in a point with a coordinate z. In this approximation, the coupling constant β is a scalar energy parameter of a random sign. The magnitude $|\beta|$ of the random coupling constant in soft-mode-acoustic interaction is subjected to a distribution density that is assumed in SMM, as apparently can be found from simulations of soft modes, and is found to be rather narrow so that the scale of $|\beta|$ can be roughly estimated as the average value of its magnitude.

The variation of excitation energy due to a weak strain can be characterized by $\delta\varepsilon(e_{ac}) \sim |\beta e|$ so in scale $|\beta| \sim \delta\varepsilon(e \sim 1)$. Let us now use the estimation $|\beta| = |d\varepsilon/de| = |(d\varepsilon/dk)(dk/de)| \approx |(\varepsilon/k)(dk/de)|$, in which for soft modes, $|dk/de| \approx |dk_0/de|$. This plausibly means that the smallness of a soft-mode spring constant can also be understood as a result of mutual cancellations of large terms in the expansion of $k = k(e)$ ($\ll k_0(e)$) in powers of a small strain e (but not in the expansion of the derivative dk/de, as any expansion term in $k(e)$ depends on e in the same way as in the expansion of $k_0 = k_0(e)$ for a standard mode [133]). Indeed, such an approach gives rise to a reasonable value $|\beta| \approx \beta_0 = \delta\varepsilon(e \sim 1) \equiv h\nu_D \approx 3 \cdot 10^{-2}$ eV for standard atomic single-well potentials with $k \approx k_0$, as expected. However, the susceptibility $\chi^{(sm)}$ of a soft mode, with respect to an imposed force, is much larger than the $\chi^{(0)} = k_0^{-1}$ of a Debye mode (Eq. (7.8)). Hence, the scale estimation of $|\beta|$ can be described as follows [125]:

$$|\beta| \approx h\nu_D(\chi^{(sm)}/\chi^{(0)}) \approx h\nu_D\eta_L^{-1} \approx 1 \text{ eV} \gg \beta_0 \approx 3 \cdot 10^{-2} \text{ eV}, \quad (7.37)$$

both for non-vibrational TLS (i.e., TS) excitations of very-low energy and for vibrational excitations of moderately low energy in soft-mode potentials, for which the magnitude of soft-mode parameters is characterized by the parameter $\eta_L \sim 10^{-2}$ (a somewhat different way of assessing $|\beta|$ for the excitations, with a similar result, is available in [127, 159]). Equation (7.37) appears to explain the occurrence of anomalously large values of the deformation potential coupling constant, $|\beta| \sim 1$ eV for TS excitations, found by comparing experimental data for acoustic phonon mean free path [106, 122], thus determining the thermal conductivity of glasses at very low $T \lesssim 1$ K, with the theoretical results [116] (see comments after Eq. (6.3)).

One can also conclude from Sec. 7.4 that the soft-mode model gives rise to a coupling constant in the interaction of soft-mode excitations with acoustic phonons, whose magnitude $|\beta|$ is anomalously large compared to that of acoustic phonons with standard local vibrations, $\beta_0 \approx h\nu_D$, because of the softness of motion modes, as first estimated in [123] (1983). Equations (7.35)–(7.37) describe interactions that, as well as the standard electron–phonon and anharmonic phonon–phonon interactions in solids [113], satisfy the van Hove criterion of dissipative interactions which also contribute to the finite width $2\gamma_{ac}(\nu)$ of acoustic phonon frequency scattered by soft-mode excitations.

The basis of soft-mode model of low-energy dynamics in glasses (SMM) contains a general description of typical, most probable single-well and non-single-well potentials for characteristic atomic "clusters", approximately of medium-range size in glassy systems, in terms of a single soft atomic motion mode coordinate. As shown in SMM (Ch. 8), typical non-single-well potentials are most probably double-well ones in which the non-phonon excitations are reduced to two-level systems, introduced in the earlier STM [116, 117], but only at very low temperatures/frequencies, $T \lesssim T_m \approx 1$ K. In fact, anomalous features of specific heat $C(T)$, thermal conductivity $\kappa(T)$ and phonon mean free path $l_{ph}(\nu)$, at very low temperatures/frequencies, appear to be in accordance with experimental data for glasses only at such temperatures frequencies. In particular, SMM appears to justify the STM suggestions and anomalous properties of two-level systems, as well as the empirically found unusually large coupling constant β for acoustic phonons and soft-mode excitations (the soft-mode model has also been formulated [128] later in terms of the excitations eigenvectors for characterizing the degree of non-localization). Since independent soft modes are implied above, a question is whether the basic properties of soft-mode excitations and the related low-energy dynamic properties of glass change due to interactions between soft modes (see Secs. 9.2 and 10.4).

8

SOFT-MODE EXCITATIONS OF VERY LOW AND "INTERMEDIATE" ENERGIES

In the present Chapter, soft-mode excitations of both very low and "intermediate" energies, as well as some related dynamic and thermal properties of glasses are discussed. Excitations of very low energies, $0 < \varepsilon \lesssim \varepsilon_m \approx (0.1-0.2)\,w$, determine dynamic properties at "external" field frequencies $\nu \lesssim \nu_m = \varepsilon_m/h$ and thermal properties at $0 < T \lesssim T_m \approx 1\text{K}$. Such excitations cannot be related to soft-mode vibrations of which the smallest energy exists, $w \sim 10^{-3} \gg \varepsilon_m \sim 10^{-4}$ eV and can be only non-vibrational excitations that are due to atomic tunneling and associated splitting of ground-state energy levels E_{0l} in two individual wells $(l = 1, 2)$ of a double-well potential with a classical energy asymmetry $\Delta_{\text{cl}} \equiv |E_{01} - E_{02}|/2 \geq 0$. On the other hand, soft-mode excitations of intermediate energies, can be associated with either classical overbarrier hopping in double-well potentials or anharmonic vibrations in the intermediate energy range, in which a smooth crossover is expected to occur from non-vibrational excitations of very low energy to almost-harmonic, vibrational excitations of moderately low energy. While for very low energies essential qualitative and scalewise analysis and results are discussed in Sec. 8.1. For intermediate energies, a tentative consideration and rough estimations are only presented in Sec. 8.2. One can conclude from what follows in the present Chapter (as well as in Chs. 9 and 10) that all soft-mode excitations, of very low and "intermediate" energies, as well as moderately low energies, can be described as Anderson non-localized states, as actually implied [129]. At the same time, the rigorous proof of this assertion still has to be given in detail elsewhere.

8.1 Soft-mode tunneling states (independent two-level systems)

For analyzing excitations in double-well potentials, it is useful to start with a symmetric potential, for which by definition the classical energy asymmetry $\Delta = 0$. A finite energy asymmetry can be accounted for by applying a perturbation theoretical approach with a small expansion parameter $\Delta/w \ll 1$. Let us first consider energy levels of a soft-mode "atom" in an individual well of a double-well potential in which the cubic anharmonicity is neglected. In this case of a symmetric double-well potential, any potential well of two types can be described as follows:

$$V_\gamma^{(0)}(x) = A_\gamma x^2 + x^4, \tag{8.1}$$

with the coefficients A_γ for $\gamma = 1, 2$ determined from Eqs. (7.3), (7.6)–(7.7) and (7.9)–(7.12). The ground-state energy level (referred as the well bottom) can be estimated as $E_\gamma^{(0)} \equiv E_\gamma^{(0)}(A_\gamma, A) \gtrsim E_\gamma^{(0)}(0, A) \approx w(\approx 5 \cdot 10^{-4}\,\text{eV})$. The harmonic term $A_\gamma x^2$ becomes essential only for sufficiently large coefficients $A_\gamma \gtrsim 5w$ at $\zeta_\gamma \gg 1$, where $\zeta_1 \equiv \eta_0/\eta_L$ and $\zeta_2 \equiv -\eta/\eta_L$. The criterion of the occurrence of at least one energy level in each potential well is that the harmonic component prevails, $E_\gamma^{(0)} \approx w[8\gamma\zeta_\gamma/(10 - \gamma)]^{1/2} \ll V_B^{(\gamma)}$ (Eq. (7.12)) at large values of $\zeta_\gamma \gg 1$ (e.g., actually at $\zeta_{1,2} \gtrsim 10$) [123] (1983):

$$E_\gamma^{(0)} \approx w[8\gamma\zeta_\gamma/(10 - \gamma)]^{1/2} \ll V_B^{(\gamma)}. \tag{8.2}$$

If the interwell barrier is high enough for the existence of several levels in the potential well, the next excited vibrational level would be $E_\gamma^{(1v)} > E_\gamma^{(0)}$, separated from the ground-state level by an interlevel gap width $\varepsilon_{1\gamma}^{(v)} \gtrsim w$. Then one can conclude that a very small width of the first interlevel energy gap, $\varepsilon'_{1\gamma} \ll w$, can be realized only if it is determined by the level splitting in a double-well potential due to interwell tunneling of the soft-mode "atom". While the next vibrational excited state of energy $E_\gamma^{(1v)}$ is separated from the ground-state energy by a gap of width $\varepsilon_{1\gamma}^{(v)} \gtrsim w$, the first nonvibrational excited state characterized by a much smaller energy gap $\varepsilon'_{1\gamma} \ll w$ by a standard definition is a TLS excitation described in STM also as a TS excitation [116] (the latter definition is mostly used in what follows). Accounting for the neglected potential cubic anharmonicity does not violate the above conclusions, as estimated corrections to the numerical coefficients are found to be relatively small (e.g., corrections to energy levels are $\sim\zeta_\gamma^{-3/2}E_\gamma^{(0)} \ll E_\gamma^{(0)}$). As usual for such systems, one can choose eigenstates in a symmetric double-potential as a zero approximation in a perturbation theoretical approach with the perturbation being the cubic anharmonicity:

$$\psi_\gamma^{(\pm)} \equiv |\gamma;\pm\rangle = 2^{-1/2}[\psi_\gamma^{(1)} \pm \psi_\gamma^{(2)}], \quad \varepsilon = \varepsilon_\gamma = \Delta_{0\gamma}^{(0)} \ll w, \tag{8.3}$$

where $\psi_\gamma^{(1)}$ and $\psi_\gamma^{(2)}$ are wave functions of the "atom" in a single well and $\Delta_{0\gamma}^{(0)}$ is a zero-order tunneling amplitude in the harmonic approximation.

Then, the standard perturbation theory (for close levels) [172] gives rise to the following first order expression accounting for a finite (small) asymmetry of the potential [125]:

$$\varepsilon = \varepsilon_\gamma = (\Delta_{0\gamma}^2 + \Delta_\gamma^2)^{1/2}. \tag{8.4}$$

The first order expression for tunneling amplitude is $\Delta_{0\gamma} = \Delta_{0\gamma}^{(0)} + \delta_\gamma$ with $\delta_\gamma = \langle \gamma;+|V'|+;\gamma\rangle$ while that for the potential asymmetry is $\Delta_\gamma = 2|\langle \gamma;+|V'|-;\gamma\rangle|$, where V' is the cubic anharmonicity perturbation operator. Since the explicit expressions of the matrix elements are well-known, the result can be expressed as follows:

$$\Delta_\gamma = w|\rho_\gamma|f_\gamma(\zeta_\gamma) \quad \text{and} \quad \Delta_{0\gamma} = \Delta_{0\gamma}^{(0)}, \quad \text{i.e., } \delta_\gamma = 0, \tag{8.5}$$

where $f_1(\zeta) \simeq |\frac{9}{\sqrt{2}}\zeta^{1/2} - \frac{891}{64}\zeta^{-1} - \frac{8}{9}\zeta^2|$ and $f_2(\zeta) \simeq \sqrt{\zeta}|2(2\zeta)^{1/2} - \frac{11}{6}\zeta^{-2} - \frac{4}{3}\zeta|$ while $\rho_1 = \eta/\eta_0 - 8/9$ and $\rho_2 = (\eta_0/\eta_L)^{1/2}$. In Δ_γ the first and second terms take into account the difference in harmonic and anharmonic components of potential wells respectively, while the last terms represent the classical asymmetry $\Delta_{\text{cl}}^{(\gamma)}$, i.e. the difference in two potential well minima.

The tunneling amplitude can be estimated by applying the semiclassical WKB approximation [172], $\Delta_{0\gamma}^{(0)} = \pi^{-1}h\nu_\gamma \exp(-S_\gamma)$ and $S_\gamma = \hbar^{-1}\text{int}_{-x_{0\gamma}}^{x_{0\gamma}}|p_\gamma(x)|dx$, where $p_\gamma(x)$ is the classical momentum, ν_γ is the classical motion frequency in an individual well and $\{-x_0, x_0\}$ are the classical turning points. In what follows, the specific dependencies of Δ_γ and $\Delta_{0\gamma}$ are not very important and the following expressions are relevant:

$$\Delta_1 \approx w\,(\eta_0/\eta_L)^2\,|\eta/\eta_0 - 8/9|, \quad \Delta_2 \approx w(\eta_0/\eta_L)^{1/2}|\eta/\eta_L|^{3/2}$$
$$\text{while } \Delta_{0\gamma} \simeq \Delta_{0\gamma}^{(0)} = \pi^{-1}h\nu_\gamma \exp[-S_\gamma(\zeta_\gamma)], \tag{8.6}$$

where the tunneling action $S_\gamma(\zeta) = b_\gamma \zeta^{1/p_\gamma}$ at $b_\gamma \approx 1$ and $1/2 \lesssim p_\gamma \lesssim 2/3$; in particular, $p_\gamma \approx 2/3$ at $E_1^{(0)} \simeq h\nu_1 \approx w(\eta_0/\eta_L)^{1/2} \ll V_B^{(1)}$ and $E_2^{(0)} \simeq h\nu_2 \approx w|\eta/\eta_L|^{1/2} \ll V_B^{(2)}$. In what follows, the approximation used is that $S_1(\zeta_1) \simeq (\eta_0/\eta_L)^{3/2} \gg 1$ and $S_2(\zeta_2) \simeq |\eta/\eta_L|^{3/2} \gg 1$ for actual values of the parameters at $|\eta| \lesssim \eta_{sm} \approx 0.1$ and $\xi^2 \lesssim \eta_{sm}$. Each S_1 or S_2 has a positive minimum value, i.e., $S_\gamma \geq S_{\min}^{(\gamma)} = \ln(2h\nu_{0\gamma}/\pi\varepsilon_\gamma) > 0$ at $\varepsilon_\gamma \geq \Delta_{0\gamma}$.

The TS excitations with energy spectrum of Eq. (8.4) can be described by the Hamiltonian $H_{TLS} = (1/2)(\sigma_z\Delta + \Delta_0\sigma_x)$ in the representation of two states, each localized in one potential well of a double-well potential, where the Pauli matrices σ_z or σ_x are related to the energy asymmetry Δ or intersite tunneling amplitude Δ_0 respectively. The criterion $\varepsilon \equiv \varepsilon_{TS} \ll w$ of the occurrence of very low energy TS excitations in the energy spectrum is satisfied at very weak atomic tunneling of a random amplitude $\Delta_0 \ll w$ and very small energy asymmetry $\Delta \ll w$ typical of soft-mode double-well potentials. The TS excitations contribute not only to specific heat, C_{TS}, but also as scatterers of acoustic phonons to thermal conductivity, χ_{TS}, as well as to other thermal and dynamic properties of the (almost) thermal equilibrium glasses under consideration. As usual for calculations of the properties, the excitation DOS $G^{(TS)}(\varepsilon)$(see $G^{(NV)}(\varepsilon)$ in (Eq. (7.33)) has to be first calculated from the standard expression (Eq. (7.32)). The calculations can be performed in more detail, for example, for $F(\eta,\xi) = F_1(\eta,\xi)$ a regular function at $\eta \to 0$ (Eq. (7.17)):

$$\begin{aligned} G_1^{(TS)}(\varepsilon) &= \sum_{\gamma=1,2} G_1^{(\gamma)}(\varepsilon) \\ &= \sum_{\gamma=1,2} \iint_{(\Omega_{TS})} d\eta d\xi F_1(\eta,\xi)\delta\{\varepsilon - [\Delta_{0\gamma}^2(\eta,\xi^2) + \Delta_\gamma^2(\eta,\xi)]^{1/2}\}, \end{aligned} \tag{8.7}$$

in which the TS excitation range Ω_{TS} can be defined as follows:

$$\Omega_{TS} = \{0 \leq |\eta/\eta_0 - 8/9| \ll 1 \text{ at } \gamma = 1, \\ \text{while } 0 \leq (\eta_0/\eta_L)^{1/2} \ll 1 \text{ at } \gamma = 2\}. \quad (8.8)$$

After passing to variables ζ_γ, integrating by accounting the δ-function and performing other rather simple calculations, the DOS can be described as $G_1^{(TS)}(\varepsilon) = \sum_{\gamma=1,2} G_1^{(\gamma)}(\varepsilon)$ with the following $G_1^{(1,2)}(\varepsilon)$ [123] (1983):

$$G_1^{(1)}(\varepsilon) = \frac{32\eta_L^{3/2}}{9w} \int_{\zeta_1}^{\infty} d\zeta F_1[8\eta_L\zeta/9, (32\eta_L\zeta/9)^{1/2}] \\ \cdot \varphi_1^{-1/2}(\zeta) f_1^{-1}(\zeta)\zeta^{1/2},$$

$$G_1^{(2)}(\varepsilon) = \frac{32\eta_L^{3/2}}{9w} \int_{\zeta_2}^{\infty} d\zeta F_1\{-\eta_L\zeta, [(32/9)\eta_L\varepsilon^2\varphi_2(\zeta) \\ \times (f_2(\zeta))^{-2} w^{-2}]^{1/2}\} \cdot \varphi_2^{-1/2}(\zeta) f_2^{-1}(\zeta), \quad (8.9)$$

where limits ζ_γ are determined from equation $\varphi_\gamma(\zeta) \equiv 1 - (\Delta_{0\gamma}^{(0)}(\zeta)/\varepsilon)^2 = 0$ (at $\varepsilon \ll w$, the main contribution to the integrals is due to $\eta \ll 1$) and $\zeta_\gamma(\varepsilon)$ is a logarithmic weak function of ε approximated for estimations as $\zeta_\gamma = [b_\gamma^{-1} \ln (E_\gamma^{(0)}(\zeta_\gamma)/\varepsilon)]^{p_\gamma}$. Since the latter is the only dependence on ε contained in the expressions of $G_1^{(1)}(\varepsilon)$ and $G_1^{(2)}(\varepsilon)$ while $p_\gamma \approx 2/3$, one can conclude that $G(\varepsilon)$ only weakly, logarithmically, depends on ε. Ignoring weak logarithmic energy dependencies, one can see that the DOS is a smooth function, in the sense that for energies $\varepsilon \ll w$, $|d \ln G^{(TS)}(\varepsilon)/d \ln \varepsilon|_{\varepsilon \ll w} \ll 1$, so the function slightly differs from a constant. Then, the scale estimation gives rise to

$$G_1^{(TS)}(\varepsilon) \approx \eta_L^{3/2} w^{-1} F_1^{(0)} \equiv g_1^{(0)} = const. \quad (8.10)$$

With $F(\eta, \xi) = F_2(\eta, \xi)$ a singular function at $\eta \to 0$ (Eq. (7.18)), a similar smooth DOS can also be obtained. In general,

$$G_\alpha^{(TS)}(\varepsilon) \approx \eta_L^{(\alpha+1/2)} w^{-1} F_\alpha^{(0)} \equiv g_\alpha^{(0)}, \quad (8.11)$$

i.e., $G_\alpha^{(TS)}(\varepsilon)$ is uniform for both possible types of the distribution densities $F(\eta, \xi) = F_{\alpha=1,2}(\eta, \xi)$ in Eqs. (7.17)–(7.19). For $F(\eta, \xi) = \sum_{\alpha=1,2} c_\alpha \; F_\alpha(\eta, \xi)$ in Eq. (7.19), a similar estimation results, with

$G_\alpha^{(TS)}(\varepsilon) = \sum_{\alpha=1,2} c_\alpha G_\alpha^{(TS)}(\varepsilon)$, at $c_1 + c_2 = 1$ and $0 \leq c_1/c_2 \leq \infty$. On the other hand, the atomic concentration $c_{sm}^{(2)}$ of soft-mode "defects" in double-well potentials can be estimated in scale in a similar way by applying Eqs. (7.6), (7.7) and (7.23)–(7.25)):

$$c_{sm}^{(2)} \approx \sum_{\alpha=1,2} \int_{-\infty}^{\eta_L} d\eta H_\alpha(\eta)\theta(-\eta) \sim \exp\left(-\eta_{sm}^{-1}\right) \sum_{\alpha=1,2} c_\alpha F_\alpha^{(0)} \eta_L^{\alpha-1}, \tag{8.12}$$

where $\theta(x) = \{1, \text{ at } x > 0; 0, \text{ at } x < 0\}$. Generally speaking, one can suggest that both types of the basic distribution densities can be important in most complex glasses in which different kinds of interactions of soft modes with neighboring local structures can coexist in general, so $F(\eta, \xi) = \sum_\alpha c_\alpha F_\alpha(\eta, \xi)$ can be relevant, and therefore the scale of $c_{sm}^{(2)}$ hardly depends on the specific value of c_1/c_2, being in this sense universal in accordance with experimental data for dynamical properties of glasses at very low temperatures and frequencies (Sec. 6). As far as this suggestion holds true and since the approximate normalization relationship (Eq. (7.23)) for the soft-mode tail can be applied to both types of functions $F_\alpha(\eta, \xi)$, the following general scale relationship between $F_0^{(1)}$ and $F_0^{(2)}$ can take place:

$$g_\alpha^{(0)} \equiv \eta_L^{(\alpha+1/2)} F_\alpha^{(0)} w^{-1} \approx g_0 = const, \quad \text{i.e., } F_1^{(0)} \approx F_2^{(0)} \eta_L \ll F_2^{(0)}, \tag{8.13}$$

i.e., $g_\alpha^{(0)}$ is approximately independent of α. In this connection, the atomic concentration c_{TS} of TS excitations in the glasses under discussion can be estimated as $c_{TS} \approx g_0 \varepsilon_m$. Since each TS excitation is located in a single soft-mode double-well potential, one can suggest that the total soft-mode concentration c_{TS} can similarly be estimated in scale from Eq. (8.12),

$$\begin{aligned} c_{TS} &\approx g_0 \varepsilon_m \approx c_{sm}^{(2)} \sim \exp\left(-\eta_{sm}^{-1}\right)\left[c_1 F_1^{(0)} + c_2 F_2^{(0)} \eta_L\right] \\ &\sim \exp\left(-\eta_{sm}^{-1}\right) F_1^{(0)}, \end{aligned} \tag{8.14}$$

in accordance with the uniformity of the DOS $G_\alpha(\varepsilon) \approx G_\alpha(0)$ in the energy range at $0 < \varepsilon \leq \varepsilon_m \approx 1$ K. On the other hand, one can estimate the TS concentration c_{TS} by comparing Eq. (6.1), describing experimental data for the thermal-equilibrium specific heat at $T \lesssim 1$ K, with the estimation of the theoretical expression in Eq. (8.18) for the total specific heat of TS, at a large enough measurement time $t_m \gg \tau_1(T)$ and typical values of related parameters. The result is that $c_{TS} \approx 10^{-6}$ and $F_1^{(0)} \approx 10^{-2}$ at $\eta_{sm}^{-1} \approx 10$ (Eq. (7.25)). Then, by applying also Eqs. (7.23)–(7.25), one can assess the total soft-mode concentration c_{sm},

$$c_{sm} \approx \int_{-\infty}^{\eta_{sm}} d\eta H(\eta) \sim c_{sm}^{(2)} \exp\left(+\frac{1}{\eta_{sm}}\right) \sim 10^4 c_{TS} \sim 10^{-2}. \qquad (8.15)$$

In this connection, such a definition of a glass as an amorphous solid, containing the highest total concentration $(c_{sm})_{\max} \sim 10^{-2}$ of soft modes does not seem to contradict the standard definition [125].

In a similar way one can estimate in this approach another distribution density

$$P(\Delta, \Delta_0) = \iint_{(\Omega_{TS})} d\eta d\xi F(\eta, \xi)\delta[\Delta - \Delta(\eta, \xi^2)]\delta[\Delta_0 - \Delta_0(\eta, \xi^2)], \qquad (8.16)$$

that is related to $G^{(TS)}(\varepsilon)$ and $P(\Delta, S)$ as follows: $G^{(TS)}(\varepsilon) = \iint d\Delta\, d\Delta_0 P(\Delta, \Delta_0)\delta\{\varepsilon - [\Delta^2 + \Delta_0]^{1/2}\}$ and $P(\Delta, \Delta_0)d\Delta d\Delta_0 = P(\Delta, S)d\Delta dS$, while $P(Z) = \int d\Delta P(\Delta, Z) = \sum_{\gamma=1,2} \iint d\eta d\xi F(\eta, \xi)\delta\{Z - Z_\gamma(\eta, \xi^2)\}$, $Z = \Delta_0$ or $Z = S$, with $S = S_1 = (\eta_0/\eta_L)^{3/2} \gg 1$ at $0 < \eta < \eta_0 \ll 1$ and $S = S_2 = |\eta/\eta_L|^{3/2} \gg 1$ at $\eta < 0$, for appropriate values of parameters: $\eta_L \ll |\eta| \lesssim \eta_{sm} \approx 0.1$ and $\eta_L \ll \eta_0 \lesssim \eta_{sm}$. At very low energies under discussion for both types of $F_{\alpha=1,2}(\eta, \xi)$ (or their superposition), the resulting estimation can be obtained with accuracy to small corrections $O(\eta_L/|\eta|; \eta_L/\eta_0)$:

$$P(\Delta, S) \approx P_0 \approx g_0 S_{\max}^{-1} \quad \text{and} \quad P(\Delta, \Delta_0) \approx P_0/\Delta_0, \qquad (8.17)$$

where P_0, as well as G_0, is a material constant of a universal scale while $\Delta_0 \simeq \pi^{-1} h\nu_0 \exp[-S]$ and $S_{\max} \approx 10$. One can interpret Eqs. (8.15) and (8.16) as follows: $G^{(TS)}(\varepsilon)$ and $P(\Delta, \Delta_0 = const)$ are smooth functions of $\{\varepsilon$ or $\Delta\} \ll w$, at $1 \gg \{|\eta|, \xi^2\} \gg \eta_L$, since the variation scale of $G^{(TS)}(\varepsilon)$ and $P(\Delta, \Delta_0 = const)$ is the soft-mode energy scale w while ε and Δ are smooth functions of η or $\eta_0 \approx \xi^2$. In contrast, Δ_0 is an exponential function of

$|\eta|/\eta_L(\gg 1)$ or $\eta_0/\eta_L(\gg 1)$ so the function $P(\Delta = const, \Delta_0)$ substantially differs from a constant. It also follows that $G_\alpha^{(TS)}(\varepsilon) \approx g_0 = const.$ at small enough $\varepsilon \ll w$ while for $\varepsilon \to w$ the DOS may considerably increase with increasing ε.

Taking into account the above-mentioned scale estimations, a plausible suggestion consistent with experimental data appears to be that the qualitative behavior and scale estimations for $G^{(TS)}(\varepsilon)$, $P(\Delta, \Delta_0)$ and total concentration c_{sm} of soft modes are rather universal for glasses for which the parameters $\eta_L \approx 10^{-2}$ and $w \sim 10\varepsilon_m \sim 10^{-3}$ eV are also universal in scale, at least in the used approximation of independent soft modes. Numerical calculations for simple structural models of soft modes (probably for both limit types of $F(\eta, \xi) = F_{1,2}(\eta, \xi)$ seem to show that $G^{(TS)}(\varepsilon) \approx const$ is valid only at $\varepsilon \lesssim (10^{-1}-10^{-2})w$, whereas $G^{(TS)}(\varepsilon)$ or $P(\Delta, \Delta_0 = const)$ at higher ε slowly increase with increasing ε or Δ, approximately like a power law ε^σ or Δ^σ at $\sigma \ll 1$, e.g., $\sigma \approx 0.1 - 0.3$ [125, 127]. This variation can give rise to additional weak dependencies T^σ or ν^σ for some dynamic and thermal properties, e.g., specific heat, observed in glasses.

For a thermal-eqilibrium glass, the total specific heat due to TS excitations can simply be calculated by applying the total DOS $G^{(TS)}(\varepsilon)$ and by taking into account that a TS is related to a two-level system of which the ground- and excited-state energy levels are separated from the higher, vibrational excited-state levels by an energy gap much larger than $\varepsilon_{TS} \equiv \varepsilon$. Then, the populations p_0 and p_1 of the ground-state and first excited-state levels E_0 and $E_1 = E_0 + \varepsilon$ are $p_0 = [1 + \exp(-\varepsilon/kT)]^{-1}$ and $p_1 \simeq 1 - p_0 = [\exp(-\varepsilon/kT) + 1]^{-1}$. The result emerging from the standard formula for total, thermal-equilibrium, specific heat is that

$$C_{TS}(T) = \int_0^\infty d\varepsilon G^{(TS)}(\varepsilon) \left(\frac{\varepsilon}{T}\right)^2 \frac{\exp(-\varepsilon/T)}{[1 + \exp(-\varepsilon/T)]} \approx \frac{\pi^2}{6} g_0 T, \tag{8.18}$$

at very low $T \lesssim T_m \approx 1\,\text{K} \ll w$. This indeed holds true for a thermal-equilibrium glass where all TS excitations are "accessible" and contribute to the specific heat, in the sense that effective times $\tau_1(\varepsilon, \Delta_0; T)$ of their relaxation to thermal equilibrium due to phonon-assisted tunneling transitions in double-well potentials, depending on both TS tunneling amplitude Δ_0

and energy $\varepsilon = (\Delta_0^2 + \Delta^2)^{1/2}$, must at least not be larger than the measurement time, $\tau_1 \leq t_m$ (cf. Sec. 9.1). However, as first shown in STM for very low $\varepsilon \ll w$, the distribution function for random tunneling amplitudes Δ_0 spans over many orders of magnitude at typical large $S \gtrsim 10$, with $\tau_1^{-1} \propto \Delta_0^2/hw$ at $\Delta_0 \approx \pi^{-1} h\nu_0 \exp(-S) \ll w$ and $S \gg 1$. The criterion of "accessibility" of an excited state is that $S \leq S_{max} = \ln(\varepsilon t_m/h)^{1/2} < \ln(w t_m/h)^{1/2}$ at actual $\varepsilon \ll w$ and $S_{max} \gg 1$. On the other hand, $\varepsilon \geq \Delta_0 \approx w \exp(-S)$ so $S \geq S_{min} \approx \ln(w/\varepsilon) > 1$ and accessible well-defined TS excitations, determining the specific heat, occur at $\ln(w/\varepsilon) \lesssim S \lesssim \ln(w t_m/h)^{1/2}$ and $t_m \gg h/\varepsilon$, the TS lifetime. Thus the specific heat was predicted to depend in general not only on T but also on t_m. For smooth enough distributions $P(S) = \int d\Delta P(\Delta, S)$ the dependence of $C_{TS}(t_m)$ of logarithmic type could be expected, in particular, $C_{TS} \propto T \ln(t_m/\tau_{min})$ at a uniform distribution $P(S) = S_{max}\, \Theta\,(S_{max} - S)$ and $\tau_{min} \propto \exp(2S_{min})$; the smallest relaxation time $\tau_{min} \propto \exp(2S_{min})$ of a TS at a given ε $(\approx T)$ corresponds to the largest tunneling amplitude $\Delta_{0,max} = \varepsilon$ at $\Delta = 0$. Indeed, in a series of experiments [171] a dependence of logarithmic type actually was observed for $C_{TS}(t_m)$, which, however, is more complicated than the simple $\ln t_m$ one. In SMM, the dependence $C_{TS}(t_m)$ can be estimated by calculating the conditional probability density $p(\tau|\varepsilon)$ of a relaxation time τ at a given TS excitation energy $\varepsilon \ll w$, which is found by accounting for the above-mentioned estimations as $p(\tau|\varepsilon) \approx \{[\ln(w^2\tau/\varepsilon^2\tau_{min})])]^{-4/3}(1 - \tau_{min}/\tau)^{-1/2}\tau^{-1}\}$. The resulting approximate expression at $\varepsilon = T \ll w$ is:

$$C_{TS}(T, t_m) = C_{TS}(T, \infty) \cdot \Psi(T, t_m), \tag{8.19}$$

where $C_{TS}(T, \infty) \approx \frac{\pi^2}{6} G^{(TS)}(\varepsilon = T)T$ at $G_{TS}(\varepsilon) = const$. In accordance with the remark below Eq. (8.17), the expression for $C_{TS}(T, \infty)$ could probably be able to describe the empirical deviation of Eq.(6.1) from the STM formula, at $G^{(TS)}(\varepsilon = T) \propto g_0 T^{\sigma}$ and $C_{TS}(T, \infty) \propto T^{1+\sigma}$ for $\sigma \ll 1$, e.g., $0.1 \lesssim \sigma \lesssim 0.3$. Moreover, $\Psi(T, t_m)$ is a logarithmic function of T and t_m, which can be approximated as follows [123, 125]:

$$\Psi(T, t_m) \approx \{1 - [1 + \ln(t_m/\tau_{min})/\ln(w/T)^2]^{-1/3}\}, \tag{8.20}$$

where τ_{min} is a minimum relaxation time corresponding to a maximum tunneling amplitude at a given energy $\varepsilon = \Delta_{0\,max} \approx w \exp(-S_{min})$

and $\Delta = \Delta_{\min} = 0$ (however, the rough estimations in Eqs. (8.19) and (8.20) have still to be thoroughly checked by direct numerical calculations, e.g., in some specific soft-mode structural models). Certainly, $\Psi(T, t_m) \simeq 1$ for large enough $t_m \gg \tau_1(T)$, with the effective relaxation time $\tau_1(T) = \tau_1(\Delta_0 = T, \varepsilon = T; T) \propto T^{-3}$ for independent, most numerous thermal TS excitations. In accordance with the remark below Eq. (8.17), the expression for $C_{TS}(T, \infty)$, with $\Psi(T, t_m) \approx 1$ at $t_m \gg \tau_1(T)$, perhaps is able to describe the empirical Eq. (6.1), $C_{TS}(T, \infty) \propto T^{1+\sigma}$ at $0.1 \lesssim \sigma \lesssim 0.3$. In addition, the function $\Psi(T, t_m)$ can also reduce at $t_m \ll \tau_{\min}(w/T)^2$ to a simple expression, $\Psi(T, t_m) \approx \ln(t_m/\tau_{\min})$, predicted in [116, 117], whereas the dependence $C_{TS}(t_m)$ is tending to a saturation not associated with a boundary of the spectrum of barrier power S at $t_m \gg \tau_{\min}(w/T)^2$. Finally, one can assume that the DOS of "accessible" TS excitations at a given $t_m < \tau_1(T)$ should be considerably smaller than the total DOS, because most excited states are separated from the occupied "accessible" ones by relatively high barriers in the soft-configuration space.

On the other hand, TS excitations of energy ε also participate in dynamic and kinetic processes, in which they can resonantly scatter acoustic phonons of energy $h\nu_{ac} \equiv h\nu \simeq \varepsilon$, giving rise to an attenuation (absorption) of acoustic waves of length $\lambda_{ac}(\nu) \simeq s_0/\nu$ and intensity I. Thus the resonant processes contribute, in particular, to the absorption coefficient $\alpha_{ac}(\nu, T; I)$, i.e., to the "internal friction" $Q^{-1}(\nu, T; I) = \alpha_{ac}\lambda_{ac}$, and to the related thermal conductivity $\kappa(T)$. As can be seen in more detail in Sec. 9.1, the resonant interaction predominates over an alternative relaxation-related one at low enough T, $T \ll h\nu \lesssim h\nu_m = T_m$ ($\ll w$) while the opposite case is realized at higher T, $h\nu \ll T \lesssim T_m$. As an example, the thermal conductivity of acoustic phonons with $h\nu \simeq \varepsilon \approx T$ in a glass at very low $T \lesssim T_m \approx 1\,\text{K}$ could perhaps be described, with $G^{(TS)}(\varepsilon = T) \propto g_0 T^{\rho}$ at $0 < \rho \ll 1$, in accordance with the empirical Eq. (6.2) as

$$\kappa(T) \approx \frac{1}{3} C(T) l_{ac}^{(res)}(\nu = T/h, T) \propto T^{2-\rho}, \quad \text{at } 0.1 \lesssim \rho \lesssim 0.3, \qquad (8.21)$$

where $l_{ac}^{(res)}(\nu, T) \simeq l_{ac,TS}^{(res)}(\nu, T) = (w/h\nu) l_0 \coth(h\nu/2T)$, $l_0 \propto g_0^{-1}$ and $C(T)$ is the Debye specific heat, $C(T) \propto T^3$ (the estimation of ρ could be made, in particular, from numerical calculations in specific soft-mode structural models, but this work is not yet accomplished). Unlike $C_{TS}(T, t_m)$, in SMM both $l_{\rm ac}^{(\rm res)}(\nu, T)$ and $\kappa(T)$ do not depend on the measurement time t_m, since the main contributions are determined by most numerous thermal TS excitations in resonant scattering, with a typical tunneling amplitude $\Delta_T \approx T$. In fact, the kinetic properties are practically independent of very small random values $\Delta_0 \ll T$ which correspond to inaccessible TS excitations with the relaxation time $\tau_1[\varepsilon, \Delta_0; T]$ exceeding a typical measurement time. The ultrasonic attenuation related to the "internal friction" $Q^{-1}(\nu, T; I)$ and determined by both resonant and relaxation-related scattering of acoustic phonons by TS excitations can also be described in SMM. Both types of processes, as well as the specific heat and thermal conductivity, can be characterized in SMM at very low $\{h\nu, T\} \lesssim T_m$ by relationships similar to, and somewhat specifying, those obtained in STM. The basic relationships of STM (at $\sigma = 0$ and $\rho = 0$) are discussed in detail in Sec. 9.1.

8.2 Soft-mode excitations of "intermediate" energies

In accordance with the comments below Eqs. (7.33), two types of soft-mode excitations are found to exist in the range of energy and temperature, at $\varepsilon_m < \varepsilon \sim T \lesssim \kappa w$ and $0 < \kappa - 1 \ll 1$, "intermediate" between very low energies (tunneling states) and moderately low energies (harmonic vibrational excitations). The excitations are non-vibrational excitations associated with thermal-activated overbarrier hopping, and related relaxation processes, as well as with anharmonic vibrational excitations of energy $\varepsilon \sim w$.

The non-vibrational excitations associated with the thermal-activated relaxation processes in double-well potentials of atomic concentration $c_{sm}^{(2)}$ are found to occur at $\widetilde{\eta} \equiv \eta/\eta_L < 0$, $\widetilde{\xi}^2 \equiv \xi^2/\eta_L$ and $\eta_L^{-1} \gg |\widetilde{\eta}|, \widetilde{\xi}^2 \gg 1$, with random heights V_B of interwell barriers subject to a temperature independent distribution density $P(V_B; \overline{V}_B, \Delta_B)$ that is a non-monotonic function with a maximum around an effective average height $\overline{V}_B$ and a relatively small distribution width $\Delta_B \ll \overline{V}_B$ (e.g., for a Gaussian function, $P(V_B) = C_0 \exp[-(V_B - \overline{V}_B)^2/2\Delta_B^2]$, at a normalization constant $C_0 \propto c_{sm}^{(2)}$). Such a distribution density might be compatible with a suggestion in [106] (1976) that an effective singularity exists in the distribution of V at a characteristic value $\overline{V}_B$ (≈ 0.05 eV for a silica glass). In fact, following Eqs. (7.10)–(7.12), a typical average interwell barrier height can be estimated as $10w \gg \overline{V}_B \sim 0.1w\{\widetilde{\eta}^2, \widetilde{\eta}_0^2\} \gtrsim w$ at $10^{-1} \gg \eta, \eta_0 \gtrsim \eta_L \approx 10^{-2}$. The excitations, by their physical origin, can be assumed to contribute largely to relaxation processes in the attenuation $\alpha_{\rm ph}^{\rm (rel)}(T, \nu)$ of acoustic (electromagnetic) waves at "intermediate" temperatures, $T_m \approx 1\,{\rm K} < T \lesssim \kappa w \sim 5\,{\rm K}$. It has to be taken into account that effective semiclassical tunneling amplitudes Δ_{0n} associated with TS excitations at discrete energy levels $\varepsilon_n (n = 1, 2, .)$ in a double-well potential rapidly increase with increasing the quantum number n, up to a "free" energy level $\varepsilon(p)$, $V_B \lesssim \varepsilon(p) = V_B + p^2/2M_{sm}^* \lesssim \varkappa w$ in the overbarrier continuum (M_{sm}^* is an overbarrier soft-mode effective mass, probably of the scale of, though perhaps smaller than, the mass M_{sm} at the ground energy level, while p is the quasiparticle momentum) [172]. Applying in this case a rather simple quantum diffusion theory [124] (1983), one can find that the total diffusion of quasiparticles in soft-mode double-well potentials can be approximated by diffusion at a single optimal energy level $\varepsilon_{\rm opt}$ that approximately depends on T as follows:

$$\varepsilon_{\rm opt}(T) \simeq \varepsilon_1 \text{ at } T < T^* \quad \text{and} \quad \varepsilon_{\rm opt}(T) \simeq \overline{V}_B \quad \text{at } \kappa w \gtrsim T \gtrsim T^*, \tag{8.22}$$

where a crossover occurs at $T = T^* \approx (\hbar^2 w/M_{sm}^* a_0^2)^{1/2}$ with realistic $T^* \equiv (\hbar^2 w/M_{sm} a_0^2)^{1/2} < T^* < w \lesssim \kappa w \ll h\nu_0$, at $1 < \kappa \lesssim 2$ (Eq. (7.30)), e.g., $T_0^* \sim w/3$. On the other hand, a typical relaxation time $\tau_{\rm rel}(T)$ of soft-mode excitations can be approximated sufficiently far below the crossover as the relaxation time for thermal TS excitations:

$$\tau_{\rm rel}(T) \approx \tau_1(T) = \tau_{TS}(\varepsilon = T, \Delta_0 = T; T) \propto T^{-3}, \quad \text{at } T < T^*, \tag{8.23}$$

whereas alternatively as the average inverse relaxation time for overbarrier hopping of soft-mode quasiparticles [173, 174]:

$$\tau_{\rm rel}^{-1}(T) \approx \tau_{\rm cl}^{-1}(T) = \nu_0 \exp(-\overline{V}_B/T) \cosh(\varepsilon/2T), \quad \text{at } \kappa w \gtrsim T > T^*, \tag{8.24}$$

where $\nu_0 \approx \nu_D \eta_{sm} \approx 0.1\nu_D$ (Eq. (7.30)), $V_B \simeq w\tilde{\eta}^2/4$ (Eq. (7.12)), $\overline{V}_B \approx w$ at $\tilde{\eta}^2/4 \approx 1$ and $\cosh(\varepsilon/2T) \approx 1$ at $\varepsilon \approx T$ for most numerous thermal excitations in double-well potentials. In other words, with increasing temperature the crossover occurs from thermal TS excitations of very low $\varepsilon \approx T \lesssim \varepsilon_m$ to classical, overbarrier, excitations associated with thermal-activated hopping at intermediate temperatures, $T^* \lesssim T \lesssim \kappa w$.

According to experimental data for SiO_2 glass in an acoustic wave of a relatively high frequency $\nu = 0.93$ THz (Fig. 6 (a)) and $\nu = 1$ THz (Fig. 6(b)), the temperature dependence of the total wave attenuation $\alpha_{ph}(\nu, T)$ exhibits a broad maximum around $T \sim w$ (Eq. (6.7)). This behavior can be in a standard way interpreted as due to the fact that the attenuation practically reduces to the relaxation related one, $\alpha_{ph}(\nu, T) \simeq \alpha_{ph}^{(rel)}(\nu, T)$. The maximum is similar to that of the acoustic attenuation determined by Debye relaxation processes described by a standard expression $D(\tau, \nu) = \nu^2\tau(1+\nu^2\tau^2)^{-1}$ with a maximum at $\nu\tau \approx 1$, while the Debye relaxation time $\tau = \tau_{cl} = \nu_0^{-1} \exp(V_B/T)$ describes the classical thermal-activated jump. For a glass, the dependencies of $\alpha_{ph}^{(rel)}(T, \nu)$ at $T > T^*$ can be expressed in terms of an integral of the function $D(\tau_{cl}, \nu)$ over random barrier heights V_B:

$$\alpha_{ph}(\nu, T) \simeq \alpha_{ph}^{(rel)}(\nu, T) \sim \int dV_B \nu^2 \tau_{cl}(V_B; T)(1 + \nu^2 \tau_{cl}^2(V_B; T))^{-1} P(V_B), \qquad (8.25)$$

or roughly

$$\alpha_{ph}^{(rel)}(\nu, T) \sim c_{sm}^{(2)} \nu^2 \tau_{cl}(\overline{V}_B; T)(1 + \nu^2 \tau_{cl}^2(\overline{V}_B; T))^{-1} \qquad (8.26)$$

The expression and estimation for the attenuation at $T > T^*$ can be extended in accordance, e.g., with [106], to lower $T < T^*$, where interwell tunneling transitions in double-well potentials account for the contributions of TS excitations, by replacing the integral of the Debye-like function $D(\tau, \nu)P(V_B)$ over the barrier heights and the expression of $\tau = \tau_{cl}(V_B, T)$ by the expression of TS relaxation time $\tau = \tau_{TS}(\varepsilon, \Delta_0; T)$ in Sec 9.1. For rough estimations reflecting the main physics of the phenomenon, the attenuation can be approximated by substituting in Eq. (8.25)

$$\tau_{\rm rel}^{-1} \approx \tau_{\rm cl}^{-1}(\overline{V}, T) + \tau_{1}^{-1}(T), \tag{8.27}$$

for $\tau_{\rm cl}^{-1}$, with $\tau_1(T) \equiv \tau_{TS}(\varepsilon = T, \Delta_0 = T; T)$, so that

$$\alpha_{\rm ph}^{(\rm rel)}(\nu, T) \sim \nu^2 \tau_{\rm rel}(T)(1 + \nu^2 \tau_{\rm rel}^2(T))^{-1}. \tag{8.28}$$

The crossover occurs at $T = T^*$ from $\tau_{\rm rel}^{-1} \approx \tau_1^{-1}(T) \propto T^3$ (Ch. 6) at $T < T^*$ to $\tau_{\rm rel}^{-1} \approx \tau_{\rm cl}^{-1}(\overline{V}, T)$ at $T > T^*$ while $\tau_1^{-1}(T^*) \approx \tau_{\rm cl}^{-1}(\overline{V}_{\rm B}, T^*)$. Then, as usual for a Debye process, the attenuation exhibits a broad maximum around the effective frequency $\nu_{\max} \approx \tau_{\rm rel}^{-1}(T)$, increasing with the increase of temperature, though in a different way below and above the crossover, in a qualitative and scalewise agreement with the data (Fig. 2).

Anharmonic vibrational excitations of energy $\varepsilon \sim w$ occur in soft-mode potentials $V_{sm}(\widetilde{x}) \simeq V_{sm}^0(\widetilde{x}) + \delta V_{sm}(\widetilde{x}; \widetilde{\eta}, \widetilde{\xi})$ (Eq. (7.27)), individual quartic potential wells $V_{sm}^0(x) = w\widetilde{x}^4$ modified by the operator $\delta V_{sm}(\widetilde{x}; \widetilde{\eta}, \widetilde{\xi}) = w(\widetilde{\eta}\widetilde{x}^2 + \widetilde{\xi}\widetilde{x}^3)$ at $0 < \widetilde{\eta}, \widetilde{\xi}^2 \lesssim 1$. Essential energy levels of such single vibrational excitations, which can resonantly interact with acoustic phonons at $h\nu \simeq \varepsilon$, are characterized by energy $\varepsilon_1^{(\rm AV)}(0{,}0)$ in a quartic anharmonic potential well and more generally by a continuous band spectrum produced by the operator $\delta V_{sm}(\widetilde{x}; \widetilde{\eta}, \widetilde{\xi})$ at important values of soft-mode parameters $\widetilde{\eta}, \widetilde{\xi}^2$. The anharmonic vibrational excitations of intermediate energies can contribute to resonant dynamic processes, e.g., to associated attenuation $\alpha_{\rm ph}^{(\rm res)}(\nu)$ of waves of frequency ν, at $\varepsilon \simeq h\nu$. At the same time, in accordance with estimations of Eq. (8.25), the effective partial DOS related to the overbarrier excitations in soft-mode double-well potentials can be determined to order of magnitude by the partial concentration of these potentials, which empirically is $c_{sm}^{(2)} \sim 10^{-6}$ and expected to be much smaller than the partial DOS $\mathrm{G}^{(\rm AV)}(\varepsilon)$ for anharmonic vibrational excitations of intermediate energies, e.g., $c_{sm}^{(\rm int)} \sim 10^{-4} \sim c_{sm}^{(\rm AV)} \gg c_{sm}^{(2)}$ (see around Eq. (7.33)). In this sense, the DOS $\mathrm{G}_{\rm int}(\varepsilon)$ for excitations of intermediate energies can be reduced to the DOS $\mathrm{G}^{(\rm AV)}(\varepsilon)$ of anharmonic vibrational excitations, which can be described by applying the general formula (Eq. (7.32)) as follows:

$$G^{(AV)}(\varepsilon) = \sum_{j=1,2} \iint_{\Omega_{\rm exc}^{(\rm anv)}} d\tilde{\eta} d\tilde{\xi} F(\tilde{\eta}, \tilde{\xi}) \delta[\varepsilon - \varepsilon_j^{(AV)}(\tilde{\eta}, \tilde{\xi}^2)], \qquad (8.29)$$

where $\Omega_{\rm exc}^{(AV)} = \{0 < \tilde{\eta}, \tilde{\xi}^2 \lesssim 1\}$. With increasing excitation energy ε, a crossover can be expected to occur at some $\varepsilon = \varepsilon_{co}$ in the range of intermediate energies, from non-vibrational TS excitations of very low energies $\varepsilon \lesssim \varepsilon_m$ to harmonic vibrational excitations of moderately low energies, $\kappa w < \varepsilon \lesssim h\nu$ (Eq. (7.33)). Since $\nu_m < \nu_{co} < \nu_0$, the crossover energy $\varepsilon_{co} \equiv h\nu_{co}$ can be estimated from equation $\alpha_{\rm ph,AV}^{(\rm res)}(\nu_{co}, T) = \alpha_{\rm ph,HV}^{(\rm res)}(\nu_{co}, T)$; here, $\alpha_{\rm ph,AV}^{(\rm res)}(\nu, T)$ or $\alpha_{\rm ph,HV}^{(\rm res)}(\nu, T)$ is the acoustic (electromagnetic) attenuation due to resonant interaction with anharmonic or harmonic vibrational excitations, the total resonant attenuation being $\alpha_{\rm ph}^{(\rm res)}(\nu, T) = \alpha_{\rm ph,AV}^{(\rm res)}(\nu, T) + \alpha_{\rm ph,HV}^{(\rm res)}(\nu, T)$.

Although there are no analytical expressions consistently describing the excitation energy spectrum of a quantum-mechanical anharmonic oscillator [175, 176], some features of the excitation DOS at intermediate energies can be found. One feature is that the dependence of excitation energies on random soft-mode parameters $\tilde{\eta}$ and $\tilde{\xi}$, determined by the operator $w(\tilde{\eta}\tilde{x}^2 + \tilde{\xi}\tilde{x}^3)$ in the Hamiltonian (7.27), is similar to that of the excitation spectrum in a 2D crystal (with wave vectors $k_{x,y}$ playing the role of $(\tilde{\eta}, \tilde{\xi})$) and contains a stationary point at some $(\tilde{\eta}_s, \tilde{\xi}_s)$ and $0 < \tilde{\eta}_s, \tilde{\xi}_s^2 \lesssim 1$, which can correspond to a saddle point at $\varepsilon = \varepsilon_s = \varepsilon(\tilde{\eta}_s, \tilde{\xi}_s^2)$ and thus to a van Hove spectral singularity described by a logarithmic divergency in the DOS (Eq. (8.29), $G^{(AV)}(\varepsilon) \propto \ln(1/|\varepsilon - \varepsilon_s|)$, at the energy ε_s [177]. The occurrence of saddle points in the lowest excitation band was shown in [178, 179] by applying explicit relationships for transformations in the $(\tilde{\eta}, \tilde{\xi})$ plane, which provide the translational invariance of the excitation spectrum at any finite displacement of the soft-mode coordinate $(\tilde{x})$. On the other hand, the lowest excitation band $\varepsilon_1^{(AV)}(\tilde{\eta}, \tilde{\xi}^2)$ is characterized in Sec. 7.3 by a characteristic energy $\varepsilon_1^{(AV)}(0,0) \simeq 1.5w$ around the upper limit κw of intermediate energies, while the second band $\varepsilon_2^{(AV)}(\tilde{\eta}, \tilde{\xi}^2)$ by $\varepsilon_2^{(AV)}(0,0) \simeq 3.5w$ is considerably higher than κw above which soft mode harmonic vibrational excitations are expected to be most essential (Ch. 10). Thus the contribution of lowest excitation band to the anharmonic vibrational spectrum is most essential in the soft-mode excitation DOS at intermediate energies $\varepsilon_m < \varepsilon \lesssim \kappa w$. In general, the expression of the lowest band energy at small $\{\widetilde{\eta}, \widetilde{\xi^2}\} \ll 1$, at which the perturbation expansion of the energy in powers of $\widetilde{\eta}$ and $\widetilde{\xi^2}$ can certainly be used, reads as follows [179]:

$$\varepsilon_1^{(AV)}(\widetilde{\eta}, \widetilde{\xi^2}) = \varepsilon_1^{(AV)}(0,0) + \Delta\varepsilon_1, \quad \text{at } \Delta\varepsilon_1(\widetilde{\eta}, \widetilde{\xi^2}) \simeq \kappa_1(\widetilde{\eta} - 3\widetilde{\xi^2}/8), \tag{8.30}$$

where $\kappa_j = 4^{-2/3}[(2j+1)^{2/3} - 1]$, so that $\varepsilon_1^{(AV)}(\widetilde{\eta}, \widetilde{\xi^2})$ decreases with increasing $\widetilde{\xi^2}$ while increases with increasing $\widetilde{\eta}$, for the lowest band at $j = 1$. It follows that $\varepsilon_1^{(AV)}(\widetilde{\eta}, \widetilde{\xi^2})$ varies non-monotonously with increasing $\widetilde{\xi^2}$ at $|\widetilde{\eta}| \ll 1$, and shows a minimum, i.e., $\partial\varepsilon_1^{(AV)}(\widetilde{\eta}, \widetilde{\xi^2})/\partial\widetilde{\xi^2} = 0$, at $\widetilde{\eta} = 0$ and $\widetilde{\xi} = \pm\widetilde{\eta}^*$, $\widetilde{\eta}^* \sim 1$, in the $(\widetilde{\eta}, \widetilde{\xi})$ plane. Since physical characteristics of the excitation spectrum, e.g., $\varepsilon_1^{(AV)}(\widetilde{\eta}, \widetilde{\xi^2})$, must exhibit a translational invariance in the sense that the values and dependencies on $\widetilde{\eta}$ and $\widetilde{\xi^2}$ must not change at a displacement of the reference point for $\widetilde{x}$, the following relationships:

$$\varepsilon_1^{(AV)}(\eta, \xi^2) = \varepsilon_1^{(AV)}(\widetilde{\eta}, \widetilde{\xi^2}), \; \partial\varepsilon_1^{(AV)}(\eta, \xi^2)/\partial\eta = \varepsilon_1^{(AV)}(\widetilde{\eta}, \widetilde{\xi^2})/\partial\widetilde{\eta}, \tag{8.31}$$

and $\partial\varepsilon_1^{(AV)}(\eta, \xi^2)/\partial\xi^2 = \varepsilon_1^{(AV)}(\widetilde{\eta}, \widetilde{\xi^2})/\partial\widetilde{\xi^2}$. Since, in general, $\partial\widetilde{\eta}/\partial\widetilde{\xi^2}$ is found to be finite, the result is that

$$\partial\varepsilon_1^{(AV)}(\widetilde{\eta}, \widetilde{\xi^2})/\partial\widetilde{\xi^2} = 0, \quad \text{at } \widetilde{\eta} = 0 \quad \text{and} \quad \widetilde{\xi} = \pm\eta^*, \eta^* \sim 1. \tag{8.32}$$

This means that the excitation spectrum $\varepsilon_1^{(AV)}(\widetilde{\eta}, \widetilde{\xi^2})$, i.e., $\Delta\varepsilon_1(\widetilde{\eta}, \widetilde{\xi^2})$, possesses analytical critical points similar to well-known van Hove spectral singularities of isoenergetic lines at $\varepsilon(q_x, q_y) = const.$ in an energy band of 2D crystals [177]. From this analogy, one can assume that saddle points only are characteristic of the lowest excitation band $\varepsilon_1^{(AV)}(\widetilde{\eta}, \widetilde{\xi^2})$.

With increasing excitation energy ε, a crossover is expected to occur at some $\varepsilon = \varepsilon_{co}$ in the range of intermediate energies, from non-vibrational TS excitations of very low energies $\varepsilon \lesssim \varepsilon_m$ to harmonic vibrational excitations of moderately low energies, $\kappa w < \varepsilon \lesssim h\nu$ (Eq. (7.32)). Since

$\nu_m < \nu_{co} < \nu_0$, the crossover energy $\varepsilon_{co} \equiv h\nu_{co}$ can be estimated from equation $\alpha^{(\mathrm{res})}_{\mathrm{ph},AV}(\nu_{co}, T) = \alpha^{(\mathrm{res})}_{\mathrm{ph},HV}(\nu_{co}, T)$. Here, $\alpha^{(\mathrm{res})}_{\mathrm{ph},AV}(\nu, T)$ or $\alpha^{(\mathrm{res})}_{\mathrm{ph},HV}(\nu, T)$ is the acoustic (electromagnetic) attenuation cross-section due to resonant interaction of excitation in question with anharmonic (AV) or harmonic (HV) vibrational excitations, while the total resonant attenuation is $\alpha^{(\mathrm{res})}_{\mathrm{ph}}(\nu, T) = \alpha^{(\mathrm{res})}_{\mathrm{ph},AV}(\nu, T) + \alpha^{(\mathrm{res})}_{\mathrm{ph},HV}(\nu, T)$. Moreover, at high enough frequency, the vibrational excitation spectra can be suggested to describe a crossover from anharmonic excitations (nonvibrational, of energy varying with increasing parameter $\widetilde{\eta}$, or vibrational ones) to harmonic vibrations. With this suggestion and Eq. (8.29) for anharmonic vibrations DOS, one can assume [179, 180] that the total anharmonic DOS $G^{(AV)}(\varepsilon)$ for appropriate energies is an increasing function of ε, which may exhibit a van Hove like singularity, e.g., a saddle point. At the same time, the contribution of such singularities to the anharmonic excitation DOS can be roughly estimated as relatively small, at least at $G^{(AV)}(\varepsilon) \propto c^{(\mathrm{int})}_{sm}$ and $G^{(AV)}(\varepsilon) \gg g_0 \propto c^{(2)}_{sm}$ for realistic values $c^{(\mathrm{int})}_{sm} \gg c^{(2)}_{sm}$. The increase of $G^{(AV)}(\varepsilon)$ with increasing anharmonic excitation energy might also give rise to a crossover to harmonic soft-mode vibrations with DOS characterized by a power law ε^n at $2 \leq n \leq 4$ (Secs. 10.1–10.2).

Unfortunately, no consistent analytical investigations of resonant scattering and relaxation processes for anharmonic soft-mode excitations interacting with acoustic phonons and for dynamic processes around inter-well barriers at inter-mediate energies (temperatures/frequencies) seem to be available so far, which could allow one not only to calculate explicitly the characteristic frequency ν_{co}, but also to describe the related soft-mode dynamics in some detail.

Let us add that the soft-mode excitations of intermediate energies with the DOS of Eq. (8.29) cannot be considered as responsible for the boson peak and related thermal anomalies in the glasses under consideration. The latter are described in Ch. 10 and found to be due rather to a qualitatively different phenomenon, Ioffe–Regel crossover from weak scattering of acoustic phonons by harmonic soft-mode excitations to strong scattering.

9

TUNNELING STATES AS VERY LOW ENERGY LIMIT CASE

The STM was the first theoretical model that described the universal, almost linear temperature dependence of specific heat and T^2 dependence of thermal conductivity observed at very low $T \lesssim 1$ K (actually not too low T, $0.1 < T \lesssim 1$ K) in glasses. The model implied that anomalous dynamic and thermal properties of glasses are characteristic of glasses in general at low $T \ll T_D$ and $\nu \ll \nu_D$ and assumed that the properties are determined by non-vibrational TS excitations of low energy $\varepsilon \ll h\nu_D$, interacting with Debye phonons. The STM also predicted an ultrasonic attenuation, with the "internal friction" $Q_{\rm ph}^{-1}$ independent of both T and ν at appropriate low values, and related changes in sound velocities. On the other hand, such universal dynamic and thermal properties at all low $T \ll T_D$ and $\nu \ll \nu_D$ can be described within the framework of the SMM in which the soft-mode excitations are TS ones only at *very low excitation energies* $\varepsilon \lesssim \varepsilon_m \approx 10^{-4}$ eV and associated $T \lesssim T_m \approx 1$ K. The relationships of SMM for other dynamic and thermal properties of glasses are also similar to those of STM at very low ε and T. In this aspect, STM can be considered as a limiting case of SMM at very low, but not too low energies. In this sense, SMM can give an additional justification, and in some respects specification, of STM at very low temperatures and frequencies.

9.1 Standard tunneling model: Independent two-level systems

STM is based on two essential hypotheses. The first hypothesis is that any glass system should contain a considerable atomic concentration of independent (not interacting with each other) local atomic groups which move in effective double-well potentials, unlike Debye phonons associated with harmonic vibrational atomic motions in standard single-well potentials. Most importantly for low $T \ll T_D$ they are non-vibrational, tunneling, atomic motions in double-well potentials. As usual in quantum mechanics, the tunneling motions give rise to a splitting of any ground state in each of two potential wells into resulting ground (0) and excited (1) eigenstates with eigenenergies E_0 and E_1, usually referred to as a two-level system excitation of energy $\varepsilon = E_1 - E_0$, as long as ε is much smaller than the energy gap to the next vibrational excitation. This has been derived in SMM only for very low energies (Ch. 8): the lowest non-vibrational excitation is a TS excitation related to a TLS one that is isomorphic to a spin 1/2, of which energy is obtained by diagonalizing the spin-like Hamiltonian $H_0 = (1/2)(\sigma_z\Delta + \sigma_x\Delta_0)$ in a standard form: $\varepsilon \equiv \varepsilon_{TS} \equiv E_1 - E_0 = (\Delta_0^2 + \Delta^2)^{1/2}$ (σ_z and σ_x are respective Pauli matrices). The effective tunneling amplitude Δ_0 and energy asymmetry Δ of a double-well potential are two basic random parameters of the model, subject to a distribution density $P(\Delta_0, \Delta)$. The second hypothesis of the model is that $P(\Delta_0, \Delta)d\Delta d\Delta_0 = (P_0/\Delta_0)d\Delta d\Delta_0$, with $P_0 = const.$, or $P(\Delta, S)d\Delta dS = P_0 d\Delta dS$, where a semi-classical atomic tunneling amplitude is $\Delta_0 \approx \pi^{-1}h\nu_0 \exp(-S) \ll h\nu_0$ with a typical large $S \gg 1$. In fact, comparing the above with the results of SMM in Sec. 8.1, one can conclude that Eqs. (8.4)–(8.6) can justify and specify the first hypothesis while Eqs. (8.15)–(8.16) can do the same for the second hypothesis.

From an experimental viewpoint, the excitation energy $\varepsilon(\equiv \varepsilon_{TS})$ and the ratio $r = (\Delta_0/\varepsilon)^2$, rather than Δ and S, are most relevant parameters of the model [106, 134]. Then, $P(S, \Delta)d\Delta dS = P_0 d\Delta dS$ is often replaced in STM by $P(\varepsilon, r)d\varepsilon dr = P_0 d\varepsilon dr/2r\sqrt{1-r}$, the distribution function diverging at boundaries of the allowed range, $r \to 0$ and $r \to 1$. A finite lower limit $r_{\min}(\varepsilon) > 0$ was introduced for getting a finite DOS of independent TS excitations $G_{TS}(\varepsilon) = \int\int d\Delta_0 d\Delta P(\Delta_0, \Delta)\delta[\varepsilon - (\Delta_0^2 + \Delta^2)^{1/2}] \simeq P_0 \ln r_{\min}$, corresponding to a finite lower limit for Δ_0 assumed to be small enough, e.g., $\Delta_{0,\min} \approx 1$ mK, or to an upper limit $S_{\max}$ for S, e.g., $S_{\max} \approx 10$. The knowledge of the DOS allows one to calculate the internal energy as well as the specific heat and other dynamic and thermal properties at very low excitation energies, temperatures or frequencies.

The famous linear temperature variation is characteristic of the specific heat of TS excitations characterized by a constant DOS. In accordance with Eqs. (8.10)–(8.11), when ignoring weak logarithmic energy dependencies, the total DOS is a finite constant, $G_{TS}(\varepsilon) \simeq G_{TS}(0) \neq 0$ at $\varepsilon \lesssim \varepsilon_m \approx 1$ K, and simple calculations give rise to the specific heat of TS excitations linear in temperature [116, 117, 125]:

$$C_{TS}(T) \simeq (\pi^2/6)Tg_0, \tag{9.1}$$

in approximate qualitative accordance with the data (Eq. (6.2)) at very low $T \lesssim 1$ K. As the tunneling transition rate τ_t^{-1} was found exponentially small, $\tau_t^{-1} \propto \Delta_0^2 \propto \exp(-2S)$ at typical large tunneling action $S \gg 1$, it was reasonably suggested in STM that, generally speaking, both g_0 and the related specific heat C_{TS} depend on the measurement time t_m. In fact, the specific heat is a thermal-equilibrium property, determined by the total DOS independent of t_m, only at a large measurement time t_m considerably exceeding a tunneling related relaxation time $\tau_1(T)$ of most numerous thermal TS excitations ($\varepsilon_{TS} \approx T$) in the (almost) thermal-equilibrium glass under discussion. In this connection, the notion of the "accessibility" of TS was introduced in [116], corresponding to excited TS with $\tau_1(T) < t$. In fact, with a standard relationship $\Delta_0 \propto \exp(-S)$ at $S \gg 1$, the approximate logarithmic dependence of the specific heat on t_m can be found from direct calculations:

$$C_{TS}(T) \approx (\pi^2/6)Tg_0 \cdot \min\{1; (1/2)\ln[4t_m/\tau_1(T)]\}. \tag{9.2}$$

Formally this means that even for a TS energy $\varepsilon = const$, the relaxation time $\tau_1(\varepsilon = const., r = (\Delta_0/\varepsilon)^2; T)$ is subject to a distribution due to the randomness of Δ_0/ε: at $r = 1$ the relaxation time is minimal, $\tau_1 = \tau_{1m} \approx \tau_1(T)$ but increases with decreasing Δ_0. Then, a corresponding effective DOS can be introduced, which increases logarithmically with increasing measurement time, $g(\varepsilon, t_m) \approx P_0 \ln\{[4t_m/\tau_1(T)]^{1/2}\}$, since more and more TS effectively interact with acoustic phonons. The logarithmic dependence of the specific heat C_{TS} on t seems to be in at least a qualitative agreement with experimental data, and a more complex dependence in Eq. (8.20) derived in SMM.

In STM, the non-vibrational TS excitations are suggested to directly interact with acoustic strain fields, and also with electric fields if TS have finite electric dipoles. The interactions are essential for dynamic and thermal properties like thermal conductivity and internal friction, at low $T \ll T_D$ and $\nu \ll \nu_D$ under discussion. Since the interactions can be described in a similar way, one can concentrate on the interaction with an acoustic strain, e.g., with longitudinal acoustic phonons, in the simplest scalar model of an isotropic elastic continuum for long-wave phonons at $h\nu \ll h\nu_D$ in a glass. The interaction was expected to be related to a change in the tunneling amplitude, $\delta\Delta_0$, and/or energy asymmetry, $\delta\Delta$, of a double-well potential in a small strain field e_{ac}, at $|e_{ac}| \ll 1$, with $\delta\Delta \simeq \beta_1 e$ and $\delta\Delta_0 \simeq \beta_2 e$ (the coupling constants β_1 and β_2 are suggested to be given constants parameters, instead of random parameters). An additional assumption applied in earlier works was that the strain fields affected the asymmetry stronger than the tunneling amplitude, i.e., $|\beta_1| \gg |\beta_2|$, the contribution associated with β_2 being neglected. Then, the TS-phonon interaction energy is approximated as follows: $\Delta\varepsilon_{TS}(e) = \varepsilon_{TS}(e) - \varepsilon_{TS}(0) \simeq (D\sigma_z + M\,\sigma_x)e \simeq D\sigma_x e$, where $D = \beta_1 \Delta/\varepsilon$ and $M = -\beta_2 \Delta_0/\varepsilon$. This approximation can also be justified in SMM, in which β_1 and β_2 are to be determined by the soft-mode-acoustic coupling energy parameter β (Eq. (7.37)). In general, the major parameter β_1 is of random sign whose scale can be estimated by comparing theoretical expressions for internal friction and thermal conductivity to appropriate experimental data, with the result $|\beta_1| \approx 1$ eV $\gg h\nu_D$ in agreement with the estimation of β in Eq. (7.37). Generally speaking, relaxation of practically localized TS excitations and their energy in macroscopic disordered systems becomes effective largely due to their interactions with non-localized excitations. The latter here are acoustic phonons, while single TS can be expected to undergo the Anderson localization due to a typical large static energy fluctuation compared to a flip-flop transition amplitude for an "updown" pair of TSs, one with an "up spin" and another with a "down spin". Then, the relaxation (decay) rate $\tau_1^{-1}(\varepsilon, r; T) \equiv \tau_1^{-1}[\varepsilon, \Delta_0; T]$ for a TS excitation with given ε, r, via typical one-phonon emission and absorption processes at low temperatures, can be estimated by applying the Golden Rule to the weak perturbation $\Delta\varepsilon_{TS}(e) \simeq (D\sigma_z + M\sigma_x)e$ [106]:

$$\tau_1^{-1}(\varepsilon, r; T) = Ar^2\varepsilon^3 \coth(\varepsilon/2T), \tag{9.3}$$

where $0 < r \equiv \Delta_0/\varepsilon \leq 1$ and parameter $A \simeq (3/2\pi\hbar)(U_0/\hbar^3 s_0^3)$ while the glass parameter $U_0 = \beta^2/\rho s_0^2$ characterizes the direct TS-phonon interaction, with ρ the mass density. The maximum value $\tau_{1,m}^{-1}$ in the spectrum of the relaxation rates $\tau_1^{-1}(\varepsilon, r; T)$ occurs for most numerous thermal TS excitations around $r = 1$ at $\varepsilon \approx T \approx \Delta_0$:

$$\tau_1^{-1}(T) = AT^3 \propto n_{\text{ph}}(T), \tag{9.4}$$

where the thermal phonon concentration $n_{\text{ph}}(T) = (T/T_D)^3$ is associated with the Debye specific heat $C_{\text{ph}}(T) \propto T^3$. The phonon mean free path $l_{\text{ph}}(\nu, T)$ of an acoustic phonon interacting with TS excitations, as well as the associated acoustic absorption coefficient $\alpha_{\text{ph}} = l_{\text{ph}}^{-1}$ and internal friction $Q_{\text{ph}}^{-1} = \lambda_{\text{ph}} l_{\text{ph}}^{-1}$, have also been calculated in STM, with a typical large mean free path $l_{\text{ph}} \gg \lambda_{\text{ph}}$ and a small $Q_{\text{ph}}^{-1} \ll 1$ at low $\nu \ll \nu_D$ and $T \ll T_D$. This was done for a resonant scattering of a phonon by a TS excitation, $h\nu \simeq \varepsilon = \varepsilon_{TS}$, with both a low acoustic wave intensity $I \ll I_c$ and a high intensity $I \gtrsim I_c$ (I_c is a critical intensity), and also for a relaxation related scattering. The expression for the phonon mean free path $l_{\text{ph,res}}$ determined by a resonant process was derived in STM [116, 117], giving rise to the following expression of $Q_{\text{ph,res}}^{-1}$ for a low-intensity wave at $I/I_c \ll 1$:

$$Q_{\text{ph,res}}^{-1}(\nu, T) = \lambda_{\text{ph}}(\nu) l_{\text{ph,res}}^{-1}(\nu, T) \approx \pi P_0 U_0 \tanh{(h\nu/2T)}. \tag{9.5}$$

The most simple derivation of $l_{\text{ph,res}}^{-1}(\nu, T)$ can be performed for "high-frequency" phonons at $h\nu > T$ for which as usual $l_{\text{ph,res}} = (\sigma_{\text{res}} n_{\text{res}})^{-1}$, where the scattering cross-section $\sigma_{\text{res}} = s_0^2/\pi\nu^2$, $n_{\text{res}} \approx (P_0/w)\langle\Gamma\rangle$ is the density of resonant scatterers and $\langle\Gamma\rangle$ is the average value of a TS excitation level width Γ. The latter was calculated by assuming that the decay of the excitation is due to emission of an acoustic phonon. The result means that the resonant process of attenuation of an acoustic wave of low intensity is proportional to the difference $p_0(T) - p_1(T) = \tanh{(h\nu/2T)}$ of thermal-equilibrium populations of lower ($p_0(T)$) and upper ($p_1(T)$) levels in the related TLS, which decreases with increasing T in accordance with the nature of the resonant process. At given T and ν, an increase of the intensity I also gives rise to a decrease of the population difference, because of growing population p_1 of the upper level, which is described by a factor $f(I/I_c)$ [180]:

$$Q^{-1}_{\text{ph,res}}(\nu, T; I) \simeq Q^{-1}_{\text{ph,res}}(\nu, T) f(I/I_c), \tag{9.6}$$

where $Q^{-1}_{\text{ph,res}}(\nu, T) \approx \pi P_0 U_0 \tanh(h\nu/2T)$ and $f(x) = (1+x)^{-1/2}$. A decrease of $Q^{-1}_{\text{ph,res}}(\nu, T; I)$ with increasing I becomes essential at $I \gg I_c$, at $f(x) \simeq x^{-1/2}$, and is referred to as a "saturation" of the resonant process. The expression for I_c is obtained to be as follows:

$$I_c = \hbar\tau_1^{-1} h\Delta\nu / 2U_0. \tag{9.7}$$

The width $\Delta\nu$ of a "hole" in the acoustic absorption spectrum, due to the population difference decrease, is proportional to a relaxation rate τ_2^{-1} generally different from τ_1^{-1} in a system of TS excitations. Taking into account the analogy between TLS and spin-1/2 excitations, one can see that τ_1^{-1} is similar by physical sense to a "longitudinal", spin-lattice, relaxation rate for independent spin excitations directly interacting with longitudinal acoustic phonons, while τ_2^{-1} is alike a "transverse", spin-spin, relaxation rate determined by interactions between spins (in what follows τ_1^{-1} and τ_2^{-1} are referred to, respectively, as "longitudinal" and "transverse" relaxation rates).

Another process for wave attenuation in STM was proposed as relaxation of independent TS excitations: since an acoustic wave modulates the TS energy, $\Delta\varepsilon \cong \beta\Delta_0 e$ and the longitudinal relaxation rate is finite, the response of an excitation is delayed with respect to the applied wave strain and the acoustic energy dissipates so the wave attenuates. This wave attenuation was found to be described by the following expression:

$$Q^{-1}_{\text{ph,rel}} = \alpha_{\text{ph,rel}}\lambda_{\text{ph}} \approx P_0 U_0 \int_0^\infty d\varepsilon \varphi(\varepsilon/2T, T)\Psi_D[\nu\tau_1(\varepsilon, r = 1)]. \tag{9.8}$$

In this formula, the internal friction is described by a superposition of Debye relaxation functions $\Psi_D(y) = \text{Im}\,(1 - iy)^{-1} = y(1 + y^2)^{-1}$ for all the participating TS excitations, each function being multiplied by the excitation thermal population function $\varphi(x, T) = 1/4k_B T \cosh^2(x)$. To this process all TS excitations of different energies ε contribute, the major contribution being due to thermal ones at energies $\varepsilon \approx T$ around $r = 1$. The integration in Eq. (9.8) can be carried out numerically but analytical solutions exist in two limiting cases, at $\nu\tau_1(T) \gg 1$ and at $\nu\tau_1(T) \ll 1$. Let us take into account that Eq. (9.4) for $\tau_1(T)$ empirically holds true only for very low, but not too low T, at $0.1 \lesssim T \lesssim T_m \approx 1$ K. Then, one can see that the condition $\nu\tau_1(T) \gg 1$ can be satisfied at "low" $T \ll h\nu$ while $\nu\tau_1(T) \ll 1$ at "high" $T \gg h\nu$ (with $(\nu\tau_1)^{1/3} \gtrless 1$ at $T \gtrless h\nu$). Then, at $\nu\tau_1(T) \gg 1$, i.e., at "low" $kT \ll h\nu$, Eq. (9.8) can be approximated as

$$Q^{-1}_{\mathrm{ph,rel}} \approx P_0U_0/\nu\tau_1(T) \propto T^3/\nu, \qquad (9.9)$$

with $\Psi_D(y) \approx y^{-1}$ in this case. Unlike the resonant process in Eq. (9.6), the relaxation process cannot be "saturated" at high $I \gg I_c$. Therefore, for "low" $T \lesssim h\nu$ at $\nu\tau_1 \gtrsim 1$,

$$Q^{-1}_{\mathrm{ph}} \simeq Q^{-1}_{\mathrm{ph,res}} + Q^{-1}_{\mathrm{ph,rel}} \simeq Q^{-1}_{\mathrm{ph,rel}} \approx P_0U_0/\nu\tau_1 \propto T^3/\nu, \qquad (9.10)$$

with $\tau_1^{-1} \propto T^3$ for high $I \gg I_c$ for which the resonant process is "saturated", whereas for low $I \ll I_c$

$$Q^{-1}_{\mathrm{ph}} \simeq Q^{-1}_{\mathrm{ph,res}} \approx \pi P_0U_0(\gg Q^{-1}_{\mathrm{ph,rel}}). \qquad (9.11)$$

On the other hand, for "high" $T \gg h\nu$ at $\nu\tau_1(T) \ll 1$, $Q^{-1}_{\mathrm{ph,rel}} \simeq \frac{\pi}{2}P_0U_0$ while $Q^{-1}_{\mathrm{ph,res}} \lesssim \frac{\pi}{2}P_0U_0h\nu/T \ll Q^{-1}_{\mathrm{ph,rel}}$, so

$$Q^{-1}_{\mathrm{ph}} \simeq Q^{-1}_{\mathrm{ph,rel}} \approx \frac{\pi}{2}P_0U_0(\gg Q^{-1}_{\mathrm{ph,res}}), \qquad (9.12)$$

even at very low $I \ll I_c$. One can conclude that Q^{-1}_{ph} is expected to exhibit a "plateau" both for "low" $T \ll h\nu$ at low $I \ll I_c$ and for "high" $T \gg h\nu$ at any $I/I_c \gtrless 1$, while increasing as $\tau_1^{-1} \propto T^3$ with growing T for "low" $T \lesssim h\nu$ and high $I \gg I_c$. Note that the occurrence of the "plateau" is a direct reflection of the postulate $P(\Delta_0, \Delta) = P_0/\Delta_0$ in STM, which appears to follow from Eqs. (8.16) for small enough $\Delta \ll w$. Then, the above-mentioned deviations in SMM with increasing $\Delta(\to w)$ can be expected to give rise to deviations from a "plateau" of Q^{-1}_{ph} in the appropriate temperature range. In principle, by measuring $Q^{-1}_{\mathrm{ph}}(\nu, T)$ over a wide enough range of T and

ν, the validity (and accuracy) of the approximation $P(\Delta, \Delta_0) = P_0/\Delta_0$ can be tested.

From above, one can also see that in STM the thermal conductivity determined by contributions of thermal acoustic phonons is basically due to resonant scattering of the phonons. The resulting expression following from STM is:

$$\kappa(T) \approx (1/3) s_0 C(T) l_{\text{ph}}^{(\text{res})}(\nu = T/h, T) \propto T^2, \tag{9.13}$$

in approximate qualitative agreement with experimental data at very low $T \lesssim 1\,\text{K}$ (Eq. (6.2)). As noted in Sec. 8.1, $l_{\text{ph}}^{(\text{res})}(\nu, T)$ and $\kappa(T)$ do not depend on the measurement time t_m, since both are determined by thermal TS excitations with a characteristic tunneling amplitude $\Delta_T \approx T$, practically independent of very small Δ_0 corresponding to inaccessible excitations.

Systematic, relatively small deviations of experimental data for the specific heat from linear temperature variation and for the thermal conductivity from T^2 dependence (Eqs. (6.1), (6.2)), which do not seem to be explained in STM, can perhaps be described by Eqs. (8.19) and (8.21), in SMM. However, this is not yet completely investigated and explained.

Experimental investigations of acoustic phonon propagation in glasses have been very important for discussing the dynamic properties at very low T and ν. As well known, a dissipative characteristic of a system, $X''(\nu, T) \equiv \operatorname{Im} X(\nu, T)$ like $Q_{\text{ph}}^{-1}(\nu, T)$ is related to a respective one $X'(\nu, T) \equiv \operatorname{Re} X(\nu, T)$ like sound velocity change $\delta s(\nu, T)/s_0$, by the Kramers–Kronig relation [135, 136], e.g., $\delta s(\nu, T)/s_0 \sim \int d\nu' Q_{\text{ph}}^{-1}(\nu', T)(\nu - \nu')^{-1}$. Hence, by accounting for $Q_{\text{ph}}^{-1}(\nu, T)$ in Eqs. (9.5) and (9.8)–(9.12), the sound velocity change can be obtained as follows [106, 134]:

$$\delta s/s_0 = (\delta s/s_0)_{\text{res}} + (\delta s/s_0)_{\text{rel}} \simeq (\delta s/s_0)_{\text{res}} \simeq P_0 U_0 \ln(T/T_0) \quad \text{or}$$

$$\delta s/s_0 = -\frac{1}{2} P_0 U_0 \ln(T/T_0), \tag{9.14}$$

at "low" T ($\nu\tau_1 \gg 1$) or at "high" T at ($\nu\tau_1(T) \ll 1$) respectively, where $(\delta s/s_0)_{\text{rel}} \simeq -\frac{\pi}{2} P_0 U_0 \ln(T/T_0)$ and T_0 is an arbitrary reference temperature. It follows that with increasing T a logarithmic increase of the velocity occurs at $\nu\tau_1(T) \gg 1$ while at $\nu\tau_1(T) \ll 1$ a logarithmic decrease with a slope twice as small as the slope at $\nu\tau_1(T) \gg 1$ occurs. Thus the temperature dependence of $\delta s/s_0$ shows a maximum at the crossover between two regimes. These theoretical results agree with Eq. (6.4) qualitatively describing the experimental data for $\delta s(\nu, T)/s_0$ at very low (but not too low, see below) $T \lesssim 1K$ and $\nu \lesssim 10^{-1}$THz with the empirical constant $B \approx P_0U_0 \sim 10^{-3}$. This equation, as well as Eqs. (9.5), (9.11) and (9.14) seem to be qualitatively correct as long as TS relax via typical one-phonon processes with $\tau_1^{-1}(T) \propto T^\alpha$ at $\alpha = 3$, the prefactor in Eq. (6.3) being $B/2$ at $\nu\tau_1(T) \gg 1$ or $-3B/2$ at $\nu\tau_1(T) \ll 1$. If Raman two-phonon or multi-phonon processes are most important, the dependence of $Q_{\text{ph}}^{-1}(\nu, T)$ on T remains the same power law as for $\tau_1^{-1}(T) \propto T^\alpha$, but with a larger α different from $\alpha = 3$ (e.g., with $\alpha = 7$ for a Raman process). Since there is a close analogy between acoustic and dielectric properties, e.g., between changes in sound velocity, $\delta s(\nu, T)/s_0$, and in light velocity, $\delta c(\nu, T)/c_0 \propto -\delta\epsilon'(\nu, T)/\epsilon_0$ (see [122] and references therein), for the sake of brevity, the acoustic properties are mainly discussed here and in what follows. Let us note that by measuring $\delta s(\nu, T)/s_0$ over a wide enough range of T and ν, the validity of the approximation $P(\Delta, \Delta_0) = P_0/\Delta_0$ postulated in STM can also be verified.

As mentioned in comments to Eq. (6.4), although at first glance this equation can be relevant for describing the experimental results, a closer inspection reveals that the ratio of the slopes in the experimental semilogarithmic plot for $\delta s(T)/s$ vs $\ln(T/T_0)$, at least at ultralow $T < 0.1$ K, is not 2: (−1) but rather 1: (−1) [134, 182]. This discrepancy is clearly seen in Fig. 1 of [134] (see also [182] (1997)), where the STM prediction is shown by a dashed curve, and it is even more evident in Fig. 2 of [134] (see [182] (1992)), at ultralow $T < 0.1$ K, of which a rather general theory is discussed in Sec. 9.2.

One may conclude that in STM the dynamic and thermal properties of glasses are determined by resonant and relaxation processes due to independent spin-like TS excitations directly interacting with acoustic phonons, of which relaxation processes are characterized by the "longitudinal" rate $\tau_1^{-1}(T) \propto T^\alpha$ at $\alpha = 3$. Although STM was assumed to hold true in general at low $T \ll T_D$ and $\nu \ll \nu_D$, experimental data (Ch. 6) demonstrated that the model was applicable only at very low (though but not too low) T, $T_l \approx 0.1 < T \lesssim T_m \approx 1$ K and respective frequencies, $\nu_l < \nu \lesssim \nu_m$.

9.2 Advanced tunneling model: Interacting two-level systems

For the lowest $T < T_l$ and $\nu < \nu_l$, the independence of spin-like TS excitations did not appear to be relevant: interactions between the excitations become important. This follows from the occurrence of many phenomena reflecting the interactions, which have been observed first in [183] and in a later series of experiments. In particular, the interactions can determine the observed hole burning in acoustic absorption spectrum due to a wave (or strong pulse) of high intensity, and the coherent echoes generated by an appropriate sequence of strong acoustic pulses (see, e.g., [122, 184]). The width $\Delta\nu$ of the "hole", which increases with increasing wave intensity, temperature and measurement time, is proportional to the above mentioned (Ch. 6) "transverse" relaxation rate τ_2^{-1} of TS excitations, which is in turn determined by universal elastic (and possible electric) interactions between excitations.

9.2.1 *Mean-field approximation: "Spectral diffusion"*

As noted in Sec. 7.4, a soft mode, as well as its excitation (in particular, a TS one) at a position l, interacts with acoustic phonons, the interaction being proportional to the acoustic strain and soft mode coordinate x_l (the latter by definition is zero at equilibrium positions of surrounding atoms). Since the soft mode excitation can be affected by the strain generated by another excitation in a neighborhood any two soft-mode excitations (l, l') can interact with each other. The mechanism of pair interaction can be understood, and the expression for the interaction energy $U_{ll'}$ can be derived, by taking into account that the energies of neighboring excitations become dependent on the quantum state of the excitation under discussion. The interactions can be described as elastic dipolar interactions between "defects", TS excitations in what follows, which are due to exchanges by virtual acoustic phonons (i.e., with no conservation of the total excitation energy [172]). The energy $U_{ll'}$ is similar to that of well-known electric dipolar interactions between defects with finite electric dipoles, due to exchanges by virtual photons, while both electromagnetic and acoustic strain fields [187] are characterized by a linear coupling with the defect displacement and by a linear dispersion law, $\nu(q) = s_0 q$.

In general, the total interaction energy E_{int} is a sum of interaction energies for pairs of excitations. The expression, describing the pair interaction at a large distance $R = R_{12} \gg a_1$ in an isotropic (glassy) elastic medium due to the "resonant" interaction (Eq. (7.37)) of each excitation with long wavelength acoustic phonons ($\lambda \gg a_1$), can be derived following, in particular, the approach used in [185] (1990) and in [122]. For the sake of simplicity, the scalar model for acoustic phonons is applied and the total Hamiltonian H of the system contains only longitudinal phonons (H_{ph}) and two excitations identified with TS ones (H_{sm}) interacting with the phonons and reads as follows:

$$H = H_{\text{ph}} + H_{sm} + V_{\text{int}},$$

where $H_{\text{ph}} = \sum_q \{[|P_q|^2/2M] + [M(2\pi\nu(q))^2|X_q|^2/2]\}$, $H_{sm} = -\Delta_1\sigma_{1z} - \Delta_2\sigma_{2z}$, $V_{\text{int}} = \sum_{l=1,2} \beta_l \sigma_{lz} \cdot e(l)$ while P and X denote respectively the momentum and coordinate displacement operators. Let one consider the system with Hamiltonian H and TS excitations displacing the original equilibrium phonon coordinates to new ones depending on the spin operators σ_{lz}, i.e. on the excitation states. What has to be done is to find the new normal modes, with their new equilibrium positions, by applying well-known unitary transformations of phonon operators [113]. Then, the total energy of the system can be described in terms of new phonon coordinates $\widetilde{X}$ and reads as follows:

$$\widetilde{H}_{\text{tot}} = \sum_q \{[|P_q|^2/2M] + [M(2\pi\nu(q))^2|\widetilde{X}_q|^2/2]$$
$$- [M(2\pi\nu(q))^2|X_q^0|^2/2]\} - \Delta_1\sigma_{1z} - \Delta_2\sigma_{2z}.$$

Here, in the parentheses the first two terms describe the phonon bath while the last term corresponds to the "defect" related energy change due to the coordinate displacements $X_q^0 = N^{-1/2}\sum_q iq(\beta_1\sigma_{1z} + \beta_2\sigma_{2z})e(q)$, and N is the total number of atoms. Straightforward calculations give rise to the following formula for the transformed term $\widetilde{H}_{sm}$ in the Hamiltonian:

$$\widetilde{H}_{sm} = H - \widetilde{H}_{\text{ph}} = -\varepsilon_1 - \varepsilon_2 + E_{\text{int}}, \quad \text{at } E_{\text{int}} = U_{12}\sigma_{1z}\sigma_{2z}, \qquad (9.15)$$

where $E_{1,2}$ describes the independent contributions of two excitations to the total energy, and $U_{12} = u_{12}R_{12}^{-3}$ is the dipolar interaction of the pair. For the spin-like (two-level system) excitation of large length $R_{12} \gg a_1$ under discussion, $u_{12} \simeq (\beta_1\beta_2/\rho s_0^2)(\Delta_1/\varepsilon_1)(\Delta_2/\varepsilon_2)$ for elastic dipolar interactions or $u_{12} \simeq (p_1p_2/4\pi\epsilon\epsilon_0)(\Delta_1/\varepsilon_1)(\Delta_2/\varepsilon_2)$ for electric dipolar ones, where p_l are characteristic electric dipoles and ϵ is the associated dielectric constant of the system (ϵ_0 is that of vacuum). As the electric interactions are actually much weaker than the elastic ones at typical $|\beta| \approx 1eV$ (Sec. 7.4) and $|p| \approx 1$ Debye, dipolar elastic interactions are mainly accounted for in what follows. In particular, the interaction gives rise to flip-flop transitions in an "updown" pair of spin-like two-level systems, one with an "up spin" and another with a "down spin". As noted in Sec. 9.1, the important resonant interaction (scattering) is characterized by energies $\varepsilon \approx \Delta_0(\gg \Delta)$ of TS excitations.

As suggested in [185], interactions between such "defects" may be responsible for universal dynamic properties of glasses observed at very low $T \lesssim 1$ K. Such a "defect" characterized by an "internal" freedom degree, a TS excitation in what follows, may be considered as an independent one as long as direct TS-phonon interaction is considerably stronger than the interaction $U_{12}(R_{12})$ in the sense that the longitudinal relaxation rate $\tau_1^{-1}(T) > |U_0|/R_{12}^3\hbar$ at given distances R_{12}. This is just the criterion of independent spin-excitations in a system of elastically interacting randomly located spins-1/2. Moreover, the solutions of Bloch equations of motion for interacting spin-1/2 **excitations** also reveal that another, "transverse", relaxation rate τ_2^{-1} exists, which is determined by the dipolar interactions $U_{12} = u_{12}/R_{12}^3$, as far as $\tau_2^{-1} > \tau_1^{-1}$ at very low $T < 1$ K. In fact, in a system of interacting spin-1/2 excitations and, by above-mentioned analogy, in a system of interacting TS excitations, $\tau_2^{-1}(T)$ determines the phase-memory or dephasing rate for coherent echoes observed in glasses at very low temperatures (Ch. 6). Coherent acoustic echoes are generated, in particular in fused silica glass, by a sequence of short acoustic pulses that interact with TS excitations. The coherent effects are largely observed at ultralow temperatures, typically $T < 0.1$ K, where the TS phase-memory time τ_2 is long compared to the acoustic pulse duration, typically of the scale 10^{-3} μs. The dependence of spontaneous (two-pulse) and stimulated (three-pulse) echo amplitude on the amplitude, width, frequency and separation time t_{12} of generating pulses has been investigated in such experiments, e.g., at typical $0.7 \lesssim \nu \lesssim 1.5$ THz and $0.02 \lesssim T \lesssim 0.08$ K. Hence, the magnitude of an average deformation-potential parameter β was estimated for fused silica glass to be $|\beta| \simeq 1.5$ eV, in agreement with its scale $\beta \approx 1$ eV found in glasses, while the two-pulse echo decay time was $t_d \simeq 15\,\mu$s at $T \simeq 0.02$ K and varied as T^{-2}. This behaviour, as well as that of the width $h\Delta\nu$ of an intense acoustic pulse generated hole in the absorption spectrum with increasing T and pulse length, may be understood in terms of so-called "spectral diffusion", due to dipolar interactions and temporal fluctuations of resonant energies at $\varepsilon_{TS} \simeq h\nu$ [138]. In thermal equilibrium, all TS with $\varepsilon_{TS} \lesssim T$ are permanently generated and destroyed, the transitions being accompanied by local strain field changes. Since ε_{TS} depends on the strain, the resonant excitations also have energies fluctuating with increasing time. Then, excitations with energy ε_{TS} close to a probing pulse energy $h\nu'$ temporarily become in resonance, $\varepsilon_{TS} \simeq h\nu'$, at finite T, so they can be generated within an energy range determined by temporary changes of ε_{TS} due to the fluctuations and statistical effect of neighboring TSs. Thus the hole width detected by the probing pulse grows with increasing time until any TS undergoes a thermal transition, the result being that the maximum hole width is $h\Delta\nu_\infty \approx P_0 U_0 T$, for most important TSs with $\langle \Delta_{0i}/\varepsilon_i \rangle \approx 1$. This estimation may explain the linear increase of the hole width with T, reflecting the rising number of thermally generated excitations. In glasses, the stationary state cannot actually be reached, since very slowly relaxing TSs occur due to the fact that a probability for a thermal transition of a TS is proportional to $[1 - \exp(-t/\tau_{\min})]$. Indeed, for $t \ll \tau_{\min} \approx \tau_1(T)$, the probability can be approximated by t/τ_1, while for $t \gg \tau_1$ the relaxation rate $\tau_1^{-1} \propto \{r \equiv (\Delta_0/\varepsilon)^2$ at $\varepsilon \approx T\}$ and hence the distribution function $P(\tau_1^{-1}) \propto P(r) \propto P_0/2r\sqrt{1-r} \to \infty$ at $r \to 0$ and $P(\tau_1^{-1}) \to \infty$ at $\tau_1^{-1} \to 0$. This actually means that with increasing time t a growing number of TSs contribute to the energy changes and the hole width is estimated to increase linearly or logarithmically correspondingly at short or long t:

$$h\Delta\nu \approx \left\{ \begin{array}{c} h\Delta\nu_\infty t/\tau_1 \text{ at } t \ll \tau_1, \\ h\Delta\nu_\infty \ln(t/\tau_1) \text{ at } t \gg \tau_1 \end{array} \right\}, \tag{9.16}$$

in a reasonable agreement with the acoustic experiments (unfortunately, a precise test in the long time limit, $t \gg t_{\rm min}$, still seems rather difficult to carry out).

In this approximation for the "spectral diffusion", a particular TS excitation of energy ε interacts with surrounding TSs assumed not to interact with each other. A specific TS excitation with an excited state of energy ε_1 is found here to be most strongly coupled with another TS in its ground state, when the excited state has a close energy $\varepsilon_2 \approx \varepsilon_1$. Such a pair, referred to in what follows as a resonant, "updown" pair with a dynamic and coherent coupling between the TSs, can undergo an Anderson delocalization at $|\varepsilon_1 - \varepsilon_2| < U_0$ [188]. Moreover, a TS excitation energy can undergo large time-dependent fluctuations due to the spectral diffusion caused by transitions of neighboring TSs. Such fluctuations can give rise to a delocalization of localized resonant pair excitations which are formed due to dipolar interactions U_{12}. This delocalization can be understood as a self-consistent process in which spectral diffusion generates transitions of TSs, which determine a spectral diffusion due to dipolar interactions (see below).

Since in a resonant pair $(i=1,2)$ of TSs the TS-phonon coupling parameter $\beta_{1,2} \equiv \beta_{1,2}^{(TS)}$ and dipole $p_{1,2}$ are of random sign, the average, of the parameter u_{12} over random locations in Eq. (9.15) is $\langle u_{12}\rangle = 0$ so the finite interaction parameter is estimated as $\langle |u_{12}|\rangle \equiv \int d\Delta \int d\Delta_0\, P(\Delta, \Delta_0)|u_{12}| \equiv U_0 \approx \beta^2/\rho s_0^2$ [122]. Then, the typical relative strength of dipolar interactions compared to energies of resonant pairs is expressed by the dimensionless parameter,

$$\chi_0/\pi \equiv P_0 U_0 \approx P_0 \beta^2/\rho s_0^2 \approx 10^{-3}, \tag{9.17}$$

that empirically is universal in scale at typical $|\beta| \approx 1\,\text{eV}$ and $P_0/\rho s_0^2 \approx 10^{-3}/\text{eV}^{-2}$ (Sec. 9.1). In this sense, only weak dipolar interactions between TSs are considered in what follows. Accounting for the interactions, the energy asymmetry Δ_i of a TS can be replaced by

$$\widetilde{\Delta}_i = \Delta_i + \sum_{j\neq i} U_{ij}\sigma_{jz}, \tag{9.18}$$

with the interaction Hamiltonian described as $H_{\rm int} = \frac{1}{2}\sum_{i,j} U_{ij}\sigma_{iz}\sigma_{jz}$ at $U_{ij} = u_{ij}/R_{ij}^3$, at $\langle u_{ij}\rangle = 0$ and $\langle |u|_{ij}\rangle \approx \beta^2/\rho s_0^2$. The second term in $\widetilde{\Delta}_i$

depends on the states of surrounding TSs approximated as independent ones. Since the TS excitations undergo chaotic transitions between their levels due to the phonon emission/absorptions processes in a thermal-equilibrium glass, the excitation energies $\varepsilon_i = [\Delta_i^2 + \Delta_{0i}^2]^{1/2}$ of TSs fluctuate with time. The changes are associated with spectral diffusion that leads to fluctuations of a TS phase $\Phi_i(t)$ in a state $\Psi_i(t) \propto \exp(i\Phi_i(t))$, $\Phi_i(t) \approx \hbar^{-1} \int_0^t dt' \varepsilon_i(t')$. This phase has a random contribution $\delta\Phi_i(t) \approx \hbar^{-1}\delta\varepsilon_i(t)t$ due to the fluctuations $\delta\varepsilon_i(t)$ in the energy $\varepsilon_i(t)$, and the phase coherence is destroyed when the random phase $\delta\Phi$ approaches unity. Thus the dephasing time τ_2 caused by spectral diffusion can be estimated from the equation

$$\hbar^{-1}\tau_2\delta\varepsilon_i(\tau_2) \approx 1, \tag{9.19}$$

where the energy fluctuation $\delta\varepsilon_i = \sum_j' U_{ij}\sigma_{jz}$ and the prime means that the summation is made over TS undergoing transitions in the time interval $(0, t)$. For assessing τ_2 one has to determine $\delta\varepsilon_i(\tau_2)$ and to assume that τ_2 is smaller than the "longitudinal" relaxation time τ_1 (Eq. (9.4)); otherwise τ_1 also participates in defining the dephasing time. The TSs largely contributing to the spectral diffusion are expected to be thermal ones at $\varepsilon \approx T$, or $\Delta \approx T \approx \Delta_0$, of which the (spatial) density $n_T \approx P_0T$ at a long measurement time $t > \tau_1(T)(\propto T^{-3})$ while $n_T(t) \approx P_0Tt/\tau_1(T)$ at $t < \tau_1(T)$. In fact, the density of TSs with energies larger than T is lower than that of thermal ones, and the TSs with smaller $\Delta_0 < T$ show very large tunneling transition times. Then, the long-time density n_T of thermal TSs (at $t \gg \tau_1$) can be found from the distribution density $P(\Delta, \Delta_0) \approx P_0/\Delta_0$ (Eq. (8.15)), $n_T \approx P_0 \int_0^T d\Delta \int_{T/2}^T d\Delta_0 \approx P_0T$, while the relaxation time of thermal TSs follows from Eq. (9.4), $\tau_1(T) \propto T^{-3}$. It is here assumed that the other TSs, nearby the TS under discussion, are still independent ones, as in general assumed in STM. Since the interaction between TSs is proportional to R^{-3} and the interaction constant has a random sign, the spatial distribution of $n_T(t)$ has to be uniform. In this case, the distribution of the energy fluctuation has been calculated in [138], and the distribution of $\delta\varepsilon(t)$ has a Lorentzian form,

$$L(\varepsilon) = (\Lambda/\pi)[(\delta\varepsilon)^2 + \Lambda^2)^{-1}, \tag{9.20}$$

where the width $\Lambda(t)$ is determined by the relevant interaction of TSs at their average distance and describes the typical value of the TS energy fluctuation during the time t : $\Lambda(t) = \frac{2\pi^2}{3} U_0 n_T(t) \approx \chi_0 Tt/\tau_1(T)$, at $t < \tau_1(T)$ or $\Lambda(t) \approx \chi_0 T$ at $t > \tau_1(T)$. Then, the dephasing rate can be estimated from $\Lambda(\tau_2)\tau_2/\hbar \approx 1$, so

$$\tau_2^{-1} \approx (\chi_0 T/\hbar\tau_1(T))^{-1/2} \propto T^2 \propto \tau_1^{-1} T^{-1}, \tag{9.21}$$

and the above condition, $\tau_2 < \tau_1$, is fulfilled at low enough T, e.g., $T <$ 0.1 K.

However, almost all recent experimental data [183, 184] show systematic deviations of $\tau_1(T)$ and $\tau_2(T)$ from the STM results at low enough $T < 0.1K$, with both $\tau_1 \propto T^{-1}$ and $\tau_2 \propto T^{-1}$, which cannot be explained by the phonon-assisted mechanism for independent TS. Another phenomenon related to the phase memory τ_2, monotonously increasing with decreasing T and observed experimentally at ultralow $T < 0.1K$, is the acoustic coherent echoes (Ch. 6). Let us consider as an example the decay time of two-pulse (spontaneous) echoes. The echo decay is also determined by the spectral diffusion processes discussed above for the hole burnt in acoustic absorption spectrum. In fact, the phase memory time can be related to the hole width $\Delta\nu_h$ as $\tau_2 \propto 1/\Delta\nu_h$, being due to dipolar interactions between TS excitations. Approximately, τ_2 is the time for which the excited-state phase change $\Delta\Phi(\tau_2) = \frac{1}{\hbar}\int_0^{\tau_2} dt \Delta\nu_h(t) \approx 1$. Two regimes can be distinguished, at $t/\tau_1 \gtrless 1$, where $t \equiv 2t_{12}$ is the pulse separation time. In the short-time limit (Eq. (9.16)) at $t \ll \tau_1$, one can get $\Delta\nu_h \approx \Delta\nu_\infty t/\tau_1$, giving rise from $\Delta\Phi(\tau_2) \approx 1$ to a dephasing of independent TS excitations with time as t^2 and the phase memory time described by the relationship $\tau_2^2 = \tau_{\min}/2(\Delta\nu_\infty)^2$. Then, if $\tau_1^{-1} \propto T^3$ and $\Delta\nu_\infty \propto T$, again $\tau_2 \propto T^{-2}$. On the other hand, an intuitive understanding of the dependence $\tau_1^{-1}(T)$ is that the "longitudinal" relaxation rate is proportional to the density $n_{\rm exc}(t)$ of thermal excitations determining the relaxation; for example, for independent TS excitations, the relaxation is due to a direct TS interaction with longitudinal phonons so that, as noted in Eq. (9.4), $\tau_1^{-1}(T) \propto n_{\rm ph}(T) \propto T^3$ [122]. However, the total specific heat of independent TS is $C_{TS}(T) \propto n_{TS}(T) \propto T$ in STM. Then, one can expect at least that a less steep dependence $\tau_1^{-1}(T) \propto T^\alpha$ at $1 \lesssim \alpha < 3$

occurs, if the TS transitions in addition to the Debye phonons contribute to the relaxation in a system of TS interacting with each other. Indeed, it was derived in [187, 188] that, in accordance with the recent experimental data, both $\tau_1^{-1}(T) \propto T$ and $\tau_2^{-1}(T) \propto T$ at ultralow $T < 0.1K$, at which the interaction between TS become more important than the TS–phonon ones. The main issues of the derivation and physical interpretation of the results are briefly presented in what follows.

9.2.2 *Many-body effects: Collective excitations*

Many-body effects due to dipolar interactions $U_{12}(R_{12})$ lead to collective excitations which mostly are coherently coupled, resonant updown pairs with one TS in its ground state while another in its excited state. As mentioned above, the excitations are associated with resonant tunneling transitions, at $\varepsilon_j = [\Delta_j^2 + \Delta_{0,j}^2]^{1/2}$, $\Delta_{0,1} \approx \Delta_{0,2}$, $\Delta_1 \approx \Delta_2$ and $\varepsilon_1 \approx \varepsilon_2$. Each collective, e.g., pair (p) excitation is shown to be itself an effective TS excitation with energy asymmetry $\Delta_p \approx |\varepsilon_1 - \varepsilon_2|$ and tunneling amplitude $\Delta_{0,p} \approx U_{12}(R_{12})\Delta_{0,1}\Delta_{0,2}/2\varepsilon_1\varepsilon_2$. At finite temperatures, the coherent coupling within the pair, determined by the amplitude $\Delta_{0,p} \propto R_{12}^{-3}$, is destroyed as usual if the time of oscillations in the pair, $\hbar/\Delta_0(R_{12})$, is longer than the life-time τ_1 of each TS, i.e., if the TS level width is large, $\hbar/\tau_1 > \Delta_0$. This criterion determines a cut-off radius R_c for the formation of coherently coupled pairs with large lengths $R_{12} \gg a_1$, and most pairs are generated by most numerous thermal TS at $\varepsilon_1 \approx T \approx \varepsilon_2$. For such pairs with $\tau_1^{-1}(T) \approx U_0(T/\hbar s_0)^3$, a coherent coupling in a pair excitation exists at $\Delta_{0,p} \approx U_{0,p}/R_{12}^3 > \hbar/\tau_1(T) \approx U_0/\lambda_T^3$, where the thermal phonon wavelength $\lambda_T \approx hs_0/T \approx a_1 T_D/T$ ($\gg a_1 \approx 3$Å at $T \ll T_D$). It follows that at finite (not too low) T, the coherent interaction occurs at distances not exceeding a characteristic value $R_c = R_c(T)(\gg a_1)$,

$$R_{12} < R_c \simeq \lambda_T, \tag{9.22}$$

which is meaningful since the coherent coupling may exist only within the thermal phonon wavelength $\lambda_T \approx hs_0/T$ by the definition of quantum coherence. In this sense, a pair of coherently coupled TSs at $R_{12} < R_c$, may really be treated as a single effective TS with $\Delta_p \approx |\varepsilon_1 - \varepsilon_2| \lesssim \Delta_{0,p} \approx U_0/R_{12}^3 \lesssim T$ for most numerous thermal TSs at $\varepsilon_1 \approx T \approx \varepsilon_2$.

The effective coupling in the pair excitations is considered weak, in accordance with Eq. (9.17). Taking into account the above, it was shown in [188] that the relatively weak coherent coupling of two resonant TS excitations cannot lead to a delocalization of such resonant pairs, because of the existence of a cut-off radius R_c for the pair length R_{12}: the probability for a single TS to find a resonant partner is much less than unity even at very low, but not too low temperatures (e.g., at ultralow $T < 0.1K$, or at $T > T_D \cdot \exp(-100)$). This statement contradicts some earlier models of "defects" interacting via dipolar interactions, in which the existence of delocalized spin-wave like excitations was assumed (see, e.g., [189]). In other words, the collective excitations generated by dipolar interactions between TS excitations at finite, not too low T are excitations localized on a finite number of centres. The Anderson delocalization of such excitations appears to be related to their ability to form an infinite system of resonant pairs. In this connection, following [188], one can use an analogy with a percolation process [191] connecting resonant pairs into an infinite cluster in which excitations become delocalized, taking into account that the average number of TSs forming a resonant pair with a given one is about $4\pi\chi_0 \ln(L/L_0)$, with L the size of the whole disordered sample and L_0 the smallest distance between TSs. Since $\chi_0 \equiv P_0U_0 \approx 10^{-3}$ is universally small for glasses and thus $4\pi\chi_0 \ln(L/a) \ll 1$ for reasonable sample sizes, the majority of TSs are localized while only a small fraction can form resonant pairs. However, true delocalized excitations may still occur due to interactions between the resonant pairs of which the (spatial) density n_2 becomes much larger than the density of single TSs with low enough excitation energy $\varepsilon < \varepsilon_0$ at $\varepsilon_0 \ll 1K$. This follows from some estimations in [187, 188]: 1) the probability of a single TS to find a resonant partner TS in the coherent range at $R_{12} < R_c$ is $W_p \approx (\pi^2/2) \cdot P_0U_0 \ln(T/T_D)$, e.g., $W_p \approx 0.1$ at $T \approx 1K$ and $T_D \approx 10^3$K, and it weakly varies with T; 2) the DOS of localized coherently coupled pair excitations is $P_2(\Delta_p, \Delta_{0,p}) \approx 2\pi P_0 W_p/12\Delta_{0,p}^2 \approx \pi^3 P_0^2 U_0/12\Delta_{0,p}^2$, for most essential $\Delta_{0p} > U_0/R_c^3$ and $T \gtrsim \Delta_{0p} \gtrsim U_0(T/\hbar s_0)^3$; 3) the pair density $n_2 \approx P_2(\Delta_p, \Delta_{0,p})\Delta_{0,p}^2 \approx (P_0U_0) \cdot (P_0T)$ in the essential range at $T \gtrsim \Delta_{0,p} > \hbar/\tau_1(T)$, for the most numerous thermal excitations, so n_2 is here independent of a specific pair tunneling amplitude Δ_{0p}. For these pair excitations, the average distance is $R_* = n_2^{-1/3}$ and the characteristic coherent coupling Δ_* between resonant pairs can be estimated as

$$\Delta_* \approx U_0/R_*^3 \approx T \cdot (4\pi P_0 U_0)^2 \ll T. \qquad (9.23)$$

Under this condition, the resonant pairs at $\hbar/\tau_1 < \Delta_{0,p} \lesssim \Delta_*$ were shown to form the "resonant cluster", an infinite coherently coupled cluster of resonant pairs (with $\Delta_{0p}/\varepsilon_p \approx 1$) that enables the occurrence of delocalized excitation consisting of coupled resonant pairs due to their dipolar interactions [187, 188]. The condition of the existence of an infinite cluster is just that a characteristic coupling energy $\Delta_* = \max(\Delta_{0,p})$ of most pairs exists, which does not depend on a specific $\Delta_{0,p}$, at $\Delta_{0,p} \approx \Delta_*$. While less strongly coupled pairs cannot exist by definition in the infinite cluster, more strongly coupled pairs remain localized, not participating in the cluster, as their interaction energy is larger in magnitude than the characteristic energy scale, $\Delta_{0,p} \approx U_0/R_{12}^3 > \Delta_*$ at $R_{12} < R_*$. That is why a weakness of coherent coupling, at $P_0 U_0 \ll 1$, is important for the existence of self-consistent dynamic infinite cluster, with pairs randomly leaving and entering it, and of its delocalized excitations. Since all resonant-pair lengths $R_{12} < R_c$, it follows for all pairs in an infinite cluster that $R_c > R_* \gtrsim R_{12}$, with $R_*^3 \approx 1/U_0 T P_0^2 \ll R_c^3 \approx (\hbar s_0/T)^3$. Thus the infinite cluster and delocalized excitations exist only at low enough T,

$$T \lesssim T_0 \approx (P_0 U_0)(\hbar^3 s_0^3/U_0)^{1/2}. \qquad (9.24)$$

This condition corresponds to sufficiently large interaction radius for excitations to delocalize. The TSs rather break up into pairs, each behaving as an effective TS, which interact with each other and give rise to the occurrence of delocalized excitations. The question is what the effects of such excitations in the spectral diffusion and in the dephasing rate are. In accordance with the above, the pairs with the energy of the scale of Δ_* change their state at the characteristic time $\tau_* \approx \hbar/\Delta_*$. The changes of their states affect the energy of other excitations, with energy change of the scale of Δ_*. Thus the resonant coupling of the pairs vanishes in a time τ_* for those with $\Delta_{0p} \leq \Delta_p < \Delta_*$ and the probability of such a real transition is $(\Delta_{0p}\tau_*)^2 < 1$. Since the transitions of TSs in the resonant cluster change the phase of any TS wave function, and for each TS the phase change is unique depending on its random surrounding, the phase coherence of the TSs will be destroyed at time $t \sim \tau_*$ and the dephasing rate in the system of TSs is determined by the single energy scale of the infinite cluster, $\tau_2^{-1} \approx \Delta_*/\hbar \sim T(P_0U_0)^2/\hbar$. As mentioned above, the dephasing time is measured in many echo experiments. Hence, both the "longitudinal" relaxation rate $\tau_1^{-1}(T)$ and the "transverse" relaxation, or "dephasing", rate $\tau_2^{-1}(T)$ indeed exhibit a linear temperature variation, in accordance with the intuitive understanding of dependencies $\tau_{1,2}^{-1}(T)$, at $\tau_1^{-1} \propto n_{TS}(T) \propto C_{TS}(T) \propto T$ and $\tau_2^{-1}\tau_1 < 1$:

$$\tau_1^{-1}(T) = b_1 T \text{ and } \tau_2^{-1}(T) = b_2 T. \tag{9.25}$$

Here b_1 and b_2 are two characteristic constants, with $b_1 < b_2$ and thus with the above-mentioned condition $\tau_1^{-1} < \tau_2^{-1}$ fulfilled:

$$b_1 \sim (P_0U_0)^3,\ b_2 \sim (P_0U_0)^2. \tag{9.26}$$

The crossover between the relaxation mechanisms is expected to occur at a characteristic temperature T' that can be found by comparing Eqs. (9.17) and (9.21) with Eqs. (9.25):

$$T' \approx T_0(P_0U_0)^{1/2} \ll T_0. \tag{9.27}$$

As noted above, the described TS related phenomena are experimentally observed at very low $T \lesssim 1K$. Then, the phonon-induced mechanism predominates at $T' < T \lesssim 1K$ while the interaction-stimulated one at lower $T < T'$, and the delocalized excitations are expected to occur not only at $T < T'$ but also at $T' \lesssim T \lesssim T_0$ and to be observable at $T < 1$ K as far as $T_0 \lesssim 1K$. In fact, tentative estimates of T_0 and T' at typical values of the model parameters, e.g., at $P_0U_0 \approx 10^{-3}$, give rise to

$$T' \sim 0.1\ T_0 \text{ and } T_0 \sim 0.1\,\text{K}, \tag{9.28}$$

while

$$b_2^{-1} \sim 10^6\ \text{K}^{-1}\text{s}^{-1} \quad \text{and} \quad b_1 \sim 10^{-3} b_2. \tag{9.29}$$

Thus, definitely $T' < 0.1$ K at plausible $T_0 < 1$ K. The factor b_2 is in a reasonable scale agreement with that obtained from measurements of spontaneous echoes in v-SiO_2 [184]. In fact, as noted above (Ch. 6), the dephasing time τ_2^{-1} has been measured in diverse echo experiments at ultralow temperatures rather similar to the spin echo experiments for nuclear spins.

In addition to a predicted new mechanism of dephasing in glasses at ultralow temperatures (Eqs. (9.25)–(9.29)) [187, 188], a new relaxation mechanism of the TS excitations is found at such temperatures by Burin and Kagan [122], which is also characterized by an observed linear temperature dependence. The main issues of the derivation and physical interpretation of the results of this theory are briefly presented in what follows while mathematical details are skipped. The effect of delocalized, collective excitations for relaxation processes in the system of interacting TSs is analyzed in detail and it is shown that the collective excitations can stimulate quicker TS transitions, exhibiting a weaker temperature dependence, than the TS-phonon interaction. In the delocalization process, TS pairs undergo transitions in each characteristic time interval $\tau_* \approx \hbar/\Delta_*$ (Eq. (9.23)). The resulting changes of local strain fields shift the energy levels of each TS so that the local environment changes, with some intra- and inter-pair resonant couplings destroyed while new ones are formed, and such rearrangement essentially affects relaxation processes in the system of TSs. Moreover, as long as the density n of TSs, undergoing the transitions, is small compared to the total density n_0, the change δ_i in asymmetry Δ_i of i-th TS determined by the transitions is estimated as $\delta_i \approx \sum_{j \neq i} U_{ij}$. Considering the evolution of TSs, determined by up-down transitions within the infinite resonant cluster of such pairs, one can conclude from the theory that, due to dipolar interactions, the asymmetry Δ_i is distributed according to a Lorentzian function (Eq. (9.20)), with the characteristic width $\Lambda[n] = \frac{2\pi^2}{3} U_0 n$. However, in this case the time dependence of the density $n(t)$ of TSs undergoing transitions is determined by the dynamics of delocalized excitations, rather than by TS–phonon relaxation. The theory also predicts that the characteristic width $\Lambda[n]$ exceeds the resonant energy Δ_* (Eq. (9.23)), which means that the majority of resonant pairs in the infinite resonant cluster have their energies shifted by more than the resonant energy and thus are broken at $t \sim \tau_* \approx \hbar/\Delta_*$. However, new resonant pairs and thus a new infinite cluster of pairs appear because the TS distribution in energy is smooth (Sec. 9.1). This self-consistent time-evolution process takes place as the total number N of TSs that undergo the transitions changes with time as $dN/dt = 2n\Delta_*/\hbar$. Assuming that the pairs that left the infinite cluster do not return while this is not the case for individual thermal TSs returning to form resonant pairs with other partners, the relaxation time τ_0 of thermal TSs can be assessed from the condition that almost all thermal TSs change their state during this time and that the solution $N(\tau_0) = 2n\tau_0\Delta_*/\hbar$ of the equation equals the density of thermal TSs, P_0T. Then, one can obtain finally that $\tau_0^{-1} = 2n\Delta_*/\hbar P_0 T \approx 10\hbar^{-1}T$ $(P_0U_0)^3$ while the distribution width $\Lambda \approx \frac{2\pi^2}{3}\tau_0\Delta_*/\tau_* \sim T\ (P_0U_0)$. In fact, the final estimations appear to be in reasonable agreement with the assumption used for their derivation.

In conclusion, the advanced tunneling model (ATM) under discussion accounts for the contributions of dipolar interactions between TS excitations and answers a basic question: why both $\tau_1^{-1}(T)$ and $\tau_2^{-1}(T)$ show a linear temperature variation at ultralow $T < 0.1$ K in glasses, whereas $\tau_1^{-1}(T) \propto T^3$ and $\tau_2^{-1}(T) \propto T^2$ at higher T, $0.1 \lesssim T \lesssim 1$ K. Moreover, Eqs. (9.25)–(9.28) agree with recent data for silica and other glasses at ultralow temperatures not only qualitatively, in temperature variation, but also in scale for parameters b_1 and b_2. Thus, in addition to the major result of STM that for very low $T \lesssim 1$ K and $\nu \lesssim 0.03$ THz the glass dynamics is related to independent TS excitations, two essential theoretical results are obtained in the tunneling model accounting for the dipolar interactions between the excitations:

(1) delocalized excitations exist in a glass at low enough $T \lesssim T_0(<1K)$, which occur in the infinite cluster of weakly coupled coherent pair TS excitations related to atomic tunneling transitions in double-well potentials; since in SMM the TS excitations are very-low-energy soft-mode ones, delocalized states appear also to be related to soft modes;
(2) the delocalized excitations stimulate a new relaxation mechanism predominant at ultralow $T < T'(\ll T_0)$, while at higher T, $T' < T(<1K)$, the standard phonon-induced relaxation mechanism predominates.

As noted in Sec. 6.1, in measurements published recently on the thermal conductivity and acoustic attenuation at very-low temperatures or frequencies ($\nu = h\nu/T$), on a total of over 60 different compositions of glasses (amorphous solids), it is found in [132] that the ratio of the acoustic phonon wavelength to mean free path is characterized by a rather universal scale, $l_{ac}^{-1}\lambda_{ac} \approx 10^{-3} - 10^{-2}$ (this experimental result is expected to be explainable in the ATM yet under discussion but for now no specific references seem to be available).

At the same time, there are still essential questions which appear to remain unanswered even in ATM yet probably can be answered in SMM:

(1) Why ATM can explain qualitatively and scalewise the experimental data (Ch. 6) for dynamic and thermal properties of glasses only at very low external-field frequencies, $\nu \lesssim \nu_m \approx 3 \cdot 10^{-2}$ THz, and temperatures, $T \lesssim T_m \approx 1K$, but not at $T_m < T \ll T_D$ and/or $\nu_m < \nu \ll \nu_D$ THz (as assumed in STM).
The restriction of energy range, in which STM is applicable, can be determined by the fact that SMM actually reduces to STM only for soft-mode TS excitations of very low energy, $\varepsilon \lesssim \varepsilon_m \equiv h\nu_m \equiv T_m$, as explained in Secs. 8.1 and 8.6.
(2) Why STM exhibits systematic "small failures" in describing the experimental data for dynamic and thermal properties of glasses at very (but not too) low T, $0.1 < T \lesssim 1$ K, and respective ν, and, in particular, gives rise to the specific heat $C(T) \propto T$ and thermal conductivity $\kappa(T) \propto T^2$, whereas the data rather show that $C(T) \propto T^{1+\alpha}$ and $\kappa(T) \propto T^{2-\rho}$ at $0.1 \lesssim \alpha, \rho \lesssim 0.3$.
The appropriate relationships of SMM may qualitatively differ from those in STM (Sec. 8.1) and be closer to the experimental data also at very (but not too) low temperatures and frequencies.
(3) Why for amorphous materials the following empirical correlation exists between dynamic properties at moderately low temperatures/frequencies and at very low ones [192]: the anomalous low-temperature thermal properties at very low temperatures, and thus TS excitations, are found to be missing in amorphous materials where no boson peak is found in their scattering spectra at moderately low-frequency shifts.

In particular, in [109], the acoustic internal friction Q_{ph}^{-1} has been measured in a-Si films hydrogenated with $\sim 1\%$ of H, in which a temperature independent part of internal friction Q_0^{-1} (plateau), rather characteristic of glasses, was found to be small (if any) and no pronounced trace of boson peak seemed to be observed (perhaps, this was the first observation of an amorphous solid without significant excess, non-Debye, low-energy excitations).

As emphasized above, the fundamental concept of STM giving rise to qualitative and scalewise theoretical description of the experimentally observed anomalous, universal, dynamic and thermal properties of almost thermal-equilibrium glasses at very low $T \lesssim T_m$ and $\nu \lesssim \nu_m$, is that the properties are due to TS excitations, their interactions with acoustic phonons and with each other. However, this theoretical concept is irrelevant for describing the properties of glasses at moderately low T and/or ν, $3T_m < T \lesssim T_M \approx (1 - 1.5) \cdot 100$ K and/ or $3\nu_m < \nu \lesssim \nu_M \approx (2 - 3)$ THz . Thus the fundamental restriction of STM is removed in SMM, in which question 3 appears to be consistently answered. Indeed, the above-mentioned empirical correlation occurs because SMM is able to describe in a unified way both the non-vibrational TS excitations, giving rise to the very-low-energy properties, and harmonic vibrational excitations, determining the moderately-low-energy properties like the boson peak, of glasses.

10

SOFT-MODE EXCITATIONS OF MODERATELY-LOW ENERGIES (BOSON PEAK)

As noted in Ch. 6, one of the most well-known anomalous, often universal, dynamic properties of glasses is the boson peak experimentally observed in photon (Raman, x-ray) and neutron inelastic scattering experiments. The BP is detected at THz scale frequency shifts, $\nu = \nu_{BP} \equiv \omega_{BP}/2\pi \sim 1$ THz, and asymmetric and broad maximum with a large width $2\gamma_{BP} \sim \nu_{BP}$ [140, 141, 142], its temperature dependence being actually described by the boson factor (Eq. (6.10)). The latter suggests that the origin of BP in a glass is in general due to inelastic scattering of photons or neutrons from *harmonic* vibrational excitations. Unlike the anomalous properties of glasses at very low $T \lesssim T_m \sim 1$ K and $\nu < \nu_m \sim 10^{-2}$ THz, which are determined by non-vibrational tunneling-state excitations, the properties at moderately low ν and T can be explained in SMM as due to soft-mode *harmonic* vibrational excitations of moderately low frequencies, $0.5 \lesssim \nu \lesssim 1.5$ THz. In the present Chapter, the SMM of the vibrational excitations is applied to glasses for which the observed THz scale vibrational dynamic properties include both BP and, above it, high frequency sound (HFS) with well-defined acoustic-like excitations [112, 146].

10.1 Ioffe–Regel crossover for acoustic phonons as origin of boson peak

No consistent microscopic approach and quantitative analytical theory of the BP and other moderately-low energy vibrational properties of the glasses under consideration seem to be available at present. Three mathematically very different theoretical approaches, mode-coupling model [193], soft-mode model (SMM) [194, 195] and random-matrices theory [196] (see also [197]), have recently been proposed for describing the origin and basic features of the low-energy vibrational properties of glasses, which exhibit both BP and HFS. These models can be considered as "mean field" theories, being based on microscopic ground, phenomenological approximations and real calculations, which together give rise to a qualitative and scale-wise description of the phenomena and their origin. All these models actually seemed to imply that the dynamic vibrational properties were determined by the occurrence of non-acoustic harmonic vibrational excitations, interacting with acoustic phonons of moderately low frequency, and might complement each other in certain aspects. Let us emphasize that in SMM both the non-vibrational tunneling excitations of very low energies, determining the glassy properties at very low T and ν, and the non-acoustic harmonic vibrational excitations of moderately low energies are correlated with each other because of their common soft-mode origin. However, unlike the completely universal properties of glasses at very low ν and T, the properties at moderately low ν and T seem to show properties that are both universal (e.g., the occurrence of BP) and non-universal (the existence or non-existence of HFS).

It appears that the origin of BP followed by HFS in the glasses can qualitatively and scalewise be described in SMM by assuming that the BP is due to the temperature independent dynamic phenomenon called Ioffe–Regel crossover (IRC) [198] for propagating wave-like, excitations of low frequency ν and large wave-length λ_{exc}. In other words, it is actually implied in the soft-mode model of excitations, particularly for harmonic vibrational excitations under discussion in the present chapter, that the excitation states can be described as Anderson non-localized (extended) states. The latter are in particular rarely scattered when characterized by their mean free path l_{exc} much larger than the wavelength, $l_{exc} \gg \lambda_{exc}$. Then, by definition, the IRC occurs with increasing excitation frequency from weak (i.e., rare) scattering to strong scattering at small $l_{exc} \approx \lambda_{exc}$ at which the wave number $q = 2\pi/\lambda_{exc}$ is no longer a good quantum number. In general, disordered systems may contain not only a static disorder providing elastic scattering, but also a "dynamic" disorder due to the existence of low-frequency, non-acoustic, harmonic vibrations of moderately-low frequencies either in certain defects of crystalline lattices [199, 200] or in soft-mode "defects" of glasses, the vibrational excitations scattering the wave-like ones inelastically.

The IRC concept was initially introduced in [198] for elastic scattering of some electronic excitations, conduction electrons or holes, in a semiconductor containing randomly located defects. In SMM, this concept can be generalized by considering the IRC as a universal temperature independent phenomenon for well-defined, wave-like excitations for which a weak, either elastic or inelastic, scattering can be described by standard perturbation theoretical expansions with terms $\propto \epsilon^n$ $(n = 0, 1, ...)$, in powers of a small temperature independent parameter $\epsilon = \lambda_{\text{exc}}(\nu_{\text{exc}})/l_{\text{exc}}(\nu_{\text{exc}}) \ll 1$ or equivalently $\epsilon = 2\gamma_{\text{exc}}(\nu_{\text{exc}})/\nu_{\text{exc}} \ll 1$ [172]. On the other hand, the crossover to strong scattering can be described by a large parameter $\epsilon \approx 1$ at which the wave-like excitation width $2\gamma_{\text{exc}}$ is not small relative to its frequency $\nu_{\text{exc}} \equiv \nu$, so the scattered excitations become "ill-defined" in the sense that they cannot be described by standard perturbation theories. In SMM, a wave-like excitation by definition is an acoustic phonon of which the width $2\gamma_{\text{exc}} = 2\gamma_{\text{ac}} \approx s_0 l_{\text{ac}}^{-1}$, at $\epsilon \ll 1$, increases with increasing frequency and is independent of temperature. For the first time, the suggestion that the BP origin in glasses can be related to the IRC phenomenon was proposed in 1992 in two limit cases of either elastic scattering of acoustic phonons ([201], Eq. (1)) or inelastic scattering of the phonons from soft-mode harmonic vibrational excitations [126], Eqs. (10) and (17); in Eq. (10) a misprint has to be corrected: the expression "$l_{\text{ph}} \approx \lambda_{\text{ph}}$ holds true for $\omega \approx \omega^* \approx \omega_b$" has to be substituted for the one "$l_{\text{ph}} \geq \lambda_{\text{ph}}$ for $\omega \geq \omega^* \approx \omega_b$". This suggestion agrees with the experimental observation that the total scattering intensity around BP is almost independent of q, $I_{ij}(\nu, q; T) \simeq I_{ij}(\nu; T)$ (Eq. (6.8)), and with the theoretical interpretation of BP in [142] (Ch. 6). It can be added that the introduced IRC appears to essentially differ from the Anderson-Mott localization occurring at a characteristic frequency $\nu_{\text{loc}}(<\nu_D)$, since the localization is due to quantum-mechanical interference processes [129] whereas the IRC actually is related to a quasiclassical scattering process [113]. Generally speaking, the effective width $\nu_{\text{loc}} - \nu_{IR}(>0)$ of the region of strongly scattered and still Anderson non-localized excitations is finite, though different for diverse excitations (the width is rather small for conduction electrons while relatively larger for acoustic phonons; see, e.g., [202]).

The general criterion of an IRC is by definition described by equation

$$\epsilon(\nu_{IR}) = 1, \quad \text{e.g.,}\ 2\gamma_{\text{exc}}(\nu_{IR}) = \nu_{IR}, \tag{10.1}$$

which determines the existence of the characteristic IRC frequency ν_{IR} that is independent of temperature; the latter, generally speaking, takes place for scattering processes giving rise to the width $2\gamma_{\text{exc}}(\nu)$ that also does not depend on T. In general, for the above-mentioned two limit types of IRC, two characteristic frequencies, $\nu_{IR} = \nu_{IR}^{(\text{el})}$ for elastic scattering and $\nu_{IR} = \nu_{IR}^{(\text{in})}$ for inelastic scattering from soft-mode vibrational excitations, is expected to exist, which can be different in magnitude (but not necessary in scale). For the glasses under discussion in SMM, the case of predominant inelastic scattering of acoustic phonons by soft-mode vibrational excitations is suggested to be realized in an IRC determining a BP at $\nu = \nu_{BP} \simeq \nu_{IR}^{(\text{in})}$ in glasses in which $\nu_{IR}^{(\text{in})}$ is lower than $\nu_{IR}^{(\text{el})}$. Vice versa, the case of predominant elastic scattering of acoustic phonons by soft-mode vibrational excitations is realized in an IRC determining a BP at $\nu = \nu_{BP} \simeq \nu_{IR}^{(\text{el})}$ in glasses in which $\nu_{IR}^{(\text{el})}$ is lower than $\nu_{IR}^{(\text{in})}$. The major mathematical problem in SMM is to develop a consistent non-perturbative analytical theory of the IRC, which could describe the ill-defined IRC excitations of which spectra would correspond to experimental BP spectra in the glasses. Since this problem is not yet solved in a consistent theory available at present, a reasonable approximation can be suggested, which allows one to estimate the BP intensity and frequency, $\nu_{BP} \simeq \nu_{IR}^{(\text{in})}$, in terms of soft-mode parameters. Recent experimental data [147] (1996, 2001) for normal and densified silica, obtained by applying the Hyper–Raman spectroscopy, appear to favor the occurrence of a BP as a maximum of an excess DOS of Einstein-like vibrational excitations, which can be correlated with anomalous thermal properties (bump of specific heat, plateau of thermal conductivity) and with dramatic changes in the glass properties at acoustic frequencies in the BP range $\nu \sim 1$ THz, providing unique information on the structure of glasses at wavelengths $\lambda \sim 50$ Å of extended length scale and possibly corresponding to the existence of a IRC . Unfortunately, the data can hardly unambiguously answer the question of whether the IRC is due to inelastic scattering for acoustic phonons of frequencies around the BP one, in agreement with the results of SMM discussed in what follows, or it is due to elastic scattering of the phonons from static disorder. It can also be noted that the assumed Einstein-like vibrational excitations actually resemble soft-mode harmonic vibrational excitations (Sec. 10.2).

One can conclude from above and from what follows in the present Chapter (see Eq. (10.8) and related comments) that the soft-mode harmonic excitations of moderately low energy can also be described as Anderson non-localized states [129].

In what follows, Secs. 10.2, 10.3, 10.5 and 10.6 describe the soft-mode model of vibrational THz scale excitations and related properties, including BP and HFS, of the glasses under discussion, in which most important is that a certain "singularity" of the excitation energy spectrum is related to the IRC for inelastic scattering. The earlier version of this model [194] approximately accounts for the IRC of acoustic phonons interacting with independent soft-mode harmonic excitations, while the "complete" model [195] takes into account the IRC for interacting soft-mode excitations. phonons. As well as in a crystalline material [113], acoustic phonons of large wave-length, $\lambda_{ac} \gg a_1$, in a glass can be approximated by Debye excitations of appropriate elastic continuum, by neglecting the specific atomic structure.

10.2 Independent soft-mode vibrational excitations

In accordance with the basic concept of SMM, two types of low energy atomic motions and related excitations in a glass, namely, propagating Debye phonons of low frequencies $\nu \ll \nu_D$, involving the vast majority of atoms, and soft-mode excitations associated with a minority of atoms of a relatively low concentration $c_{sm}(\ll 1)$ in soft local configurations, coexist and interact with each other. For moderately-low-energy excitations, the SMM was developed [194, 195] as an analytical "mean field" theory that described the glass as an isotropic elastic $3D$ continuum containing randomly immersed localized "defects", soft modes, characterized by harmonic vibrations of THz scale frequencies $\nu'(\lesssim \nu_{max} \ll \nu_D)$, where the upper limit can be estimated as $\nu_{max} \approx 2$–3 THz. The purpose of the earlier SMM of harmonic vibrational excitations [194] was to calculate the total single-excitation DOS $J(\nu^2)$ for a macroscopic $3D$ disordered system. For the sake of simplicity, the scalar model is used in this analysis, which does not distinguish longitudinal and transverse acoustic phonons. On the other hand, soft-mode harmonic vibrational excitations in a glass are similar to "resonant" [199], or "quasilocal" [200], low-frequency ($\nu \ll \nu_D$) vibrational states of defects in crystals, characterized by large amplitudes $\Delta x \approx (\hbar/4\pi M_{sm}\nu'_{sm})^{1/2} \gg \Delta_{ac} \approx (\hbar/4\pi M\nu_D)^{1/2}$, the typical acoustic amplitude [113]. As an example, let us consider in the Green's function approach resonant vibrations in the case of independent, randomly located heavy-isotope point defects of low concentration $c \ll 1$, in an isotropic elastic continuum [203]. The averaged Green's function matrix G of a disordered system can be described as a function of site vector differences $r_n - r'_{n'}$, as well as the Green's function G^0 of a perfect crystal, because of the macroscopic homogeneity over distances larger than the average distance $\sim a_1 c^{-1/3}$. Then, the long-wave vector q representation for the averaged Green function can be found in a canonical form that is actually not restricted to a linear approximation in c:

$$G(\varepsilon = \nu^2, \mathbf{q}) = \{\varepsilon - \nu_{ac}^2(\mathbf{q}) - cP(\varepsilon)\}^{-1}, \tag{10.2}$$

where $\nu_{ac}(q) \approx s_0 q$ and the function $P(\varepsilon)$ can be presented also in a familiar form:

$$P(\varepsilon) = I_0/D(\varepsilon). \tag{10.3}$$

The parameter I_0 characterizes the interaction, $I_0\delta(r - r_d)$, of long-wave phonons with a point defect located at $r = r_d$, and $D(\varepsilon) = 1 - I_0 J^0(\varepsilon)$, where $J^0(\varepsilon) = \int dz J_0(z)(\varepsilon - z + i\eta)^{-1}$ (at $\eta \to +0$) is the Green's function and $J_0(z)$ is the total vibrational DOS for acoustic phonons in the medium, so $P(\varepsilon)$ is a complex function. Equations (10.2) and (10.3) can be analyzed for finding the resulting vibration eigenvalues and eigenstates, e.g., plane waves of averaged displacements describing collective excitations characterized by wave vector $\mathbf{q}$ and frequency ν. As usual, the dispersion law for the excitation eigenvalues, $\nu(\mathbf{q})$, is determined by the poles of the Green's function $G(\varepsilon, \mathbf{q})$, i.e., by the roots of dispersion equation $\varepsilon - cP(\varepsilon) \simeq \nu_{ac}^2(\mathbf{q})$.

For real wave vectors $\mathbf{q}$, the solutions are complex ones for damping plane-wave excitations, $\mathbf{u}(\mathbf{r},t) \approx const.\ \exp(-2\gamma t)\exp[i\mathbf{qr} - i2\pi\nu t]$, for which the finite vibration width 2γ is determined by Im $P(\varepsilon)$. The well-defined wave-like excitations occur when the damping is weak, at $2\gamma \ll \nu$. In this case, in the used scalar model of vibrations it is easy to estimate the eigenvalues $\varepsilon(q) = [\nu(q) - i2\gamma]^2 \simeq \nu^2(q) - i4\gamma\nu(q)$ in the lowest order of $2\gamma/\nu(q) \ll 1$ from Eqs. (10.2) and (10.3), where $\nu(q)$ is the dispersion law of the resulting vibrational excitations with a finite width $2\gamma \sim \nu_D^2 \nu \delta J(\nu^2)$. The "excess" total vibration DOS $\delta J(\nu^2)$ is proportional to the concentration c of defects with quasilocal vibrations described by the term $cP(\varepsilon)$. The lifetime $(2\gamma)^{-1}$ of quasilocal vibration is finite because the energy of the defect-related vibration can effectively be transferred to the (almost) resonant levels of continuous phonon spectrum. In the Anderson classification of excitations [129], the harmonic vibrational excitations of quasilocal defects and of similar soft-mode "defects" are extended, i.e., non-localized ones, which are propagating wave-like when their damping is weak, at $2\gamma \ll \nu$, far enough from the corresponding IRC.

For independent soft modes, the harmonic vibrational excitations (Eqs. (7.29)–(7.30)) occur mostly in single-well potentials at $\eta_0 < \eta \lesssim \eta_{sm} \approx 0.1$ with $\eta_0 = 9\xi^2/32$ (Eq. (7.5)), and can be characterized by the general expression of (normalized) total vibrational DOS in Eq. (7.32). In general, the total vibrational DOS $J^{(HV)}(\nu^2)$ and related frequency DOS $g^{(HV)}(\nu)$ can be calculated by applying the following expression:

$$J^{(HV)}(\varepsilon \equiv \nu^2) = g^{(HV)}(\nu)/2\nu = \sum_\alpha c_\alpha J_\alpha^{(HV)}(\nu^2)$$
$$= \sum_\alpha c_\alpha \iint_{\Omega_{exc}} d\eta d\xi F_\alpha(\eta,\xi)\delta[\nu^2 - \varepsilon_1^{(HV)}(\eta,\xi^2)], \tag{10.4}$$

where $J^{(HV)}(\nu^2)d\nu^2 = g^{(HV)}(\nu)d\nu$, Ω_{exc} denotes the appropriate range of eigenvalues $\varepsilon = \varepsilon_1(\eta,\ \xi^2)$ for soft-mode excitations in their lowest band, $\sum_{\alpha=1,2} c_\alpha = 1$ and $0 \leq c_\alpha \leq 1$. For the soft-mode harmonic vibrations $h\nu(\eta,\xi^2) \simeq h\nu(\eta) \simeq h\nu_D\eta^{1/2} \propto k^{1/2}$ in the vast majority of related, actually single-well potentials, $V(x) \simeq kx^2/2$ with $k \approx k_0\eta \ll k_0$, at $9\xi^2/32 < \eta \ll 1$ (Eq. (7.5)). Then, the following result can straightforwardly be obtained:

$$J_\alpha^{(HV)}(\varepsilon) = \int\int_{\Omega_{exc}} d\eta d\xi F_\alpha(\eta,\xi)\delta[\nu^2 - \varepsilon_1^{(HV)}(\eta,\xi^2)]$$

$$\approx 2\int_0^{\eta_{sm}} d\eta \int_0^{2\eta^{0.5}} d\xi F_\alpha^{(0)}\eta^{\alpha-1}\delta(\nu^2 - \nu_D^2\eta)$$

$$= 0.5 \cdot C_\alpha \nu_D^{-2}(\nu^2/\nu_D^2)^{(\varkappa-1)/2}, \qquad (10.5)$$

where $C_\alpha \equiv 8F_\alpha^{(0)} = const. \propto c_{sm}$, and $\varkappa = 2\alpha = 2$ at $\alpha = 1$ or $\varkappa = 4$ at $\alpha = 2$ in the limit cases (7.17) or (7.18) of the distribution density $F_\alpha(\eta,\xi)$, while the exponent $\varkappa$ varies in the range $2 \leq \varkappa \leq 4$ in the general case (7.19). Thus the vibrational DOS increases as a power law with increasing frequency ν' and contains two material parameters: exponent $\varkappa$ and typical soft-mode frequency $\nu_0 \approx 1$ THz (Eq. (7.30)).

The problem under discussion is similar to that of the theory of vibrational spectra in a disordered lattice containing a finite atomic concentration c_d of randomly located defects with quasilocal vibrational excitations; the concentration is low in the usual sense, $c_d \ll 1$, but can be "high" in a sense defined below [203, 204]. Then, SMM for harmonic vibrations can be considered as an appropriate extension of the theory for a crystalline lattice to an elastic continuum with immersed randomly located soft-modes of concentration c_{sm}. The problem is not only to derive the dispersion law of the frequency spectrum for given realizations of the random system, depending on defect (e.g., soft-mode) parameters, but even more to find basic spectral properties, averaged over all realizations in macroscopic ensembles of defect locations, which do not depend on random microscopic fluctuations. The fundamental result of the theory of vibrational spectra in disordered lattices is that the spectral properties described, e.g., by the total vibrational DOS $J(\varepsilon = \nu^2) = g(\nu)/2\nu$ and frequency DOS $g(\nu)$ (as well as by the dynamic structure factor $S(\mathbf{q},\nu)$) are so-called self-averaging properties, or measurable macroscopic characteristics of the systems, which can be calculated by averaging the respective microscopic characteristics over all random realizations. The calculations of self-averaging DOS can be performed only for "long-wave" acoustic phonons, of which wavelengths overlap large numbers ($\gg 1$) of defects in mesoscopic sub-volumes characterized on average by spatial homogeneity. The range of validity of the theory is determined by the criterion of "long-wave" acoustic excitations: the typical wavelength $\lambda_{ac}(\nu_0) = s_0\nu_0^{-1} \gg r_{av} = a_1 c_d^{-1/3}$, the mean separation of defects. In SMM, as well as in the vibrational dynamics of disordered lattices, one can calculate the single-excitation Green's function (at continuum points r, r') of the system, averaged over all random realizations, $G_{av}(\mathbf{r} - \mathbf{r}'; t) \equiv \langle G_1(\mathbf{r}, \mathbf{r}'; t)\rangle$, its time Fourier transform, $G_{av}(\mathbf{q}, \varepsilon)$, and the total vibrational DOS $J(\varepsilon) = \pi^{-1}\,\mathrm{Im}\,Tr\;G_{av}(\mathbf{q}, \varepsilon)$. The averaged Green's function is related in a well-known simple way to averaged space-time correlators of acoustic and soft-mode displacements ($\mathbf{u}_{ac}, x$), particularly to $\langle \mathbf{u}(\mathbf{r}, t)u(\mathbf{r}', 0)\rangle$ and $\langle x(\mathbf{r}, t)x(\mathbf{r}', 0)\rangle$, and hence to products of averaged displacements, $\mathbf{u}_{av}(r, t)$ and $x_{av}(r, t)$, in a reasonable approximation (in the same way as $G_1(\mathbf{r}, \mathbf{r}'; t)$ is related to the products of the dynamic variables $\mathbf{u}(\mathbf{r}, t)$ and $x(\mathbf{r}, t)$) [203, 204]. For often studied types of defects (e.g., heavy-isotope defects or interstitial-like ones) in a lattice, the resulting spectra of vibrational excitations are explicitly found to be essentially different in two diverse ranges of c_d, "low" $c_d \ll c_0$ and "high" c_d, $c_0 \ll c_d$ ($\ll 1$). The characteristic value c_0 of c_d can be defined from the equation $r_{av}(c_0) = \lambda_{ac}(\nu_0)$, so that $c_0 = (\nu_0/\nu_D)^3$ at $\nu_D = s_0 a_1^{-1}$. Moreover, the limit case of "high" concentrations defines the range of validity of the theory that holds true for any kind of quasilocal defects, in particular, for soft-mode defects. For the latter, only the limit case of "high" concentrations is important, since the actual soft-mode concentration is $c_d = c_{sm} \approx 10^{-2} \gg c_0 \approx 10^{-3}$, at typical $\nu_0 \approx 0.1\nu_D$. Since the "bare" soft-mode vibrational excitations are characterized by a direct harmonic interaction with acoustic phonons, an inelastic (resonant) scattering of acoustic phonons from soft-mode vibrational excitations occurs at $\nu_{ac}(q) = s_0 q/2\pi \simeq \nu'$, at $q \ll q_D \equiv 2\pi\nu_D/s_0$. In the limit case of "high" $c_d = c_{sm} \gg c_0$, the Green function approach can be applied for analytical calculations of $J(\varepsilon)$ as an expansion in powers of a small parameter, $\zeta \equiv R_{av}^3(c_{sm})\lambda_{ac}^{-3}(\nu_0) = c_0/c_{sm} \ll 1$, in which the basic contribution to calculated quantities is determined by the main finite, lowest order term $\sim \zeta^{-1} \propto c_{sm}$ (as usual, the approach is a consistent perturbation-theoretical one valid for $c_0 \ll c_{sm} \ll 1$, with accuracy to small corrections $\sim \zeta^n$ at $n = 0, 1, \ldots$).

From this viewpoint, the defining equations of SMM consist of both the equations of motion for averaged acoustic ($\mathbf{u}_{av}$) and soft-mode displacements (x_{av}) and the general relationship describing the total vibrational DOS $J(\varepsilon)$ for the spectrum of vibrational eigenvalues $\varepsilon = \nu^2(q)$. In the Green's function approach, the equations of motion for microscopic displacements ($\mathbf{u}(\mathbf{r}, t)$, $x(\mathbf{r}, t)$) can be derived by applying standard dynamic equations [203] to the system Hamiltonian $\widehat{H} = \widehat{K}_{ac} + \widehat{K}_{sm} + \widehat{V}_{ac} + \widehat{V}_{sm} + \widehat{V}_{int}$ or Lagrangian $L = \widehat{K}_{ac} + \widehat{K}_{sm} - V_{ac} - V_{sm} - \widehat{V}_{int}$, expressions for acoustic and soft-mode kinetic energies, $\widehat{K}_{ac} \propto \int d^3r\ (de_{ac}(\mathbf{r})/dt)^2$ and $\widehat{K}_{sm}$, potential energies, $\widehat{V}_{ac} \propto \int (d^3\mathbf{r})\ (e(\mathbf{r}))^2$, $\widehat{V}_{sm}$, and for interaction energy operator $\widehat{V}_{int}$ of a soft mode randomly located ($\mathbf{r} = \mathbf{R}$) in the continuum with the elastic strain magnitude $|e(\mathbf{R}, t)|$, e.g., $|e(\mathbf{R}, t)| = |\partial \mathbf{u}_z(\mathbf{R}, t)/\partial z| \ll 1$, of acoustic vibrations (Ch 7). In accordance with Eq. (7.36), the interaction energy can most simply be approximated by $\widehat{V}_{int} = \sum_{\mathbf{R}} \beta x(\mathbf{R},t) e(\mathbf{R},t)$, in which one can replace the sum over random locations $\mathbf{R}$ of soft modes by the term $c_{sm}\beta x e_{ac}$ that results from averaging such sums over spatially homogeneous sub-volumes. Taking into account the above comments, one can readily obtain the equations of motion (retaining the main terms in the small parameter ζ) as follows:

$$\rho \frac{\partial^2 \mathbf{u}_{av}}{\partial t^2} \approx \rho s_0^2 \nabla_{\mathbf{r}}^2 \mathbf{u}_{av} + c_{sm}\beta \frac{\partial x_{av}}{\partial \mathbf{r}} \quad \text{and}$$

$$M_{sm} \frac{\partial^2 x_{av}}{\partial t^2} \approx -M_{sm}\nu'^2 x_{av} - \beta\ \partial \mathbf{u}_{av,z}/\partial z, \tag{10.6}$$

where ρ is the average mass density of the elastic continuum. On the other hand, the DOS $J(\varepsilon)$ can be calculated by applying its general formula and taking into account the spectrum of eigenvalues of vibrational excitations (see below). As also follows from above, Eqs. (10.5), (10.6) are linear in c_{sm} and, as well as the resulting relationships, hold true at $c_0 \approx 10^{-3} \ll c_{sm} \ll 1$.

The vibrational spectrum is found in a standard way for propagating vibrational excitations, e.g., acoustic ones, by substituting into Eq. (10.6) extended, plane-wave states $\exp\{i(\mathbf{qr} - 2\pi\nu t\})$ for $\{x_{av}(\mathbf{r}, t), \mathbf{u}_{av}(\mathbf{r}, t)\}$, which by standard definition are well-defined in the usual sense that their width $2\gamma_{exc}$ is much smaller than the frequency $\nu_{exc} \to 0$, $2\gamma_{exc}/\nu_{exc} \ll 1$. The spectrum of eigenvalues $\varepsilon \equiv \nu^2$, at a given value of the soft mode parameter $\varepsilon' \equiv \nu'^2$, is a two-branch polariton-like one with the dispersion law $\varepsilon_j(q, \varepsilon'), j = 1, 2$ [194]:

$$\begin{aligned}\varepsilon &= \epsilon_j(q, \varepsilon') \equiv \nu_j^2(q, \varepsilon') \\ &= (1/2)(\varepsilon' + s_j^2 q^2/4\pi^2 + \Delta) \\ &\quad \times\{1 + (-1)^j[1 - \pi^{-2}\varepsilon' s_j^2 q^2(\varepsilon' + s_j^2 q^2/4\pi^2 + \Delta)^{-2}]^{1/2}\},\end{aligned} \tag{10.7}$$

at $q \equiv |\mathbf{q}| = 2\pi/\lambda_{ac}$. Here, $\Delta \simeq c_{sm}Q^2\varepsilon'$ and a typical Q^2, e.g., $Q^2 = \beta^2\ (\mu_{av}s_0^2\mu_{sm}\nu_0^2a_0^2)^{-1} \approx 10$ for a typical soft-mode coupling parameter, $\beta^2 \approx 1\,\text{eV}$, and $\mu_{av}s_0^2 \approx 10\,\text{eV}$, $\mu_{sm}\varepsilon_0 a_0^2 \sim 0.01\,\text{eV}$, $\varepsilon_0 \equiv \nu_0^2, a_0 = 1\,\text{Å}$. For each soft-mode vibrational eigenvalue ε', the branches are separated by a narrow gap of a width $\Delta = \epsilon_{2,\min} - \epsilon_{1,\max} \equiv \delta \cdot \varepsilon' \ll \varepsilon'$, at $\delta \approx 0.1$, $\epsilon_{2,\min} = \varepsilon' \approx \varepsilon_0$, $\epsilon_{1,\max} = \varepsilon'(1 - c_{sm}Q^2/\varepsilon_0)^{1/2}$ and $\varepsilon_0 \equiv \nu_0^2$ (Fig. 10). The existence of the gap is a remarkable new feature, effective "singularity", of the excitation spectrum of which the well-defined states, weakly hybridized acoustic vibrations ($\exp(i\mathbf{qr})$) and vibrational soft-mode excitations, due to their interaction (Sec. 7.5) and related weak inelastic scattering (Sec. 10.1), can be described by the approximate wave function $\phi_{\alpha\mathbf{q}}(x, \mathbf{r}) \approx \{c_{\alpha\alpha}(\mathbf{q})\varphi_\alpha(x, \mathbf{r}) \exp(i\mathbf{qr})$

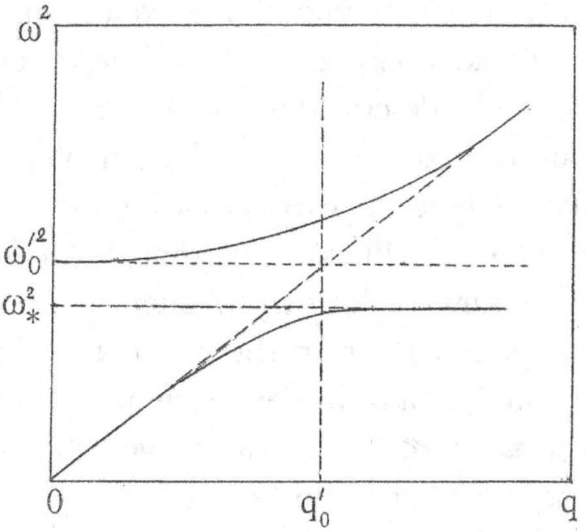

Figure 10: Spectrum of harmonic vibrational excitations due to hybridization of acoustic phonons and soft-mode vibrations.

$+\sum_{\mu\neq\alpha,\mathbf{q}'\neq\mathbf{q}} c_{\mu\alpha}(\mathbf{q}')\varphi_\mu(x,\mathbf{r})\exp(i\mathbf{q}'\mathbf{r})\}$ with $c_{\mu\alpha}(\mathbf{q})$ having random phases, at $\sum'_{\mathbf{q},\mu\alpha}|c_{\mu\alpha}(\mathbf{q})|^2 \ll 1$ and $|c_{\alpha,\alpha}(\mathbf{q})|^2 \simeq 1$ (the prime means $\mu \neq \alpha$). The gaps overlap each other at $\varepsilon' \approx \varepsilon_0$, merging into a narrow pseudogap of an effective width $\Delta_{\mathrm{eff}} \sim 0.1 \cdot \varepsilon_0$. The existence of such gaps is a manifestation of the general quantum-mechanical effect of splitting and repelling off each other two bare frequency "levels" of sound wave and soft-mode vibrations, which intersect at $s_0 q \approx 2\pi\nu_0$ and interact with each other [172] (as expected from physical considerations, if the acoustic-soft mode coupling parameter $\beta = 0$, as well as if $c_{sm} = 0$, the gap disappears, $\Delta = 0$, and two separate branches occur for non-interacting soft-mode excitations, $\nu = \nu'$, and acoustic phonons, $\nu \simeq s_0 q/2\pi$). For frequencies relatively far from the gap, the dispersion relations $\nu_j^2(q,\varepsilon')$ of well-defined excitations are similar to that of acoustic phonons, $\epsilon_1(q,\varepsilon') \simeq s_1^2 q^2/4\pi^2$, with a sound $l_{\mathrm{ac}}^{(\mathrm{in})}(\nu)$ velocity $s_1 \simeq s_0(1-\Delta\omega_0^{-2})^{1/2} < s_0$ at $q < q_0 = 2\pi\nu_0/s_0$ in the lower branch, and are similar to $\epsilon_2(q,\varepsilon') \approx \nu_0^2 + s_2^2 q^2/4\pi^2$ with a slightly larger velocity $s_2 \simeq s_0 > s_1$ at $q_0 < q < q_M = \nu_M/s_0$ ($\ll q_D = \nu_D/s_0$) in the upper branch. The latter seems to resemble the recently observed HFS (Ch 6), with a finite background in the limit at $q \to 0$ and an upper limit frequency $\nu_M \equiv p_M \nu_0 < \nu_D/3$ at an empirical $p_M \approx (2 \div 3)$ [112].

Another remarkable new feature of the excitation spectrum is that quite different, not well-defined vibrational excitations of THz scale frequencies can occur in and around such a gap as an effective spectral "singularity", which cannot be described by the spectrum of Eq. (10.7). Since these excitations are related to a region of eigenvalues located between those for two different types of extended excitations, one can assume that the gap excitations are still extended though "ill defined" states, i.e. their widths are of the same scale as the frequencies (this assumption has still to be rigorously verified within the concept of Anderson localization) [129]. Taking into account the definition of a well-defined state with half-width $\gamma_{\mathrm{ac}}^{(\mathrm{in})}(\nu_{\mathrm{ac}} \equiv \nu) \ll \nu$, one can suggest that the width of "ill-defined" excitations is not small and reaches the physically reasonable limit value of acoustic phonon width, $2\gamma_{\mathrm{ac}}^{(\mathrm{in})}(\nu) \approx \nu$, that is characteristic for strong inelastic scattering and related IRC in Eq. (10.1), at $\nu \approx \nu_{IR}^{(\mathrm{in})}$.

In principle, the IRC frequency can explicitly be obtained from estimations of width, or the equivalent mean-free-path $l_{\rm ac}^{(\rm in)}(\nu) \approx s_0/2\gamma_{\rm ac}^{(\rm in)}(\nu)$, determined by an appropriate interaction of a low frequency phonon with harmonic soft-mode vibrations, so that

$$l_{\rm ac}^{(\rm in)}(\nu_{IR}^{(\rm in)}) \approx \lambda_{\rm ac}(\nu_{IR}^{(\rm in)}) \text{ or equivalently } \gamma_{\rm ac}^{(\rm in)}(\nu_{IR}^{(\rm in)}) \approx \nu_{IR}^{(\rm in)}. \qquad (10.8)$$

It seems reasonable to assume that $\nu_{IR}^{(\rm in)} \approx \nu_0$, since ν_0 is the only characteristic frequency of harmonic soft-mode vibrations under discussion (Eq. (7.30)). The "ill-defined" IRC excitations can be characterized phenomenologically by a wave function $\phi_\alpha(x, r)$, similar to the above-mentioned function but at probabilities $|c_{\mu\alpha}(q')|^2$ comparable with each other. Then, it seems plausible for estimates to suggest that the IRC eigenstates can be characterized by diffusive motions with random phases, effective velocity $v_{\rm eff}$ and diffusion coefficient $D_{\rm eff} \sim a_1^2 \nu_{\rm eff}$ (rather than by a well-defined localized state with $v_{\rm eff} = 0$). In other words, the IRC eigenstates can be considered as Anderson ill-defined extended (i.e., not well-defined) non-localized states. The eigenstates seem to be qualitatively not very much different from the phenomenologically introduced "diffuson" excitations in [205]. By physical sense, one can estimate in scale $v_{\rm eff} \sim (D_{\rm eff}\gamma_{\rm eff})^{1/2} \sim (a_1^2 \nu_{\rm eff}\gamma_{\rm eff})^{1/2} \lesssim a_1 \nu_{BP} \approx 3 \cdot 10^4\,\text{cm/s} \approx 0.1 s_0$, where the effective frequency is $\nu_{\rm eff} \approx \nu_{BP} \approx 1$ THz while the effective excitation width is $\gamma_{\rm eff} \approx \gamma_{IR}^{(\rm in)} \approx \nu_{IR}^{(\rm in)}$. It seems worth emphasizing that the propagating acoustic-like excitations resembling HFS ones, in the spectrum of Eq. (10.7), occur for frequencies higher than the IRC frequency for inelastic scattering.

Based on Eqs. (10.7) and (10.8), a theory of the total vibrational DOS of independent soft-mode vibrations (Sec. 10.3)) and of interacting ones (Sec. 10.5) can be presented in what follows. It is able to describe both BP and HFS in glasses exhibiting these correlated phenomena by applying also a phenomenological description [194, 195] and by accounting recent results that concern interacting soft-mode "defects" [206]. What is most important here and should be taken into account is that by definition the IRC and its frequency $\nu_{IR}^{(\rm in)}$, as well as the related inelastic scattering, should be independent of T.

10.3 Total vibrational density of independent soft-mode states

The major spectral characteristic for the soft-mode harmonic excitations under discussion is the total vibrational DOS $J(\varepsilon)$ for the eigenvalues $\varepsilon \equiv \varepsilon(q) = \nu^2(q)$, analyzed here for the simplest case of independent soft-mode harmonic-vibration excitations interacting with acoustic phonons. The main points of the calculations and the results are only described and discussed here (for more details see the Appendix). Moreover, by definition, $J(\varepsilon)d\varepsilon = g(\nu)d\nu$ where $g(\nu) = 2\nu J(\varepsilon = \nu^2)$ is the equivalent total DOS for the excitation frequency, and the total vibrational DOS can be described by the following formula (Eqs. (A.1), (A.2), in Appendix A)

$$J(\varepsilon) = g(\nu)/2\nu = \int_{\varepsilon'_1}^{\varepsilon'_2} d\varepsilon' I(\varepsilon;\varepsilon') J^{(HV)}(\varepsilon'), \qquad (10.9)$$

which is a convolution of the soft-mode vibrational DOS $J^{(HV)}(\varepsilon' \equiv \nu'^2)$ (Eq. (10.5)) and the kernel $I(\varepsilon;\varepsilon')$ transforming soft-mode excitations at eigenvalues ε' into vibrational excitations at eigenvalues ε, resulting from the soft-mode–acoustic interaction. The lower limit of the integral is $\varepsilon'_1 = (3w/h)^2 \approx 0.01\nu_0^2$ and its upper limit is $\varepsilon'_2 \approx \nu_0^2$ ($\ll \nu_D^2$) (Sec. 7.4), while $J^{(HV)}(\varepsilon')$ is the vibrational DOS for independent soft-mode vibrational excitations, both $J(\varepsilon)$ and $J_{HV}(\varepsilon')$ being normalized to unity, $\int J(\varepsilon)d\varepsilon = 1 = \int J^{(HV)}(\varepsilon')d\varepsilon'$. The calculations are performed by applying the power-law soft-mode vibrational DOS $J^{(HV)}(\varepsilon')$ of Eq. (10.5) and by taking into account that for a given soft-mode frequency ν' the spectrum of well-defined excitations consists of two branches separated by a narrow (pseudo) gap. For the acoustic-like parts of each branch, with small excitation widths $2\gamma_{ac}(\nu) \ll \nu$, the standard approximation for the transforming kernel is (Eq. (A.3)):

$$I(\varepsilon;\varepsilon') \simeq I_0(\varepsilon;\varepsilon') = (\pi/2) \cdot \Sigma_{j=1,2} \int (q^2/q_D^3)dq\, \delta[\varepsilon - \epsilon_j(q;\varepsilon')], \qquad (10.10)$$

where $q_D = \pi/a_1$. With new variables $u = \varepsilon/\varepsilon'$ and $y = \varepsilon'/\varepsilon_0$, one can obtain that $I_0(\varepsilon;\varepsilon') \propto \sqrt{\varepsilon}\Phi_0(u,\delta)$ and that the function $\Phi_0(u,\delta) = (u - 1/u - 1 + \delta)^{1/2}\{1 + [\delta(1-\delta)/(u - 1 + \delta)^2]\}$ describes the lower spectral branch ($j = 1$) at $0 \leq u \leq 1 - \delta$ while the upper branch at $1 \leq u \leq u_M \equiv p_M^2$ corresponds to the above-mentioned empirical upper limit of HFS frequencies. Moreover, the asymptotic limit for this function in the upper branch in the vicinity of its lower edge (ε_u), at $0 < u - 1 \lesssim \delta \sim 0.1$, reads: $\Phi_0(u,\delta) \sim \sqrt{u-1}/\delta^{3/2}$. It follows from this expression that, at $\varepsilon' = const$, $I_0(\varepsilon;\varepsilon')$ exhibits near the upper edge of the narrow gap a singularity, $I_0(\varepsilon;\varepsilon') \propto \sqrt{\varepsilon - \epsilon_{2,\min}}$ at $\varepsilon - \varepsilon_{2,\min} \to +0$, and above it a narrow peak at $\epsilon_{2,\min} < \varepsilon \lesssim \epsilon_{2,\min} + \varepsilon_0\delta \equiv \varepsilon_u(\ll \varepsilon_0)$; a similar feature, $I_0(\varepsilon;\varepsilon') \propto \sqrt{\epsilon_{1,\max} - \varepsilon}$, is also characteristic of the lower edge (ε_l) of the gap at $\epsilon_{1,\max} - \varepsilon \lesssim \varepsilon_l \ll \varepsilon_0$ (the singularities are similar to appropriate van Hove's band-edge ones in spectra of $2D$ vibrational eigenvalues [177]). The occurrence of the singularities corresponds to the fact that the narrow gap in the spectrum (10.4) can be considered as a spectral quasi-singularity. For numerical calculations of $J(\varepsilon)$ from Eqs (10.9) and (10.10), the δ-functions in $I_0(\varepsilon;\varepsilon')$ can be approximated by a "pre-limit" regular function $D[X]$ at $X \equiv \varepsilon - \epsilon_j(q;\varepsilon')$ and $\kappa_1^2 \ll 1$, the result being weakly sensitive to the specific form of the function and the specific value of the small parameter $\kappa_1^2(\longrightarrow 0)$.

As explained in Sec. 10.1 and Eq. (10.8), the vibrational excitations are ill-defined, i.e. are characterized by a large width $2\gamma_{ac}(\varepsilon^{1/2}) \equiv \kappa_2^2\nu_0 \approx \nu_0$, in the IRC region of frequencies, where the perturbation-theoretical approach for the excitation description fails. Therefore, a phenomenological ansatz has to be introduced for describing the transformation kernel $I(\varepsilon;\varepsilon')$, which suggests that a regular function $D[\varepsilon - \epsilon_j;\kappa_2^2]$ for the excitations has to be substituted for the δ- function,

$$I(\varepsilon;\varepsilon') = (\pi/2)\cdot\Sigma_{j=1,2}\int_{\Omega(q)} dq q_D^{-3}\, q^2 \cdot D[\varepsilon - \epsilon_j(q;\varepsilon')], \qquad (10.11)$$

where $D[X] \equiv D[X;\kappa^2]$ is the "pre-limit" function for the δ-function $\delta(X)$ at small $\kappa^2 \equiv \kappa_1^2 = 2\gamma_{ac}(\varepsilon^{1/2})/\varepsilon^{1/2} \ll 1$, i.e., $\delta(X) = \lim D(X;\kappa_1^2)$ at $\kappa_1^2 \to +0$ for well-defined acoustic phonons, far enough from the pseudogap in the spectrum $\epsilon(q) = \nu^2(q)$, and at small width $\Gamma(\varepsilon = \nu^2) = 2\gamma_{ac}(\varepsilon^{1/2})\varepsilon^{1/2} \equiv \kappa_1^2\varepsilon \ll \varepsilon$. In principle, the function $D[X]$ could be derived from a Green's function analysis of the IRC problem. Since such a derivation is not yet completely available (see comments to Eq. (10.8)), the function $D[X;\kappa^2]$ **preferably** has to be chosen in terms of variables associated with the vibrational eigenvalues $\varepsilon = \nu^2$ and can be approximated by an often-used Lorentzian function $D_1[X]$:

$$D_1[X] \equiv D_1(X;\kappa^2) = \pi^{-1}\{\Gamma/[X^2 + \Gamma^2]\} \approx \pi^{-1}\{\kappa^2\nu_0^2/[X^2 + \kappa^4\nu_0^4]\}, \tag{10.12}$$

where the characteristic width is $\Gamma(\varepsilon = \nu^2) = 2\gamma_{ac}(\varepsilon^{1/2})\varepsilon^{1/2} \equiv \kappa^2\varepsilon$ for typical eigenvalues $\varepsilon \approx \nu_0^2$. At $\kappa^2 = \kappa_1^2 \ll 1$, the results of calculations of Eq. (10.11) is rather insensitive to the specific form of $D(X;\kappa_1^2)$ and value of κ_1^2, as the contribution of the Debye acoustic phonons to $I(\varepsilon;\varepsilon')$ is well approximated by an expression similar to $I_0(\varepsilon;\varepsilon')$ in Eq. (10.10). In the alternative region of frequencies around the IRC pseudogap, the width $\Gamma(\varepsilon = \nu^2) = 2\gamma_{ac}(\varepsilon^{1/2})\varepsilon^{1/2}$ is large, e.g., $\Gamma(\varepsilon) = \kappa_2^2\varepsilon \approx \varepsilon$ at $\kappa^2 = \kappa_2^2 \approx 1$. Then, Eqs. (10.9) and (10.12) can also be calculated by applying the suggested ansatz, as long as the results are in scale and are weakly sensitive to the specific expression of $D[X;\kappa^2 = \kappa_2^2]$ at $\kappa_2^2 \approx 1$. The latter condition can indeed be fulfilled in numerical calculations with the function of Eq. (10.12). In fact, one can find from Eqs. (A.4) the kernel $I(\varepsilon;\varepsilon') \equiv I_1(\varepsilon;\varepsilon')$, which accounts for both contributions of well-defined acoustic-like excitations, $I_{(ac)}(\varepsilon;\varepsilon')$, and, more roughly, of ill-defined states around the pseudogap, $I_{(IRC)}(\varepsilon;\varepsilon')$:

$$I_1(\varepsilon;\varepsilon') = I_{(ac)}(\varepsilon;\varepsilon') + I_{(IRC)}(\varepsilon;\varepsilon'), \tag{10.13}$$

where the term $I_{(IRC)}(\varepsilon;\varepsilon')$ contains the phenomenological regular function $D[\varepsilon - \epsilon_j;\kappa_2^2 \approx 1]$ instead of the δ-function.

Let us now introduce new, dimensionless, variables and parameters:

$$u \equiv \varepsilon/\nu_0^2 \equiv \nu^2/\nu_0^2,\ t \equiv \varepsilon'/\nu_0^2,\ \delta \equiv \Delta/\nu_0^2 \text{ and } z \equiv \epsilon/\nu_0^2, \tag{10.14}$$

and, for certainty, choose $D(X;\kappa_2^2) = D_1(X;\kappa_2^2)$ as the Lorentzian-like function. Then, the resulting total vibrational DOS can be described by the expressions of Eqs. (A.7)–(A.11) and straightforwardly be expressed in terms of the dimensionless function $I(u)$:

$$J(\varepsilon) = 0.5 \cdot C_\alpha \cdot (\nu_0/\nu_D^3)(\nu_0/\nu_D)^{2\alpha+1} \cdot I(u) \tag{10.15}$$

where

$$
\begin{aligned}
I(u) &= g(\sqrt{u})/2\sqrt{u} \\
&= \int_{\Omega(z,t)} dt \cdot t^{(2\alpha-1)/2} \int dz\sqrt{z}p(z,t,\delta) \cdot D_1[z-u;\kappa^2] \\
&= I_{(\mathrm{ac})}(u) + I_{(IRC)}(u), \qquad (10.16)\\
p(z,t;\delta) &= (z-t-\delta)^{1/2}(z-t)^{-1/2} \cdot [1 + t\delta \cdot (z-t)^{-2}], \quad D_1[z-u;\kappa^2] \\
&= \kappa^2[(z-u)^2 + \kappa^4]^{-1} \qquad (10.17)
\end{aligned}
$$

at $\kappa^2 = \kappa_1^2 \ll 1$ in $I_{(\mathrm{ac})}(u)$ while at $\kappa^2 = \kappa_2^2 = 1$ in $I_{(IRC)}(u)$, and the constant C_α is defined in Eq. (10.5). Then, one can readily obtain from Eqs. (10.16), (A.15) and (A.16) that the dimensionless DOS $I(u)$ can be expressed as a sum of two double integrals,

$$
\begin{aligned}
I(u) &= [F(u) + G(u)], \qquad (10.18)\\
F(u) &= \int_{t_1}^{t_2} dt \cdot t^{(2\alpha-1)/2} \\
&\quad \times \left[\int_0^{t_1-\varsigma} dz \cdot D_1[z-u;\kappa_1^2] + \int_{t_2+\delta}^{\infty} dz \cdot D_1[z-u;\kappa_1^2]\right] \\
&\quad \cdot \sqrt{z} \cdot p(z,t;\delta), \qquad (10.19)
\end{aligned}
$$

at $\kappa^2 = \kappa_1^2 \ll 1$, and

$$
\begin{aligned}
G(u) &= \int_{t_1-\varsigma}^{t_2-\varsigma} dz\sqrt{z}D_1(z-u;\kappa_2^2) \int_{z+\varsigma}^{t_2} dt \cdot t^{(2\alpha-1)/2}p(z,t;\delta) \\
&\quad + \int_{t_1+\delta}^{t_2+\delta} dz\sqrt{z}D_1(z-u;\kappa_2^2) \int_{t_1}^{z-\delta} dt \cdot t^{(2\alpha-1)/2}p(z,t;\delta), \qquad (10.20)
\end{aligned}
$$

at $\kappa^2 = \kappa_2^2 \approx 1$, where $\sqrt{u} \equiv \nu/\nu_0$.

As the characteristic soft-mode vibration frequency is ν_0, the typical values of parameters in the basic equations, calculations and results

(Eqs. (10.5)–(10.20)) are chosen to be

$$\nu_{IR}^{(\text{in})} = \nu_0,\ 2\gamma_{IR}^{(\text{in})} = \nu_0, \quad t_i = \varepsilon_i' \nu_0^{-2} (i = 1, 2),$$
$$\varsigma = |z - t|_{\min} = 0.05, \quad \delta \equiv \Delta / h\nu_0 = 0.1, \qquad (10.21)$$

where the lower limit for the variable t is estimated as $t_1 \simeq (3w/h\nu_0)^2 \approx 0.1$ while the upper limit is $t_2 \simeq (\nu_{sm}/\nu_0)^2 \approx 1$, by taking into account that the upper limit of soft-mode vibrational frequencies is $\nu_{sm} \approx \nu_0$ at $\nu_{sm} > \nu_0$ (Eq.(7.30)). The additional parameter ς is introduced as the lowest finite value of $(t - z)$ for cutting off the divergence of the integral at $t - z \longrightarrow +0$. The cut-off provides the long-wave approximation (Sec. 10.2), at $\lambda_{ac} = 2\pi q^{-1}(\epsilon; \varepsilon') = (s/\nu_0)[(t - z + \delta)(t - z)^3]^{1/2} \cdot [\sqrt{z}(t - z)^2 + t\delta]^{-1} > a_1 c_{sm}^{-1/3}$. For typical values of parameters in (10.21), including $a_1 = 3$ Å, $s_0 = 3 \cdot 10^5$ cm/s and $c_{sm} = 3 \cdot 10^{-2}$, one can estimate $t - z \gtrsim (t - z)_{\min} = 0.05$. The integration over t in the double integrals $I_\varrho(u; \delta, \kappa)$ can be carried out analytically by substituting $y = [(t - z - \delta)/(t - z)]^{1/2}$ for t, for example, giving rise to the following expression: $I_2(u; \delta, \kappa) = (2/3)z x_1^3 + 2\delta \cdot x_1 - \delta \cdot x_1 (x_1^2 - 1)^{-1} + (3\delta/2) \ln |(x_1 - 1)(x_1 + 1)^{-1}|$, where $x_1 = [(z - t_1 - \delta)^{1/2} (z - t_1)^{-1/2}]$. In general, the calculations of the double integrals were carried out numerically by applying the MATHCAD program OMIT, with $\gamma = 0.05, \delta = 0.1, t_1 = 0.1, t_2 = 0.95$ and $t_2 = 1.05$. In particular, it is found at $u \gg 1$ that $I(u) \simeq I_0(u) = (\pi/2)(t_2 - t_1)[(\kappa^4 + u^2)^{1/2} + u]^{1/2}$ corresponds to the gapless spectrum in Eq. (10.7), as can be expected.

Similar calculations can be carried out by choosing the suggested regular function $D(X; \kappa^2)$ to be the well-known "damped harmonic oscillator" spectral function $D_2[X; \kappa^2] = \pi^{-1}\tilde{\Gamma}(q)/[X^2 + \tilde{\Gamma}^2(q)\nu^2]$, with $\tilde{\Gamma}(q) \simeq 2\gamma_{ac}[\nu(q)]\nu(q) \equiv \kappa^2 \nu(q)^2$ for typical eigenvalues $\nu(q) \approx \nu_0$.

The results of numerical calculations for $J(\varepsilon) = g(\varepsilon^{1/2})/2\varepsilon^{1/2}$ are practically indistinguishable (within the accuracy available) with both regular functions $D(X; \kappa^2) = D_1[X; \kappa^2]$ and $D(X; \kappa^2) = D_2[X; \kappa^2]$, and are presented in Fig. 11 for a typical width of the pseudogap, $\delta = 0.1$. Ill-defined excitations in the IRC region, $2\gamma_{IR}^{(\text{in})} \simeq \kappa^2 \nu_0$, with $\kappa^2 = \pi \cdot (0.4, 0.3$ and $0.2)$ (curves 1, 2 and 3), and an "intermediate" $\kappa^2 = \pi \cdot 0.1$ (curve 4) and non-realistic, very small $\kappa^2 = \pi \cdot (0.05$ and $0.01)$ (curves 5, 6) are presented for comparsion. In curves 1, 2, 3, two narrow peak singularities of $I_0(\varepsilon; \varepsilon')$ near the gap edges merge in $J(\varepsilon)$ into a single broad peak around the IRC

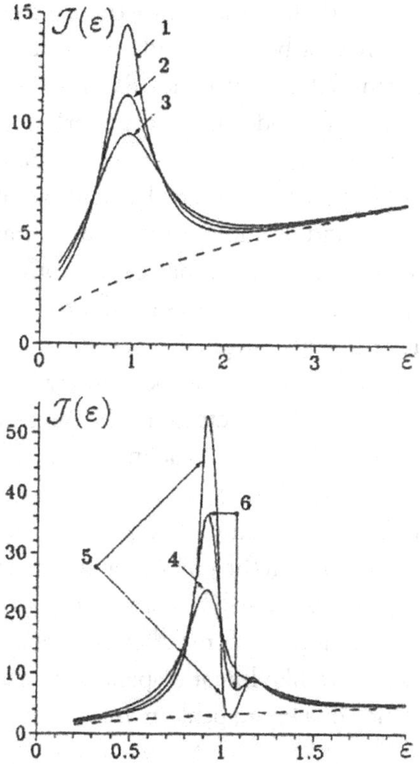

Figure 11: Evolution of total vibrational DOS $J(\varepsilon)$ of harmonic excitations (in a.u., at $\varepsilon_0 = 1$ and typical $\delta = 0.1$), at the transformation kernel $I(\varepsilon, \varepsilon') \simeq I_0(\varepsilon, \varepsilon_0')$, with the variation of parameter $\varkappa^2$ from very small $\varkappa^2 = \pi \cdot (0.05, 0.01)$ (curves 5 and 6) through "intermediate" $\varkappa^2 = \pi \cdot 0.1$ (curve 4) to relatively large $\varkappa^2 = \pi \cdot (0.4, 0.3, 0.2)$ (curves 1, 2 and 3) in the range at $\varkappa^2 \sim 1$.

frequency $\nu_{IR}^{(in)} \approx \nu_0 \approx 1$ THz. The large width of ill-defined excitations, $2\gamma_{IR}^{(in)} \sim 0.3 \cdot \nu_{IR}^{(in)}$ at $\nu_{IR}^{(in)} \approx \nu_0$, is similar to that of a BP with the empirical $\nu_{BP} \approx 1$ THz and $2\gamma_{BP} \approx 0.3 \cdot \nu_{BP}$. On the other hand, two narrow peaks do not merge into a single broad peak in curves 5 and 6 at small widths, while an intermediate behavior characterizes the curve 4 with an intermediate width. Furthermore, both $J(\nu^2) = g(\nu)/2\nu$ and the reduced DOS $g(\nu)/\nu^2$ exhibit above the broad peak acoustic-like behavior similar to that of HFS (while similar to a standard Debye-like behavior below the peak), with $g(\nu) \propto \nu^2$ at $0 < \nu/\nu_0 \lesssim 0.9$ and $1.2 \lesssim \nu/\nu_0 \lesssim 2 - 3$.

It follows from above that in the SMM theory based on spectrum of Eq. (10.7) the occurrence of both BP and, above it, HFS in glasses can be explained as determined by both the IRC for inelastic scattering of acoustic phonons from soft-mode harmonic vibrations and the spectral "singularity" of vibrational eigenvalues $\varepsilon = \nu^2$ in the narrow pseudogap. The spectrum, derived from the averaged equations of motion of interacting acoustic phonons and harmonic soft-mode vibrations, corresponds to extended and propagating vibrational excitations above the pseudogap characterized by spectral singularities similar to appropriate van Hove ones. In the whole spectrum range, the extended acoustic-like excitations can be characterized by a linear dispersion law and can coexist with non-propagating, ill-defined IRC excitations that seem also to be extended because at higher frequencies the well-defined wave-like HFS occur (see above Eq. (10.8)). This conclusion appears to qualitatively agree with recent experimental data, described at the end of Sec. 6, which assert that a typical BP frequency is located in the range in which the dispersion relation for Brillouin peak frequencies can still be considered as an acoustic-like one, $\nu_B(q) \propto q$. In this connection, no contradiction seems to appear between the existence of an acoustic-like linear dispersion law for the HFS excitations and the occurrence of a broad peak in an effective pseudogap around the IRC.

10.4 Generalization for interacting harmonic excitations

As follows from Sec. 9.2 (Eqs. (9.15), (9.17)), interactions between very-low-energy soft-mode excitations, i.e., TLSs, and acoustic phonons generate universal weak elastic dipolar ($\propto 1/r^3$) interactions between the excitations, which give rise to new properties in both coherent and relaxational processes in the advanced tunneling model of [187, 188, 191]. Quite similarly, interactions between moderately-low-energy soft-mode harmonic vibrational excitations and the phonons is expected to generate universal weak elastic dipolar interactions between the vibrational excitations that can give rise to new properties in the processes and related excitation DOS.

In fact, recently a "mean-field" model of weakly interacting soft-mode vibrational excitations and their DOS has been proposed [206], which accounted for a renormalization of the original quasi-local harmonic excitations of low enough frequency $\nu' \ll \nu_c$, due to their elastic pair interactions of energy $E_{12} = U_{12}x_1x_2$, with much more numerous vibrations of higher frequencies $\nu' \approx \nu_c$ ($x_{1,2}$ are dimensionless soft-mode coordinates). Here, $|U_{12}| \approx U_0 R_{12}^{-3}$ in a pair of soft modes, at a large distance $R_{12} \gg a_1$, and $U_0 = \beta^2/\rho s_0^2$ is the effective strength parameter, while the frequency parameter ν_c is determined by the typical interaction strength and assumed to be substantially lower than the Debye frequency, $\nu_c \ll \nu_D$, for the interactions which are weak in the sense that $P_0 U_0 \ll 1$ in Eq. (9.17) (the many-body problem seems here to be effectively reduced to that for separate pairs).

The problem for a pair of weakly interacting "bare" excitations with original frequencies ν_1' and ν_2', effective soft-mode masses $M_{sm}^{(1,2)}$ and interaction energy U_{12} , is resolved in the "mean-field" approximation that gives rise to the spectral eigenvalues $(\nu_{\mp}')^2 = (1/2)[(\nu_1'^2 + \nu_2'^2) \mp \sqrt{(\nu_1'^2 - \nu_2'^2)^2 + 4E_{12}^2[M_{sm}^{(1)}M_{sm}^{(2)}]^{-1}}]$ which, as well as the parameters U_{12} and ν_c, are random variables subject to an appropriate distribution density. It follows that a mechanical, vibrational, instability occurs for such low frequencies $\nu_{i=1,2}'$ that the eigenvalues are negative, $\nu_-'^2 < 0$, at $|E_{12}| > E_c = h\nu_c \sim \nu_1'\nu_2'\sqrt{M_{sm}^{(1)}M_{sm}^{(2)}}$. The stable vibrational states characterized by positive renormalized eigenvalues are found to be restored in a similar way as in the original soft-mode model (see below Eqs. (7.3) and (7.7)) due to finite, not small, quartic anharmonic interaction introduced in the soft-mode potentials. The important result of the model under discussion appears to be that the DOS $\widetilde{g}(\nu_r')$ for eigenfrequencies of stabilized harmonic modes, at $\nu_r'^2 > 0$ and low enough frequencies $\nu_r' \ll \nu_c \ll \nu_D$, becomes linear in the eigenfrequencies, $\widetilde{g}(\nu_r') \propto \nu_r'$. This dependence follows from the theoretical result that the stabilized low-frequency soft-mode oscillators also interact with each other and the interaction is equivalent to the occurrence of random forces f acting on the soft-mode vibrations of which the generalized susceptibility to forces is high (Eq. (7.8)). In a purely harmonic case, these linear forces would not affect the frequencies. However, the quartic anharmonicity provides an additional renormalization of the low-frequency spectrum. As found from appropriate calculations, the resulting vibration frequency DOS can be described as $g^*(\nu_r') \propto (\nu_r')^4$ for the low $\nu_r' \ll \nu_c$, which qualitatively resembles the DOS of vibrational excitations in Eq. (10.5) at $\alpha = 2$ in the case of Eq. (10.18). The respective reduced DOS at low frequencies $\nu_r' \ll \nu_c$ of the vibrations in this model has been calculated and found to contain a single material parameter that is the characteristic vibration frequency ν_0^* determined by the typical strength parameter U_0 of pair interactions, or by the equivalent typical random force parameter $f_0 \propto (\nu_0^*)^3$. Then, the reduced DOS is shown at $w \ll h\nu_0^* \ll h\nu_D$ to be described by the following expression:

$$g^*(\nu_r')/\nu_r'^2 \propto (1/\nu_0^{*3})(\nu_r'/\nu_0^*)^2 \int_0^1 \frac{dz}{[1 + (\nu_r'/\nu_0^*)^6 z^2(3 - 2z^2)]}, \tag{10.22}$$

which increases with increasing ν_r' like $\nu_r'^2$ at low $\nu_r' < \nu_0^*$ and decreases like $(\nu_r')^{-1}$ at high $\nu_r' > \nu_0^*$, exhibiting a broad (rather weak) peak at $\nu_r' \simeq \nu_0^*$. It was assumed in [206] that the broad peak of $g^*(\nu_r')/\nu_r'^2$ could be identified with the empirical boson peak, a broad peak in the reduced inelastic scattering intensity (Eq. (10.9)), at $\nu \simeq \nu_0^*$ of THz scale. The theory hardly seems to allow one to calculate quantitatively ν_0^* for a specific glass, so one can consider this frequency as a model parameter. The parameter ν_0^* plays here the role similar to that of the typical soft-mode vibration frequency $\nu_0 \approx 1$ THz in the SMM for independent soft modes in Secs. 10.2 and 10.3. Let us add that at increasing ν_r' the total vibrational DOS $J^*(\nu_r'^2) = g^*(\nu_r')/2\nu_r'$ exhibits a plateau-like behaviour instead of a peak.

In most recent works [207], it appears to be argued that the physical mechanism of stabilization of interacting soft-mode excitations, which is related to the quartic anharmonicity in soft-mode potentials, is fundamental not only for interacting harmonic vibrational excitations of moderately low frequencies, with the peak feature in the reduced DOS, but also for a seemingly different phenomenon: the formation of interacting TLS excitations of very low energies in double-well potentials with a wide distribution of barrier heights (Chs. 7–9). The mechanism is the vibrational instability of weakly interacting harmonic modes, which is controlled by the anharmonicity and creates below the characteristic frequency $\nu_c (\ll \nu_D)$ a new stable harmonic spectrum with both a peak feature and double-well potentials. In this sense the vibrational instability model can be considered as a generalization of the soft-mode model with independent soft modes and their excitations, considered in Ch. 7, by taking into account the universal weak elastic interactions between soft-mode excitations. In particular, one should suggest that the model has to describe new properties predicted in the advanced tunneling model (Sec. 9.2) for both coherent and relaxation processes in a system of weakly interacting TLS excitations at ultralow temperatures, $T < 0.1$ K, which have been observed experimentally [122]. It appears that in this model the vibrational modes cease to exist as plane waves at frequencies higher than the BP one [208], which is not the case in the glasses under discussion exhibiting the earlier observed high-frequency sound excitations above the BP [112]. This fact may be interpreted as an indication that the vibrational instability model, which does not take into account the spectrum (10.7) of harmonic vibrational excitations of moderately low energy with its pseudogap and related singularities, is not related to the IRC for inelastic scattering of acoustic phonons by soft-mode excitations discussed in Secs. 10.2 and 10.3. In particular, the model does not describe the BP tail above ν_{BP} (Eq. (10.24)) in the glasses. One can conclude from the above comments that the vibrational instability model is not able to provide a complete qualitative and scalewise description of the major features of the BP intensity for the glasses under consideration, and that the broad peak of $g^*(\nu_r')/\nu_r'^2$ in Eq. (10.22) is a "bare" peak which has to be transformed to the measurable BP around $\nu_{BP} \approx \nu_0^*$ after its renormalization due to the soft-mode–acoustic interaction in the IRC region (Sec. 10.5).

Let us note that an earlier model of the BP [209] took into account the elastic dipolar interactions between soft-mode harmonic vibrations in a rather different way. The main idea seemed to be that the dipolar interactions could essentially renormalize an original DOS like $g_2^{(HV)}(\nu') \propto \nu'^4$ for frequencies ν' of independent vibrations and that a broad peak identified as a BP would appear in the reduced renormalized DOS $g^*(\nu')/\nu'^2$ at a characteristic frequency identified as the BP frequency ν_{BP}. It was argued that the reason for the occurrence of the peak was the general quantum-mechanical effect of strong level repulsion due to dipolar interactions which is characterized by non-diagonal transition matrix elements between the vibrational states, $\Delta J \equiv \langle 1, 2|E_{12}|2, 1\rangle$, estimated as $\Delta J \simeq \hbar\beta^2 g_2^{(HV)}(\nu)\delta\nu/4\pi M\rho s_0^2$ for vibrations in a small frequency interval $(\nu' + \delta\nu', \nu')$. The resulting DOS was found to become linear in frequency, $g^*(\nu') \simeq g_2^{(HV)}(\nu')\hbar\delta\nu'/\Delta J \simeq (M\rho s_0^2/\beta^2)\ \nu'$, and to be essentially independent of the original DOS shape, when the level repulsion is large, $\Delta J > \hbar\delta\nu'$, while still $g^*(\nu') \simeq g_2^{(HV)}(\nu')$ when $\Delta J < \hbar\delta\nu'$. On the other hand, the mean free path of acoustic phonons in the case under discussion, at weak resonant scattering from soft-mode vibrations, was estimated as $l_{ac}^{-1}(\nu') \simeq (\pi\beta^2/2M\rho s_0^3)g_2^{(HV)}(\nu')$, so an IRC could occur at $\nu' = \nu'_{IRC}$ and $l_{ac}(\nu'_{IRC}) \approx s_0/\nu'_{IRC}$. However, such a mechanism is hardly compatible with the mechanism of BP described in Eq. (10.22), which does not appear to be associated with a IRC.

10.5 Total vibrational density of states: dynamic properties

One can now propose a "complete" theory of harmonic vibrational properties of the glasses under consideration, describing the general features of correlated BP and HFS in the photon and neutron inelastic scattering intensities, which can be obtained within the framework of SMM by accounting also for the contribution of interactions between soft mode and their harmonic vibrations. The main idea of the complete SMM is that the total vibrational DOS $J^*(\nu'^2) = g^*(\nu')/2\nu'$, with $g^*(\nu')$ of Eq. (10.22), has to be substituted for the original vibrational DOS $J_\alpha^{(HV)}(\nu'^2)$ (Eq. (10.5)) containing two parameters, ν_0 and α. In contrast, for scale estimations, the complete model can contain a single scale parameter ν_0^* in the integral of $J^*(\nu'^2)$, with $\varepsilon'_1 \approx 0.1\nu_0^{*2}$ the lower limit and $\varepsilon'_2 \approx \nu_0^{*2}(\ll \nu_D^2)$ the upper limit. This takes into account that the soft-mode frequencies ν_0 of Eq. (10.7) and ν_0^* of Eq. (10.22) are hardly distinguishable in scale, with $\nu_0^* \equiv \sigma\nu_0$ at $\sigma \approx 1$, in the function

$$f(\sigma t) = (\sigma t)^{3/2} \int_0^1 dz[1 + (\sigma t)^3 z^2(3 - 2z^2)]^{-1}. \tag{10.23}$$

Thus for estimations of Eqs. (10.18)–(10.20) the dimensionless function $t^{(2\alpha-1)/2}$ has to be replaced in the complete theory by the function $f(\sigma t)$, where σ can be considered as another parameter of the theory, at $\sigma \equiv (\nu_0/\nu_0^*)^2 \approx 1$, with a typical variation range like $0.5 \leq \sigma \leq 2$.

Now, the following principal question has to be answered in what follows: is the IRC for the inelastic scattering of acoustic and soft-mode vibrations the most important phenomenon, or is the renormalization of soft-mode vibrations due to their elastic interactions of comparable importance for determining the qualitative and scalewise properties of dynamic vibrational properties for the glasses under discussion [195], particularly of the BP position (ν_{BP}) and the scale of its width ($2\gamma_{BP}$). The answer can be found from the results of numerical calculations of (dimensionless) reduced DOS $g_r(u^{1/2}) \equiv g(u^{1/2})/u$ and total vibrational DOS $I(u) \equiv g(u^{1/2})/2u^{1/2}$ as functions of $u^{1/2} \equiv \nu/\nu_0$, which are carried out from Eqs. (10.18)–(10.20) by applying a MATHCAD programme OMIT. Both the original soft-mode DOS with $t^{(2\alpha-1)/2}$ at $\alpha = 1$ and $\alpha = 2$, and the renormalized soft-mode DOS with $f(\sigma t)$ at $\sigma = 0.5, 1, 2$, are calculated and the results are shown in Figs. 12 and 13.

The results for $g_r(u^{1/2})$ as a function of $u^{1/2} = \nu/\nu_0$ are presented in Fig. 12 with $\sigma = 0.5$ and 2 in $f(\sigma t)$, as well as with $\alpha = 2$ in $t^{(2\alpha-1)/2}$, and show a broad asymmetric peak that, by the standard definition, is a boson peak in the reduced scattering intensity $I_R^{(r)}(\nu)$ (Eq. (6.9)) at $u^{1/2} = 1$, i.e., $\nu/\nu_0 = 1$. A weak increase of $g_r(u^{1/2}) \propto \nu^\theta$ with increasing ν, at $0 \lesssim \theta < 1$ (cf. the Debye behavior at $\theta = 0$), is also seen above the peak in Fig. 12. One can see from the figure that the qualitative features, including the peak position, of $g_r(u^{1/2})$ are weakly sensitive to the difference in the soft-mode DOS itself, i.e., of $t^{(2\alpha-1)/2}$ at different $\alpha = 1, 2$ or $f(\sigma t)$ at different $\sigma = 0.5; 2$. The behavior of $g_r(u^{1/2})$ above the BP is similar to, though relatively weakly deviates from, the Debye behavior. The similarity may be interpreted as a manifestation of the existence of high-frequency acoustic-like excitations in the upper spectral branch ($j = 2$) of Eq. (10.7), similar to the experimentally observed HFS one. The origin of the deviation is not yet quantitatively explained, though it can be qualitatively understood, partly at least, as due to an appropriate weak dispersion of HFS velocity $s_2(q) \simeq 2\pi\nu_2(q)/q > s_1(q) \simeq s_0$, at $q_0 = 2\pi\nu_0/s_0 < q < q_M = 2\pi\nu_M/s_0$.

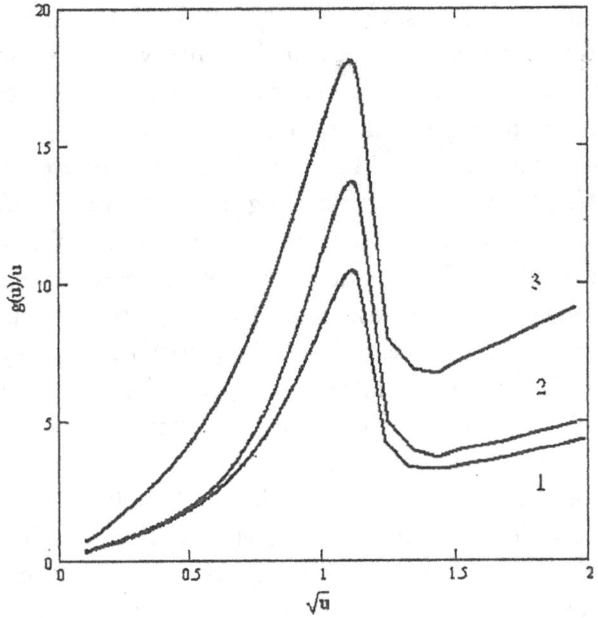

Figure 12: Evolution of reduced frequency DOS $g_r(u^{1/2}) = g(u^{1/2})/u$ (in a.u.), as a function of $u^{1/2} \equiv \nu\nu_0^{-1}$, with the variation in $J_{sm}(t)$ from $J_{sm}(t) = t^{(\alpha-1)/2}$ at $\alpha = 2$ (curve 1) to $J_{sm}(t) \equiv J^*(\sigma t)$ at $\sigma = 0.5$ (curve 2) and $\sigma = 2$ (curve 3); the values of main parameters in this and other figures are defined in Eq. (6.21).

The results for total vibrational DOS $I(u) = g(u^{1/2})/$ vs $u^{1/2} \equiv \nu/\nu_0$ obtained at different $\alpha(=1;2)$ and $\sigma(=0.5;2)$, as well as at different values of the gap width $\delta = 0.05; 0.10; 0.15$ around a typical $\delta = 0.10$, are shown in Ref. [195]. As seen from the results, $I(u)$ exhibits a broad peak of asymmetric shape and large half-width $\gamma_{IR}^{(\text{in})} \approx \nu_{IR}^{(\text{in})} \simeq \nu_0$. This is qualitatively similar to BP in reduced DOS and thus may be identified as BP in the scattering intensity related to Raman dynamic susceptibility χ''_R (Eqs. (6.10)), practically at the same position $\nu = \nu_{BP} \simeq \nu_{IR}^{(\text{in})} \simeq \nu_0$. Above the peak, $I(u)$ exhibits a Debye-like increase with increasing ν, $I(u) \propto \nu^{1+\theta}$ at $\theta \ll 1$, which is characteristic of high-frequency sound.

The qualitative features, including the peak position, of $I(u)$ are also weakly sensitive to the difference in soft-mode DOS itself, i.e., to specific values of α or σ (at typical value $\delta = 0.1$), as well as to specific values of δ at given $\sigma = 1$, in Fig. 13.

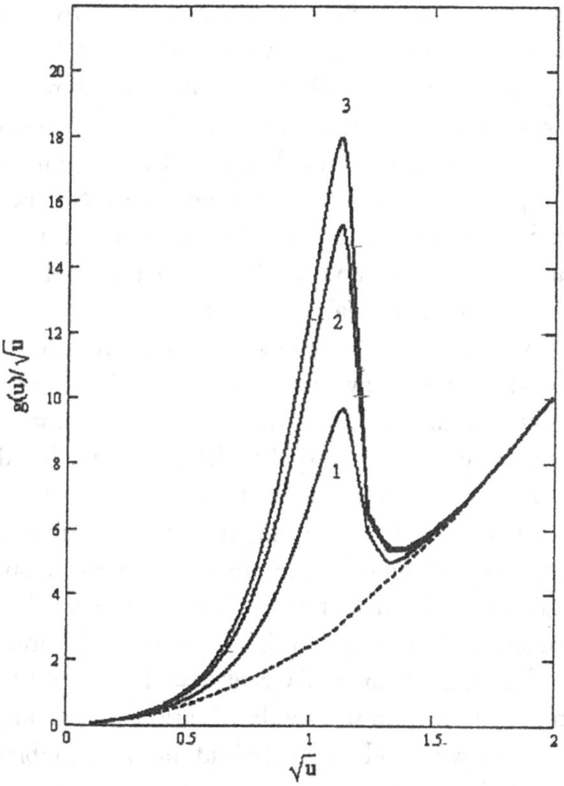

Figure 13: Evolution of total vibrational DOS (in a.u.), as a function of $u^{1/2}$, with the variation of $\delta \equiv \Delta/\nu_0^2 \propto \beta^2$, where $J_{sm}(t) \equiv J^*(\sigma t)$ at $\sigma = 1$. Curve 1 is obtained at $\delta = 0.05$, curve 2 at $\delta = 0.1$ and curve 3 at $\delta = 0.15$. The dashed curve is given for the hypothetical case at $\beta = 0 = \delta = \varsigma$ and describes the Debye (gapless) spectrum.

For a hypothetical case at $\beta = 0 = \delta$, the gap between the two separate spectral branches (Eq. (10.7)) disappears, for non-interacting acoustic phonons ($\nu \simeq s_0 q/2\pi$) and soft-mode excitations ($\nu = \nu'$), so the IRC does not occur. In this case, in Eqs. (10.13)–(10.15), the function $p(z, t; \delta = 0) = 1$ and $D(z - u; \eta)$ is a δ-function, $D(z - u; \eta) = \delta(z - u)$, at both $\kappa^2 = \kappa_1^2 \longrightarrow +0$ and $\kappa^2 = \kappa_1^2 \longrightarrow +0$, and the integration range reduces to $t_1 \leq t \leq t_2$ and $z_{min} \leq z \leq z_{max}$ at $\varsigma = 0$, as shown in the Appendix. The resulting basic DOS $I(u)$ becomes a product of the normalization integral for soft-mode DOS and the Debye DOS $I_D(u)$ for gapless acoustic spectrum, so that eventually $I(u) = I_D(u)$. In fact, this follows from calculations and can be shown from Fig. 13, by comparing the dashed curve with curves 1, 2, 3. The results discussed above allow one also to answer a basic question noted below Eq. (10.23). In accordance with above considerations, in the soft-mode model the IRC of inelastic scattering of acoustic phonons from soft-mode vibrations is the most important phenomenon determining the origin of BP, the scale of its characteristic frequency $\nu_{BP} \approx \nu_0 \approx 1$ THz and width, as well as the related anomalous dynamic and thermal vibrational properties.

Then, in SMM the broad asymmetric peak of IRC for inelastic scattering of acoustic phonons by harmonic soft-mode vibrations and the acoustic-like behavior above the peak in both total vibrational DOS $I(u)$ of the hybridized excitations (Eqs. (10.15)–(10.20)) and reduced vibrational DOS $g_r(u^{1/2}) \equiv g(u^{1/2})/u$ are identified as the correlated BP and HFS in the glasses under discussion. At the same time, the renormalization of soft-mode vibrations due to their elastic interactions appears to give rise rather to a quantitative effect in characteristics like BP intensity height. In fact, one can see quantitative changes in the properties of BP intensity height in Fig. 12 for different soft-mode DOS at $\alpha = 1$ or $\alpha = 1.5$ with strong hybridization of acoustic and soft-mode vibrational states important, and at $\sigma = 0.5$ or $\sigma = 2$ with both strong hybridization and vibrational instability effects contributing to the function. On the other hand, in the HFS part of the curves, noticeable quantitative changes can be seen in Fig. 12 at different α or σ but smaller ones in Fig. 13. The difference between dependencies of the properties on α or σ appears to be due to the fact that the gap states (Eq. (10.7)) are essentially linked to IRC (i.e., BP), whereas no relation between the gap states and HFS excitations and thus no substantial dependence of the HFS part of the curves on δ are available. It is worth emphasizing also that the BP frequency is located in the frequency range, intermediate between ranges of standard low frequency acoustic phonons and of HFS excitations for which the dispersion relation of propagating, wave-like excitations is still almost linear, $\nu(q) \propto q$, in accordance with experimental data (Ch. 6).

Let us now briefly compare the results of complete SMM for total vibrational DOS $I(u)$ and reduced DOS $g_r(u) \equiv g(u^{1/2})/u$ with the experimental data for BP and associated effects in scattering intensities.

For neutron scattering the data is usually described in terms of the reduced DOS, $g(\nu)\nu^{-2}$, of harmonic vibrational excitations, since the coupling coefficient of neutrons with the vibrations does not depend on ν. The latter does not hold for the Raman coupling coefficient $c_R(\nu)$ of light with vibrations, which may have diverse dependencies on ν for various groups of glasses. Then, generally speaking, the BP in Raman scattering cannot necessarily be described as a peak of only the reduced DOS. It is preferable to concentrate on these scattering data, of which comparison with different theoretical models seems to be most helpful for finding relevant models of vibrational dynamics in glasses. As follows from Eqs. (6.9)–(6.13), in glasses in which BP is found in $I_R(\nu, T)$ at high T, it can be practically the same as in $I_R^{(r)}(\nu)$, i.e., in $g(\nu)\nu^{-2}$ for one group of glasses with $c_R(\nu) = c_1(\nu) \approx const$, or in $g(\nu)\nu^{-1}$ for another group with $c_R(\nu) = c_2(\nu) \propto \nu$. On the other hand, in glasses in which BP is found in $I_R(\nu, T)$ at low T, it can be the same as in the dynamic susceptibility $\chi_R''(\nu)$, i.e., in $g(\nu)\nu^{-1}$ for the first noted group of glasses or in $g(\nu)$ for the second group. As noted in Ch 6, BP appears to be observed also in spectra of $\chi_R''(\nu)$ of a number of different glasses like diglycidyl bisphenol A (DGEBA), polymethyl methacrylate (PMMA) and calcium potassium nitrate (CKN), polysterene, polycarbonate, all those with the characteristic frequency $\nu_{BP} \sim 1\,\text{THz}$ and width $\gamma_{BP} \sim 0.3\ \nu_{BP}$. These data qualitatively and scalewise can be interpreted within SMM in which all these scattering intensities exhibit a broad asymmetric boson peak and, above it, a "wing" corresponding to HFS excitations characterized by a group velocity s_0. In contrast to SMM, Eqs. (10.22) and (6.10)–(6.13) give a (weak) peak in the reduced Raman scattering intensity $I_R^{(r)}(\nu) \propto g^*(\nu)\nu^{-1}[1 + B \cdot (\nu_{BP}/\nu)]$ for one group of glasses with $B \approx 0.5$, whereas it does not for another group with $B \approx 0$ and for both groups of glasses in $\chi_R''(\nu)$. Moreover, Eq. (10.22) does not give rise to a pronounced HFS wing observed above the BP in Raman scattering intensity, because this model neglects direct contributions of acoustic excitations and their interactions with soft-mode vibrations accounted for in the SMM. In addition, the reduced DOS $g^*(\nu)/\nu^2$ in Eq. (10.22) practically decreases like ν^{-1} at $\nu > \nu_0^*$ and contradicts the experimentally observed temperature independent exponential decrease of this DOS at $\nu > \nu_{BP}$ (Eq. (6.14)). For the latter, two dashed curves in Fig. 14 approximate (with accuracy to

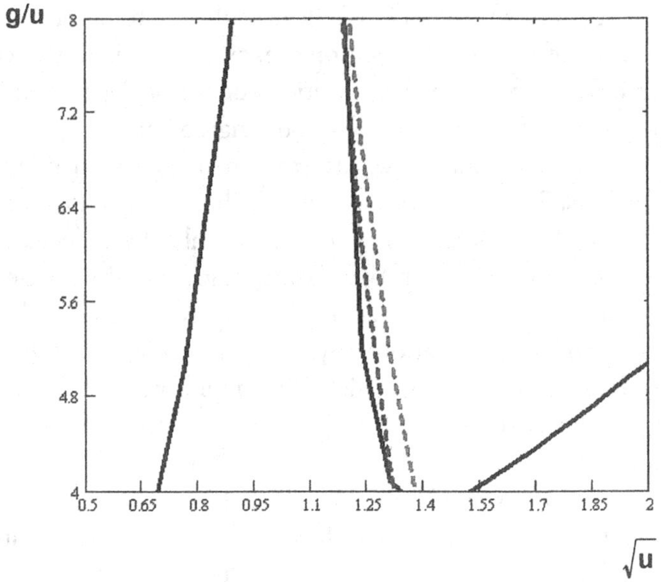

Figure 14: Theoretical approximation of the temperature independent BP tail at $\nu_{BP} < \nu < \nu_{min}$ (HFS), where the dashed or dashed-dotted curves approximate the part of essential interest in curve 2 of Fig. 12, with accuracy to corrections $\lesssim 5\%$ at $\nu = 0.20$ or $\lesssim 10\%$ at $\nu = 0.25$, respectively.

corrections $\lesssim 10\%$) the part of essential interest in curve 2 of Fig. 12 at $\nu_{BP} < \nu < \nu_{min}$(HFS) [210],

$$g(\nu)/\nu^2 \equiv g/u \propto \exp\left(-\nu/\nu_t\right) \text{ at } \nu_t = \upsilon\nu_{BP}, \tag{10.24}$$

and are similar to the empirical exponential decrease of reduced DOS described by Eq. (6.14), where $\upsilon \approx 0.20 - 0.25$ agrees with the empirical $\upsilon \simeq 0.5$ to a factor of about 2 (for the glasses under discussion here, ν_{min}(HFS) denotes the effective lowest frequency of HFS). In other words, the exponential function in Eq. (10.24) appears to describe qualitatively the empirical frequency dependence of the reduced DOS g/u. Moreover, the correlation length L_c, characteristic of the harmonic vibrational excitations of Eq. (10.7), can roughly be estimated in terms of the IRC wavelength, $L_c \sim \lambda_{ac}(\nu_{IR}^{(in)}) = s_0/\nu_{IR}^{(in)} \sim 30$ Å at typical $\nu_{IR}^{(in)} \approx \nu_0 \approx 1$ THz, and the estimation also agrees in scale with the empirical assessment.

In conclusion, the observed [143] universality of frequency dependence of the Raman coupling coefficient, for each of the two groups of glasses in Eqs. (6.12)–(6.13), appears to support the idea that the extended vibrational excitations can hardly be separated into propagating and non-propagating ones in the BP region. This idea seems to be in accordance with a conclusion of the complete SMM that a strong hybridization of propagating (acoustic) and non-propagating (soft-mode) excitations occurs in the IRC range of frequencies.

10.6 Width (attenuation) of acoustic phonons

As seen from Eq. (10.8), an essential question in SMM theory of the BP accompanied by HFS is what frequency dependence is typical of the width $2\gamma_{ac}^{(in)}$ of THz scale frequency acoustic phonons inelastically scattered by harmonic vibrational soft-mode excitations, at phonon frequencies ν considerably lower than $\nu_{BP} \simeq \nu_{IR}^{(in)} \simeq \nu_0$. In what follows, the qualitative behaviour of the acoustic width is discussed, by taking into account a simple mechanism of inelastic scattering. The latter can be characterized within the framework of a continuum approximation based on the elastic strain tensor e_{ik} and applied to describing long-wave acoustic phonons. One can see from Eq. (7.36) that the scattering can be due to almost "resonant" interaction between a harmonic vibrational soft mode excitation and an acoustic phonon described by a standard operator $\widehat{V}_{int}^{(res)}(x, e_{ik}) \simeq \beta_{ik} e_{ik} x$, accounting for the first nonvanishing term in the power expansion of both soft-mode harmonic vibrational displacement x and acoustic strain e_{ik} at realistic $|x| \lesssim 1$ and $|e_{ik}| \ll 1$ (β_{ik} is the deformation-potential coupling tensor). This interaction is implied to be a weak perturbation for both types of well-defined bosonic excitations far below the IRC, at a small width $2\gamma_{ac}^{(in)}(\nu) \simeq 2\gamma_{ac}^{(res)}(\nu) \ll \nu_{ac}(q)$. Neglecting for simplicity the tensorial character of $\widehat{V}_{int}^{(res)}(x, e)$, one can use the scalar model of phonons, in which $\widehat{V}_{int}^{(res)}(x, e) \simeq \beta x e$ and the acoustic strain $e \to du(q)/dq$. Indeed, the interaction gives rise to a "resonant", inelastic, scattering: one phonon of energy $h\nu$ is absorbed ($n_{ac}^{(i)}(\nu) = 1 \to n_{ac}^{(f)}(\nu) = 0$) while one soft-mode excitation of energy $h\nu'$ is emitted ($n_{HV}^{(i)}(\nu') = 0 \to n_{HV}^{(f)}(\nu) = 1$) or vice versa, at $\nu' \simeq \nu$, in the population number representation ($(n_{ac}(\nu) = 0, 1;\ n_{HV}(\nu') = 0, 1)$ for the operators of soft-mode (x) and acoustic (u) displacements. Applying the Golden Rule to the interaction matrix element $V^{(res)}_{int}(\nu', \nu)$ for acoustic phonons of frequencies $\nu < \nu_{BP}/2 \approx \nu_0/2$, one can estimate the acoustic-phonon width by a temperature independent expression [211] that is described in terms of a sum $\sum_{\nu'}\{|V^{(res)}_{int}(\nu', \nu)|^2\}$ over continuous soft-mode excitation frequencies ν. After transforming the sum into integral, the following expressions for the small width result:

$$2\gamma^{(res)}_{ac}(\nu) = \sum_{\alpha=1,2} c_\alpha 2\gamma^{(res)}_{ac,\alpha}(\nu), \tag{10.25}$$

and

$$2\gamma^{(res)}_{ac}(\nu) = 4\pi^2 h^{-1} \sum_{\alpha=1,2} c_\alpha \int d\nu' |V^{(res)}_{int}(\nu', \nu)|^2 g^{(HV)}_\alpha(\nu')\delta(\nu' - \nu) \ll \nu, \tag{10.26}$$

where $g^{(HV)}_\alpha(\nu') \approx C_\alpha \nu_D^{-1}(\nu'/\nu_D)^{2\alpha}$ is the frequency DOS of single vibrational soft-mode excitations (Eq. (10.5)), at $\alpha = 1, 2$. A standard calculation of the expression gives rise to a simple formula that qualitatively and to scale describes the acoustic-phonon width at moderately low ν, $3w \lesssim h\nu < h\nu_0/2$:

$$2\gamma^{(res)}_{ac}(\nu) = (\beta^2/hMs_0^2) \cdot [g^{(HV)}_\alpha(\nu)wh^{-1}] \equiv 2\gamma_0 \cdot [g^{(HV)}_\alpha(\nu)wh^{-1}] \ll \nu. \tag{10.27}$$

The related formula for inverse mean free path,

$$(l^{(res)}_{ac,\alpha})^{-1}(\nu) = 2\gamma^{(in)}_{ac,\alpha}(\nu)/s_0 = (2\gamma_0/s_0) \cdot [g^{(HV)}_\alpha(\nu)wh^{-1}], \tag{10.28}$$

is similar to that obtained in a different, coordinate, representation [212] at $\alpha = 2$, $c_2 = 1$, at $g^{(HV)}(\nu) = g^{(HV)}_2(\nu) \propto w^{-1}(h\nu/w)^4$. This particular power-law function is hardly a general expression even in the simplest case of independent soft modes, in which the universal dipolar interactions between soft-mode "defects" are neglected, when the soft-mode parameters η and ξ are subject to a basic distribution density $F(\eta, \xi)$ which, generally speaking, can be described in two limit cases at $\eta \to 0$ (Eqs. (7.17), (7.18), or in general with Eq. (7.19)). In fact, one can find from Eqs. (10.4), (10.5) and (10.26)–(10.28) that two limit types of $\gamma_{\mathrm{ac},\alpha}^{(\mathrm{in})}(\nu)$ can be distinguished at $\nu < \nu_0/2$:

$$2\gamma_{\mathrm{ac},\alpha}^{(\mathrm{res})}(\nu) \approx 2\gamma_0 \cdot D_\alpha(\nu/\nu_0)^{2\alpha} \ll \nu, \tag{10.29}$$

where $2\gamma_0 \sim 10^2$ THz, $\nu_0 \approx 1$ THz, D_α is independent of T, $D_\alpha \equiv C_\alpha \cdot 10^{-(1+2\alpha)}$, $C_\alpha \equiv q_0 F_\alpha^{(0)} \eta_L^{\alpha-1}$ at $q_0 \approx 10$ and $\nu_D \approx 10\nu_0$ (Eq. (7.30)). The dimensionless material parameters $C_\alpha \approx C \approx 10F_1^{(0)} = const.$ can be considered as independent of α, in accordance with the estimate $F_1^{(0)} \approx F_2^{(0)}\eta_L$ in Eq. (8.13) (the independence of the parameters of α is in accordance with experimentally found universality of dynamic properties at very low $\nu \lesssim \nu_m$ and $T \lesssim T_m$). Since DOS $g_0 \approx g_\alpha^{(0)} \equiv \eta_L^{\alpha+1/2} F_\alpha^{(0)} w^{-1}$ and parameter $F_\alpha^{(0)} \approx 10^{(2\alpha+1)} g_0 w$ at empirical $g_0 w \equiv G^{(TS)}(0)w \sim 10^{-5}$, one can assume that $F_1^{(0)} \sim 10^{-2}$, $F_2^{(0)} \sim 1$ and

$$2\gamma_{\mathrm{ac},\alpha}^{(\mathrm{in})}(\nu) \approx 2\gamma_0 \cdot 10^{-2(1+\alpha)} \cdot (\nu/\nu_0)^{2\alpha} \ll \nu. \tag{10.30}$$

Then, for independent soft modes, two limit types of frequency dependence of temperature independent phonon width exist, with either $c_1 \simeq 1$ and $\gamma_{\mathrm{ac}}^{(\mathrm{in})}(\nu) \simeq \gamma_{\mathrm{ac},1}^{(\mathrm{res})}(\nu) \propto \nu^2$ or $c_2 \simeq 1$ and $\gamma_{\mathrm{ac}}^{(\mathrm{in})}(\nu) \simeq \gamma_{\mathrm{ac},2}^{(\mathrm{res})}(\nu) \propto \nu^4$, while in general

$$2\gamma_{\mathrm{ac}}^{(\mathrm{res})}(\nu) = \sum_{\alpha=1,2} c_\alpha 2\gamma_{\mathrm{ac},\alpha}^{(\mathrm{res})}(\nu) \propto \nu^\kappa \tag{10.31}$$

at $0 \leq c_\alpha \leq 1$ and $c_1 + c_2 = 1$ so that $\varkappa \approx 2$ at $c_1 \approx 1$ or $\varkappa \approx 4$ at $c_2 \approx 1$.

On the other hand, the experimentally measured Brillouin line widths in the glasses at low $\nu \ll \nu_{BP}$ are small, $\gamma_{BL}(q = \nu/s_0) \ll \nu$, and is described by a power-law frequency dependence $\gamma_{BL}(\nu/s_0) \approx 2\gamma_{\mathrm{ac}}^{(\mathrm{res})}(\nu) \propto \nu^\varkappa$ at $\varkappa \approx 2$, rather than at $\varkappa \approx 4$. In the approximation of Eqs. (10.26)–(10.31), this behaviour can be qualitatively understood at $\nu_{BP} \simeq \nu_0$ and $\nu/\nu_0 \ll 1$ in case at $c_1 \simeq 1$. In this aspect, the theory contains a phenomenological constituent concerning the value of the ratio c_1/c_2 at $c_1 + c_2 = 1$, of which the study, as well as that of the contribution of universal weak dipolar interactions between soft modes [206], is not yet accomplished. As apparently follows from the complete SMM (Sec. 10.5), the interactions are not expected to substantially change the BP intensity shape and HFS spectrum DOS (as usual, DOS describes the real part of excitation self-energy while the "phonon width" is linked to the imaginary part, the two parts of self-energy being related by Kramers–Kronig-like formula [135, 136]). The main result that can be obtained in such an approximation appears to be that the frequency dependence of soft-mode vibrational DOS $g^{(HV)}(\nu)$ and of the acoustic phonon width at moderately low ν can be described by a power-law function, $f(\nu) \propto \nu^{\varkappa}$, where $\varkappa$ can be either $\varkappa \approx 2$ or $\varkappa \approx 4$, out of the BP region at $\nu \approx \nu_{BP}$. However, the phonon width at $\varkappa = 4$ cannot even qualitatively explain recent experimental data with $\varkappa \simeq 2$ (see the end of Ch 6) for such $\nu \ll \nu_{BP}$ for which the data are observed. Although the derivation of the experimental frequency dependence is not yet accomplished in SMM, the behaviour of phonon width,

$$2\gamma_{\rm ac}^{\rm (res)}(\nu) \propto \nu^{\varkappa}, \tag{10.32}$$

at $\varkappa \approx 2$ at not too low $\nu \ll \nu_{BP} \approx \nu_0 \sim 1\,\text{THz}$, still cannot be excluded as relevant for a qualitative interpretation of recent data in this model for glasses exhibiting HFS above BP. An argument in favour of such an interpretation seems to be that a mathematically different theoretical model of BP [193] (Sec.12.1) can give rise to acoustic phonon width characterized at $\nu < \nu_0$ by dependencies $\gamma_{\rm ac}(\nu, T) \propto \nu^{\varkappa} T^{\kappa}$ at $\varkappa \approx 2$ and $\kappa = 0$. These frequency and temperature dependencies may be due to contributions of weak anharmonic interactions (for harmonic vibrations alike soft-mode harmonic vibrations) to frequency dependence of the phonon width $\gamma_{\rm ac}(\nu) \propto \nu^2 T^0$.

In conclusion, the resonant mechanism of acoustic phonon broadening in glasses (Eq. (10.26)) may give rise to a temperature independent and small width $2\gamma_{\rm ac}^{\rm (res)}(\nu) \ll \nu$, with $2\gamma_{\rm ac}^{\rm (res)}(\nu) \propto \nu^2$ considerably below IRC , e.g., at $\nu < \nu_0/2 \sim (1/2)\,\text{THz}$. Equations (10.26)–(10.32) can determine (or considerably contribute to) the width due to resonant phonon interaction with harmonic vibrational soft-mode excitations in such glasses. A quantitative problem is still available, concerning the estimation of phonon width close to the IRC for inelastic scattering, $2\gamma_{\rm ac,\alpha}^{\rm (in)}(\nu) \approx 10^{-2\alpha}\cdot(\nu/\nu_0)^{2\alpha}\,\text{THz} \ll \nu_0 \sim 1\,\text{THz}$ (Eq. (10.30)) at $\nu < \nu_0/2$, so the width is too small to directly derive Eq. (10.8) by extrapolating the estimation to $\nu \approx \nu_0$. Perhaps, however, this question is not too important for the soft-mode model since the latter is a qualitative (and apparently scalewise) mean-field theory, but not a quantitative theory.

10.7 Thermal vibrational properties of glasses

Anomalous thermal properties at moderately low temperatures, $5 \lesssim T \lesssim T_M \sim 10^2$ K, can also be described within the framework of SMM for non-metallic glasses like v-SiO_2, exhibiting both BP and HFS in their dynamic properties. The standard assumption applied, which seems to be experimentally supported up to $T \sim T_M$, is that the heat carriers are acoustic phonons (Ch. 6) [131]. The anomalous properties include the "bump", broad asymmetric peak, in the reduced specific heat $C(T)/T^3$, and "plateau" followed with increasing T by quasi-linear increase, in the thermal conductivity $\kappa(T) \propto T^\alpha$ at $\alpha \simeq 1$ [104, 107, 111, 131]. In SMM, the properties have a common origin with that of BP followed at higher frequencies by HFS.

Let us now consider the temperature dependence of the reduced specific heat at moderately low T corresponding to THz scale frequency dynamics, including "low" frequencies ν, $\nu_{\rm min}^{(HV)} \equiv T_{\rm min}^{(HV)}/h \approx 0.5$ THz $\lesssim \nu \lesssim \nu_-$, "high" ν, $\nu_+ \lesssim \nu < \nu_M \approx 2-3$ THz, and "intermediate" ν, $\nu_- < \nu < \nu_+$, between the BP edges at $\nu_+ - \nu_- \simeq 2\gamma_{BP}$. The specific heat $C(T)$ related to harmonic vibrational soft-mode excitations can be estimated as that of thermal excitations of energy $h\nu \approx T$. For calculating $C(T)$, the basic vibrational DOS $J(\varepsilon \equiv \nu^2)$ for soft-mode excitations can be *harmonic soft-mode vibrations* approximated by a Debye term, $J(\varepsilon) \simeq J_D(\varepsilon) \simeq A_D\varepsilon^{1/2}/2$, at "low" and "high" $\varepsilon \equiv \nu^2$, and by $J(\varepsilon) - J_D(\varepsilon) \simeq \pi^{-1}A_L\Gamma_{BP}[(\varepsilon - \varepsilon_{BP})^2 + \Gamma_{BP}^2]^{-1}$, a Lorentzian-like function, at "intermediate" ε. Here A_L and $A_D = 2\pi\nu_D^{-3}$, as well as $\nu_{BP} \equiv \varepsilon_{BP}^{1/2}$ and $\Gamma_{BP} \simeq \gamma_{BP}\nu_{BP}$, are material constants. Then, at the intermediate frequencies,

$$g(\nu)/g_D(\nu) \simeq 1 + H_{BP}[1 + (\nu^2 - \nu_{BP}^2)^2/\Gamma_{BP}^2]^{-1} \times (\nu_{BP}/\nu)\Theta(\nu - \nu_-)\Theta(\nu_+ - \nu), \qquad (10.34)$$

where $\Theta(Z) = \{1, \text{if } Z > 0; 0, \text{if } Z < 0\}$ and the BP height $H_{BP} = 2A_L/\pi A_D\Gamma_{BP}\omega_{BP}$. Then, the reduced specific heat $c(T)/c_D(T)$, with a standard $c_D(T) \approx 2\pi(\nu_T/\nu_D)^3$ at $\nu_T = T/h$, can be described by the following approximate expressions [213]:

$$C(T) = \beta^2(-\partial/\partial\beta)\int_{\nu_{h\nu}}^{\nu_M} d\nu N(\beta\nu)\nu g(\nu) \simeq C^{(ac)}(T) + C^{(BP)}(T), \tag{10.35}$$

where

$$C^{(ac)}(T)/C_D(T) \approx \Sigma_{\pm} I_{\pm}(T) \quad \text{and} \quad C^{(BP)}(T)/C_D(T) \approx I_{BP}, \tag{10.36}$$

$$I_-(T) = \int_{x_{h\nu}}^{x_-} dx\, x^2 f(x), \quad I_+(T) = \int_{x_+}^{x_M} dx\, x^2 f(x),$$

$$I_{BP} = (2\pi)^{-1}\nu_D^3\nu_T^{-2}\int_{x_-}^{x_+} dx h(x,T). \tag{10.37}$$

In Eq. (10.37): $h(x,T) = f(x)g(xT/\hbar)$, $f(x) = (x/2)^2(\sinh(x/2))^{-2}$ and $x_\alpha \equiv h\nu_\alpha/T \equiv T_\alpha/T$, where ν_α are material constants. The reduced specific heat is numerically calculated by applying Eqs. (10.35)–(10.37), and the results are presented in curve (a) in Fig. 15. As seen, $C(T)/T^3$ exhibits a rather broad asymmetric peak, "bump", around a maximum at $T = T_{max}$ between $T_- \equiv h\nu_-$ and $T_+ \equiv h\nu_+$. Moreover, the reduced specific heat also exhibits a HFS wing at $T_+ \lesssim T < T_M \equiv h\nu_M$. The existence of a Debye-like wing, weakly varying around a constant with increasing T, appears to agree qualitatively and in scale with asymptotic estimates of $C(T)/C_D(T)$, in which the function $x^2 f(x)$ is approximated by

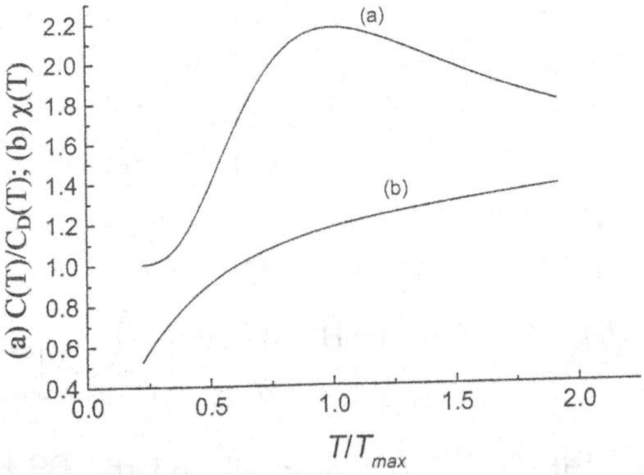

Figure 15: Theoretical approximation of reduced specific heat (a) and thermal conductivity (b) (in a. u.) as functions of T/T_{max}.

x^2 for $x_+ < x < 1$ or by $x^4 \exp(-x)$ for $1 < x < x_M$, and by $g(\nu) \simeq g_D(\nu)$ for acoustic-like excitations at $g(\nu)/g_D(\nu) \simeq 1$ for $\nu_M > \nu_+$. The bump of reduced specific heat qualitatively agrees with the data [104, 107] for $C(T)/T^3$ vs T/T_{max} with $T_{max} \equiv h\nu_{max}$ and $\nu_{max} \approx \nu_{BP}$, e.g., $T_{max} \sim 30$ K$\sim T_{max}^{(exp)}$ for v-SiO_2. The shape of the bump is similar to that of the reduced DOS $g(\nu)/\nu^2$ vs ν/ν_{BP}, in the BP region around $\nu/\nu_{BP} = 1$, being determined by thermal excitations ($h\nu \approx T \sim T_{max}$). Thus, the bump and BP indeed appear to have a common origin due to the IRC for resonant, inelastic, and scattering of acoustic phonons from the *harmonic soft-mode vibrations*.

Let us now discuss the temperature dependence of the thermal conductivity $\kappa(T)$ at moderately low temperatures at which the scatterers of acoustic phonons are harmonic soft-mode vibration excitations. Within the framework of SMM, the general expression of this characteristic (Eq. (2.3)) can be approximated as follows [213]:

$$\kappa(T) = \int_0^\infty d\nu C(\nu, T) s_0 l_{ac}(\nu), \tag{10.38}$$

$$\kappa(T) = \kappa_0 + \Delta\kappa \text{ at } \kappa_0(T) \equiv \left[\int_{\nu_{h\nu}}^{\nu_-} d\nu + \int_{\nu_+}^{\nu_M} d\nu\right] C(\nu, T) s_0 l_{ac}(\nu), \tag{10.39}$$

and

$$\Delta\kappa(T) = \int_{\nu_-}^{\nu_+} d\nu\, C(\nu, T) s_0 l_{ac}(\nu), \tag{10.40}$$

where $C(\nu, T) \simeq (\nu/2\nu_T)^2[\sinh(\hbar\nu/2T)]^{-2} g_D(\nu)$, $g_D(\nu) \simeq A_D\nu^2$, $l_{ac}(\nu) \simeq s_0/2\gamma_{ac}(\nu)$, and $2\gamma_{ac}(\nu)$ is the width of acoustic phonons. This corresponds to the fact that, in accordance with the above mentioned assumption, thermal acoustic excitations are heat carriers determining the major term, $\kappa_0(T)$, of the thermal conductivity in the Boltzmann heat transport, whereas thermal non-propagating excitations in the BP region can give rise only to a small correction $\Delta\kappa(T) \ll \kappa_0(T)$. The accuracy of this approximation in Eq. (10.38) can roughly be assessed by taking into account the estimations below Eq. (10.8) for non-propagating states in the BP region at $\omega_- < \omega < \omega_+$. In fact, the non-propagating states can be characterized by a small effective group velocity V_{eff} associated with random-walk type diffusive motions in the IRC and by a related low diffusion coefficient D_{eff}, compared to a much larger sound velocity s_0 for propagating sound-like states:

$$\Delta\kappa(T)/\kappa_0(T) \sim (V_{\text{eff}}/s_0)^2[g(\nu_{BP})/g_{ac}(\nu_{BP})][\gamma_{ac}(\nu_{BP})/\gamma_{BP}] \lesssim 10^{-2}. \tag{10.41}$$

where $V_{\text{eff}} \sim (D_{\text{eff}}\gamma_{\text{eff}})^{1/2} \sim (a_1^2\nu_{\text{eff}}\gamma_{\text{eff}})^{1/2} \lesssim a_1\nu_{BP} \sim 0.1s_0$, $\nu_{\text{eff}} \approx \nu_{BP} \sim 1\,\text{THz}$, $g(\nu_{BP})/g_{ac}(\nu_{BP}) \gtrsim 10$ and the IRC excitation width is at least one order of magnitude larger than the sound-like state width at $\nu \approx \nu_{BP}$, $\gamma_{BP} \gtrsim 10\gamma_{ac}(\nu_{BP})$.

Hence, the main thermal conductivity $\kappa_0(T)$ is determined by thermal excitations at "low" and "high" T, whereas for "intermediate" T, $T_- < T < T_+$, $\chi_0(T)$ is rather determined not by thermal *BP* excitations but by much more mobile "sub-thermal" ($\nu \lesssim \nu_- < \nu_T$) and "super-thermal" ($\nu \gtrsim \nu_+ > \nu_T$) acoustic excitations. Here, the acoustic phonon width is assumed to be due to inelastic scattering from soft-mode vibrational excitations, $\gamma_{ac}(\nu) = c_1\gamma^{(1)}(\nu) + c_2\gamma^{(2)}(\nu)$, with $\gamma^{(1)}(\nu) \propto \nu^2$ and $\gamma^{(2)}(\nu) \propto \nu^4$ (Sec. 10.6), while the ratio c_2/c_1 does not seem to be well-known. Therefore, we can suggest to use for calculations of $\kappa_0(T)$ recent experimental data [112] for the glasses under discussion: $2\gamma_{ac}(\nu) \simeq 2\gamma_{\exp} = P_{\exp}\nu^2$ for $\nu_- \lesssim \nu \lesssim \nu_M$, where $P_{\exp}$ is a material dependent constant ($\gamma_{\exp}(\nu)$ may contain also a contribution of a different type of phonon scattering). The thermal conductivity is numerically calculated from Eq. (10.38), the results being presented in curve (b) of Fig. 15. As can be seen from this curve, the behavior of $\kappa_0(T)$ vs $T/T_{\max}$ with its "plateau" and quasi-linear increase, is qualitatively in accord with experimental data (Eq. (6.2)), e.g., for v-SiO_2, around $T = T_{\max}$. Then, simple analytical estimations of $\kappa_0(T)$ at $T_- \lesssim T < T_M$ can be summarized in the following relationship:

$$\kappa(T) \simeq \kappa_0(T) \propto P_A(\mu_0 + \mu_1\nu_T), \tag{10.42}$$

which describe both the plateau, $\mu_0 \simeq \nu_-$, and the quasi-linear increase with increasing T, at $\mu_1 \approx 1$, also in a qualitative agreement with curve (b) of Fig. 15. Hence, the plateau of thermal conductivity and its quasi-linear increase with increasing T occur in a rather natural way and have a common origin with both the bump of reduced specific heat and BP accompanied by HFS.

One can conclude that the bump of reduced specific heat at a characteristic $T = T_{max} \simeq T_{BP} \equiv h\nu_{BP}/2 \sim 20$ K and the plateau of thermal conductivity at $T \approx T_{max}$ and quasi-linear increase with increasing T, at $T_{max} \lesssim T \lesssim T_M \equiv h\nu_M$, are correlated with each other and can be described by Eqs. (10.35)–(10.37) and (10.38)–(10.41), in terms of the DOS of harmonic vibrational soft-mode excitations. In other words, the anomalous thermal properties of glasses are determined by harmonic vibrational soft-mode excitations of moderately low energies, interacting with acoustic phonons.

Earlier assumptions introduced for explaining the anomalous thermal properties of glasses at moderately low temperatures do not seem to clearly show such a common origin with that of BP, at moderately low temperatures. In particular, the theoretical efforts to explain the plateau and following quasi-linear rise of thermal conductivity with increasing temperature included fractons in the fractal model [121], soft-mode *nonvibrational* TLS excitations or a kind of localized phonons [214] as the excitations, responsible for inelastic scattering of the heat carriers (acoustic phonons). It is not clear, however, how one can map a glass with its medium-range order local atomic configurations onto a self-similar percolating network in the fractal model, while in the other two models how to simply explain the rise of thermal conductivity with increasing temperature.

Let us now compare the main results of some earlier theoretical papers on the specific heat and thermal conductivity of non-metallic glasses, at moderately low temperatures, with Eqs. (10.34)–(10.42). It was suggested in [168], within the original soft-mode model [123], that scattering of low-frequency phonons by TLS excitations could limit thermal transport above the plateau, and the general formula (10.38) of thermal conductivity χ was investigated in two limit cases of the soft-mode model at $T \ll w$ and $w \ll T$. In the first case, the main contribution to thermal conductivity was ascribed to acoustic phonons and their resonant scattering from TLSs and the result was that $\kappa \propto T^2$, in accordance with STM (Sec. 9.1) described the observed plateau and quasi-linear increase in $\kappa(T)$ by increasing T at moderately low temperatures T around $T_{max} \simeq T_{BP} \sim 10$ K. In other words, in this temperature region, the approach discussed (as noted also in

[127]) is not sufficient to provide the explanation of experimental data. On the other hand, in the first paper of [214], the plateau was assumed to be linked with a crossover from a low-frequency region ($\nu \lesssim \nu_* \sim 0.1$ THz) with an acoustic phonon long mean free path $l_{ac} \sim 10^2\ \lambda_{ac}$ to a high-frequency region ($\nu \gtrsim 1$ THz) with a short mean free path $l_{ac} \sim \lambda_{ac}$. To describe the decrease of l_{ac} at $\nu > \nu_*$, an assumption was introduced that a sharp increase of DOS $g(h\nu) = g_0[1 + S \cdot \Theta(h\nu - h\nu_*)]$, where $S = const. \gg 1$ and $\Theta(x) = \{1, x > 0; 0, x < 0\}$, occurred for some localized harmonic vibrational excitations described as (unspecified) Einstein ones (S is a fitting parameter and g_0 is the constant DOS of TLS excitations of low frequencies, $\nu < \nu_*$, predominant at very low energies in STM). The enhancement of DOS was argued to account for the plateau of thermal conductivity and the bump of reduced specific heat, while the specific heat was assumed to be due to the contribution of TLS excitations, Einstein ones and Debye phonons. Then, the bump was suggested to correspond to a crossover from TLS excitations to the Einstein ones: For both types, the specific heat was proportional to T but the coefficients were essentially different at $S \gg 1$, while the frequency ν_* might be the same for $\kappa_0(T)$ and $C(T)$ only if the Rayleigh scattering was assumed to dominate in the crossover region, with $l_{ac} \propto \lambda_{ac}^{-4}$. As can be seen from comparing the results of the first paper in [214] with Eqs. (10.35)–(10.42) in SMM, the models could be in a qualitative accordance, only if the function $g_0 S \cdot \Theta(h\nu - h\nu_0)$ at large $S \gg 1$ was substituted at $\nu_0 \simeq \nu_{BP}$ for the asymmetric broad maximum (BP) in the DOS of harmonic vibrational soft-mode excitations. This seems to be a rather rough approximation compared to SMM in which the IRC is the decisive phenomenon while the contribution of the Rayleigh scattering is not. In the second paper of [214], an explanation for the plateau in thermal conductivity of orientational glasses like KBr:KCN was provided. In fact, it was shown that librations, or angular harmonic vibrations of individual CN ions moving in effective double-well potentials (in the libration coordinate), have a sharply peaked DOS in the THz range of frequencies. The related excitations were found to give rise both to the bump in specific heat (which empirically was considerably weaker than in conventional glasses) and to a strong resonant scattering of thermal acoustic phonons in the THz frequency range, which together with the assumed essential Rayleigh scattering can indeed provide a plateau in thermal conductivity. Moreover,

it was suggested that the plateau could be due to acoustic phonon localization which was assumed to be correlated with the occurrence of IRC for the scattering of acoustic phonons off the libration modes. This picture for the plateau of orientational glasses is essentially different from SMM picture for BP (Secs. 6.2 and 6.5) and for the specific heat bump and thermal conductivity plateau in conventional, structural, glasses, in which the BP states are hardly either propagating, wave-like excitations or localized ones, but rather correspond to an "intermediate" diffusive motion in the sample. By the way, the authors of the paper under discussion disagreed with [168] on the origin of the thermal conductivity increase above the plateau, arguing that the TLS relaxational scattering dominated under these circumstances, and thus no linear temperature dependence would be seen above the plateau, in contrast to the experimental data.

In conclusion, according to the soft-mode model, TLS excitations as scatterers of phonons are actually important only at very low frequencies $\nu \lesssim \nu_m \ll \nu_0$ and temperatures $T \lesssim T_m \sim 1$ K while harmonic vibrational soft-mode excitations are essential at moderately low frequencies/temperatures for the reduced specific heat bump and thermal conductivity plateau. Thus in the glasses under discussion, the origin of thermal conductivity increase above the plateau hardly can be ascribed to resonant scattering of very-low-frequency phonons from TLS excitations.

11

ON UNIVERSAL AND NON-UNIVERSAL DYNAMIC PROPERTIES OF GLASSES

As noted above, at present no consistent quantitative microscopic theory of the low-energy dynamics of glasses appears to be available, only a number of mean-field theories (models) of the dynamics, including SMM, are developed. Generally speaking, in the framework of SMM, the glassy dynamic and thermal properties under discussion at low temperatures and frequencies may exhibit both universal and non-universal features. The situation seems to be essentially different at very low (Chs. 8, 9) and moderately low (Ch. 10) temperatures. At very low T and ν, the properties appear to be universal at least in their qualitative behaviour and scales, whereas at moderately low T and ν the properties exhibit both universal and non-universal features in their qualitative behaviour, though apparently important scales remain rather universal.

11.1 Very low temperatures and frequencies

As noted in Ch. 6, the experimentally observed scale of dynamic properties (e.g., the "saturation" of resonant processes and related "hole" in absorption at high wave or pulse intensity) and thermal properties (e.g., the behaviour of specific heat and thermal conductivity) of thermal-equilibrium glasses at very low ν and T does not depend on their chemical nature and specific short-range and medium-range orders, the properties being in this sense universal. In STM, as well as in SMM, this means that the basic parameter P_0 and the related atomic TS concentration c_{TS} are in scale universal, fitting the experimental data at $c_{TS} \sim 10^{-6}$. In this connection, a basic question in STM, as well as in ATM (Advanced Tunneling Model, Sec. 9.2) accounting for the universal dipolar elastic (and possible electric) interactions between double-well potentials and related TS excitations, is why the postulated probability distribution density $P(\Delta, \Delta_0) = P_0/\Delta_0$ with $P_0 = const.$ and the related dynamic properties of glasses are universal. At present, two approaches for answering this basic question appear to be available: one approach developed in essential detail follows from ATM [187, 188] and is relevant only at very low ν and T, while another approach straightforwardly follows from SMM for independent (not interacting with each other) soft-mode "defects" and their excitations, both at very low and moderately low ν and T. From the viewpoint of SMM, these two approaches can be complementary to each other, so their synthesis could be of importance. Since the synthesis is not yet available, the main points of two approaches can be characterized separately. In the ATM approach, the universality of very-low-energy spectral properties of amorphous systems can be a result of the above-mentioned (elastic, electric) dipolar interactions $U(R) \propto 1/R^3$ between "defects", with each "defect" having an internal degree of freedom and being similar to a soft mode. For SMM approach, this might mean that the interactions between soft-mode TS excitations do not violate the universality of the properties at very low ν and T.

11.1.1 *On universality of basic distributions in ATM*

In the ATM (Advanced Tunneling Model) approach, the occurrence of "defect" centres can be ascribed to the removal of "spatial degeneracy" in crystal structures due to local stresses. An example is a crystalline chain compressed in its length: instead of one orientation in space it has at least two orientations with similar energies and an internal degree of freedom for tunneling transitions of random amplitude Δ_0 between two atomic configurations, each in its well of a double-well potential, at very low $T \to 0$. If such a double-well model is applied to describe primary "defects", not interacting with each other, the original distribution density function $P(\Delta_0) = P_0(\Delta_0)$ may be rather diverse for different "defects" and the system of such "bare" centres is not expected to have universal spectral properties independent of the form of $P_0(\Delta_0)$. However, the dipolar interactions between the centres can give rise to the occurrence of an increasingly large number of multi-centre clusters of which the relative role increases with decreasing excitation energy, while the the relative role of the primary "defects" centres. Then, a universal structure of low-energy spectral properties can be established, which is practically insensitive to the specific form of $P_0(\Delta_0)$, with the amplitudes of multi-centre tunneling transitions having a universal, almost uniform, logarithmic distribution $P_0(\Delta_0) \propto \ln \Delta_0$. For systems with $1/R^3$ interactions, an essential point is that the effective interaction region logarithmically increases with the system site (see below).This suggests that a renormalization group procedure can be used. Such a procedure was first demonstrated in [216] for localization–delocalization of excitations in a system with dipole–dipole interaction. The use of a similar approach can give rise to a general renormalization group equations of which the solutions determine the main results for $P_0(\Delta_0)$. When low-energy excitations are found in a system of interacting centres at $T \longrightarrow 0$, correct determination of the ground state is of fundamental importance. Its stability requires positive values for energies of any multi-centre excitations, thereby introducing a hierarchy of constraints. The latter affect the DOS, the influence being similar to the Coulomb gap in semiconductors [190], although in a different form. In the problem under discussion, the stability constraints are very significant and must be taken into account from the very beginning. In the present section, we follow the main points of the essential work [215] and, for the sake of physical clarity, we begin with an analysis of the spectral properties of the system in the limit of weak interaction, concentrating on two-centres interactions.

Let us consider an amorphous medium with randomly distributed double-well "defects" with an arbitrary distribution of the parameters. The Hamiltonian in standard spin representation is (Sec. 9.2):

$$H = H_0 + U + V, \tag{11.1}$$

$$H_0 = \sum_i h\nu_i \sigma_{iz}, \quad U = \frac{1}{2}\sum_i u_{ij}\sigma_{iz}\sigma_{jz}, \quad V = \sum_i \Delta_{0i}\sigma_{ix}. \tag{11.2}$$

As well as in Sec. 9.2, it was originally assumed that u_{ij} for different (i,j) pairs were uncorrelated and had random signs, so $\langle u_{ij}\rangle = 0$ while $\langle |u_{ij}|\rangle = U_0 = const.$, and also that the dispersion W of the centres excitation energies $(h\nu_i)$ was large, $W \gg nU_0$, with n the centres density, and that $|\Delta_{0i}| \ll h\nu_i, |u_{ij}|$, i.e., in this sense tunneling amplitudes $\Delta_{0i} \simeq 0$. In this case, the distribution of the energies of single-centre excitations, $\varepsilon_i = h\nu_i + \sum_i u_{ij}\sigma_{jz}$, was essentially given by the original dispersion W. Moreover, for the sake of simplicity, it is assumed that correlations between ε_i and Δ_{0i} are negligible, so

$$P(\Delta, \Delta_0) \simeq P_0(\Delta)P_0'(\Delta_0), \quad \int P_0'(\Delta_0)d\Delta_0 = 1,$$

$$P_0(\Delta) = \frac{1}{\Omega}\left\langle \sum_i \delta(\Delta - \varepsilon_i) \right\rangle, \tag{11.3}$$

with Ω the volume of the system, and it is taken into account that $\int P_0(\Delta)d\Delta = n$ since $P(\Delta, \Delta_0)$ is calculated per unit volume. It is also assumed that Δ_{0i} lie in an energy range with the characteristic value $\Delta_{0*} \ll W$. Let us now consider the DOS at low energies ε:

$$\Delta_{0*} < \varepsilon \ll W, \tag{11.4}$$

so that V can be omitted in Eq. (11.1). The stability of the ground state requires that $\varepsilon_i > 0$, this also holds for many-particle excitations. In the case of weak interaction under discussion, it is relevant to restrict the analysis to that of the role of 2- and 3-particle excitations, with the stability conditions, $\varepsilon_{ij} \equiv \varepsilon_i + \varepsilon_j - U_{ij} > 0$ and $\varepsilon_{ijk} \equiv \varepsilon_i + \varepsilon_j + \varepsilon_k - U_{ij} - U_{ik} - U_{jk} > 0$. Then, to the first order in the interaction, the contribution of pair excitations to the DOS can be found as follows:

$$\begin{aligned}
P_2(\Delta) &= \frac{1}{2\Omega}\left\langle \sum_{ij} \delta\left(\Delta - \varepsilon_{ij}\right) \right\rangle \\
&= \frac{1}{2\Omega}\left\langle \sum_i \delta\left(\Delta - \varepsilon_i\right) \right\rangle \left\langle \sum_j \delta\left(\Delta - \varepsilon_j\right) \right\rangle \\
&\quad \times \frac{\left\langle \sum_{ij} \delta\left(\Delta - \varepsilon_{ij}\right) \right\rangle}{\left\langle \sum_i \delta\left(\Delta - \varepsilon_i\right) \right\rangle \left\langle \sum_i \delta\left(\Delta - \varepsilon_j\right) \right\rangle} \\
&= \frac{1}{2}P_0^2(0) \int dR_{ij} \int d\varepsilon_i \int d\varepsilon_j \left\langle \delta\left(\Delta - \varepsilon_i - \varepsilon_j + \frac{u_{ij}}{R_{ij}^3}\right) \right\rangle_u .
\end{aligned} \tag{11.5}$$

Since small $\varepsilon_i, \varepsilon_j \ll W$ contribute mainly to the integral, $P_0(\Delta)$ is replaced by the limiting value $P_0(0) \approx n/W$, while the denotation $\langle \ldots \rangle_u$ means averaging over u_{ij}. Then, with $\varepsilon_i > 0$,

$$P_2(\Delta) \approx \frac{1}{2} P_0^2(0) \int dR_{ij} \left\langle \left(\Delta + \frac{u_{ij}}{R_{ij}^3} \right) \Theta \left(\Delta + \frac{u_{ij}}{R_{ij}^3} \right) \right\rangle_u , \tag{11.6}$$

where $\Theta(x)$ is a standard unit step function. Let the region of abovementioned coherent interaction, e.g., of coherent pairs, be limited by a radius $R_{\max}$, at which

$$U_0/R_{\max}^3 > \Delta_{0*}, \tag{11.7}$$

and which is significantly different from the interaction radius, i.e., for finite temperatures $R_{\max} = R_T \approx hs_0/kT$ ($R_T \to \infty$ at $T \to 0$); in the energy range $\Delta < U_0/R_{\max}^3$ only positive u_{ij} contribute to the integral ($\Theta(\Delta + u_{ij}/R_{ij}^3) = 0$ for $u_{ij} < 0$). Assuming that the distribution of u_{ij} is symmetric relative to $u_{ij} = 0$, one can get

$$P_2(\Delta) \approx \pi P_0(0) \chi_0 \cdot \left(\xi + \frac{1}{3} \frac{\Delta}{U_0 R_{\max}^3} \right), \tag{11.8}$$

where

$$\chi_0 = P_0(0) U_0 \ll 1, \quad \xi = \ln\left(R_{\max}/R_{\min}\right). \tag{11.9}$$

To logarithmic accuracy, $R_{\min}$ can be determined from the relation $U_0/R_{\min}^3 = W$, and the contribution of incidental nearby pairs, with the level shift exceeding W, can be neglected. Since actually $\xi \gg \{1; \Delta \cdot R_{\max}^3/3U_0\}$, the main contribution to $P_2(\Delta)$ at $\chi_0 \xi \ll 1$, with accuracy to (including) terms $\propto (\chi_0 \xi)^2$, eventually is

$$P_2(\Delta) \approx P_0(0) \chi_0 \xi. \tag{11.10}$$

In the same approximation, the DOS $P_1(\Delta)$ of single-centre excitations is also renormalized due to the stability condition for two-centre excitations:

$$P_1(\Delta) = \Omega^{-1} \left\langle \sum_i \delta(\Delta - \varepsilon_i) \prod_j \Theta(\varepsilon_{ij}) \right\rangle, \tag{11.11}$$

with $\Theta(\varepsilon_{ij}) = 1 - \Theta(u_{ij}/R_{ij}^3 - \varepsilon_i - \varepsilon_j)$ different from 1 only in a small part of the phase space. Then, by again replacing $P_0(\Delta) \approx P_0(0)$ one can get that, at $\chi_0\xi \ll 1$,

$$P_1(\Delta) \approx P_0(0)(1 - 2\chi_0\xi). \tag{11.12}$$

A trend to a decrease of $P_1(\Delta)$ and $P_2(\Delta)$ with increasing $\chi_0\xi$ is referred to as an argument for the existence of a quasigap ("dipole quasigap") in the spectrum, due to dipolar interactions. One can conclude for the calculated low-energy DOS $P(\Delta)$, at $\Delta \ll W$ and $\chi_0\xi \ll 1$, that the DOS $P(\Delta)$ depends on a single dimensionless small parameter $\chi_0\xi$. Moreover, $P(\Delta)$ practically does not depend on Δ at small enough energy, $P(\Delta) \approx P(0)$, and with increasing $\chi_0\xi$, the role of multi-centre excitations increases while that of single-centre excitations diminishes; since $P_\nu(\Delta) \approx P_\nu(0) \propto (\chi_0\xi)^{\nu-1}$ holds also for larger ν, multi-centre excitations can dominate at $\chi_0\xi \gtrsim 1$.

Now, the perturbation-theoretical approach with the small expansion parameter $\chi_0\xi \ll 1$ is applied to the complete Hamiltonian (11.1) of interacting tunneling centres with finite small intra-centre tunneling amplitudes $\Delta_{0i} < \varepsilon_i = h\nu_i + \sum_j U_{ij}\sigma_{iz}$. In this approach, one has to calculate and analyse the contributions of ν-centre excitations ($\nu \geq 2$) to the pair excitation DOS $P_2(\Delta, \Delta_0)$ and other spectral characteristics. Already in this approach, in which the main contribution is due to pair excitations, $P_2(\Delta, \Delta_0)$ is found to be uniform in Δ and proportional to Δ_0^{-1}, an arbitrary "bare" $P_0(\Delta, \Delta_0) \approx P_0(0)P'(\Delta_0)$ at assumed uncorrelated Δ and Δ_0. At $\chi_0\xi \ll 1$, the main contribution to $P_2(\Delta, \Delta_0)$ is due to independent pair excitations, each with the Hamiltonian

$$\begin{aligned} H_p &= \varepsilon_1\sigma_{1z} + \varepsilon_2\sigma_{2z} + U_{12}\sigma_{1z}\sigma_{2z} - U_{12}(\sigma_{1z} + \sigma_{2z})/2 \\ &\quad + \Delta_{01}\sigma_{1x} + \Delta_{02}\sigma_{2x}. \end{aligned} \tag{11.13}$$

The latter is obtained by adding the last two tunneling terms to the "static" Hamiltonian $H_p^0 = \varepsilon_1\sigma_{1z} + \varepsilon_2\sigma_{2z} + U_{12}\sigma_{1z}\sigma_{2z} - U_{12}(\sigma_{1z} + \sigma_{2z})/2$. Transforming the spin operators $(\sigma_{1x,z})$ to those $(\widetilde{\sigma}_{1x,z})$ corresponding to the eigenvalues $E_i(\varepsilon_i^2 + \Delta_{0i}^2)^{1/2}$ of single excitations, by rotating the quantization axes in the spin space, at $\sigma_{iz} = \widetilde{\sigma}_{iz} \cdot (\varepsilon_i/E_i) - \widetilde{\sigma}_{ix} \cdot (\Delta_{0i}/E_i)$ and $\sigma_{ix} = \widetilde{\sigma}_{iz} \cdot (\Delta_{0i}/E_i) + \widetilde{\sigma}_{ix} \cdot (\varepsilon_i/E_i)$, one can get the pair Hamiltonian

$H_p = h_0 + h'$ in the new representation, also assuming that $\Delta_{0i} < \varepsilon_i$. This allows one to set $(\varepsilon_i/E_i) \approx 1$ in the operators h_0 and h'. With small $\Delta_{oi} \ll \varepsilon_i$ above, for an up–up pair excitation ($\widetilde{\sigma}_{iz} = 1$), the effective energy asymmetry determined by the operator h_0 is $\Delta_p = E_1 + E_2 - U_{12}$, while the effective tunneling amplitude is

$$\Delta_{0p} = \frac{\Delta_{01}\Delta_{02}}{2E_1E_2}U_{12} \ll U_{12} = u_{12}/R_{12}^3. \tag{11.14}$$

As well as above, it is assumed that $\Delta_{0i} \ll \varepsilon_i \ll W$ (i.e., $E_i \simeq \varepsilon_i$). Then, the pair excitation DOS is calculated from the formula $P_2(\Delta_p, \Delta_{0p}) \propto \frac{1}{2}P_0^2(0) \int d\Delta_{01}P'(\Delta_{01}) \int d\Delta_{02}P'(\Delta_{02}) \int d\bar{R}_{12} \int\int d\varepsilon_1 d\varepsilon_2 \langle\delta(\Delta_p - \varepsilon_1 - \varepsilon_2 + U_{12})\delta(\Delta_{0p} - \Delta_{01}\Delta_{02}|u|/2\varepsilon_1\varepsilon_2 R_{12}^3)\rangle_u$, with $\langle K\rangle_u \equiv duP(u)K(u)$ and $|u|$ substituted for $u(\gtrless 0)$ because by definition of $P_2(\Delta_p, \Delta_{0p})$ both $\Delta_p > 0$ and $\Delta_{0p} > 0$ and so the sign is unimportant. Here $P_2(\Delta_p, \Delta_{0p})$ and other low-energy spectral properties are analysed for $\Delta_p < U_0/R_{\max}^3(\ll W)$, but for simplicity $\Delta_p > \Delta_{0*}$, which is consistent as $\Delta_{0*} < U_0/R_{\max}^3$. Under these conditions, $\delta(\Delta_p - \varepsilon_1 - \varepsilon_2 + U_{12}) \neq 0$ only for pairs with $u_{12} > 0$ at $\varepsilon_i > 0$. Here $P_2(\Delta_p, \Delta_{0p})$ and other low-energy spectral properties are analysed for $\Delta_p < u/R_{\max}^3(\ll W)$, but for simplicity $\Delta_p > \Delta_{0*}$, which is consistent as $\Delta_{0*} < U_0/R_{\max}^3$. Assuming again that $\Delta_0, \Delta \ll \Delta_{0*}$ and that $P'(\Delta_{0i})$ does not have a singularity at $\Delta_{0i} = 0$, one can find that after rather cumbersome calculations, a universal dependence of $P_2(\Delta, \Delta_0)$ on Δ_0 is indeed obtained,

$$P_2(\Delta, \Delta_0) \approx \frac{1}{3\Delta_0}P_0(0)\chi_0, \tag{11.15}$$

at $\int d\Delta_0 P'(\Delta_0) = 1$ and $n = \int d\Delta P_0(\Delta) \approx P_0(0)W$, for the smallest Δ_0, $\Delta \ll \Delta_{0*}$. The inequality $\Delta \ll \Delta_{0*}$ is easily satisfied, the lower bound for Δ_0, $\Delta_0 > 2\Delta_{01}\Delta_{02}R_{\min}^3 U_0^{-1} \sim \Delta_{0*}^2 W^{-1}$ at $\Delta_{0*} \ll W$, is consistent with $\Delta_0 \ll \Delta_{0*}$ and marks a broad range for Δ_0, $\Delta_{0*} \gg \Delta_0 > \Delta_{0*}^2/W$. The conclusion is that at least up–up pair excitations of interacting double-well centres exist, at an arbitrary non-singular $P_0'(\Delta_0')$, for which $P_2(\Delta, \Delta_0)$ is consistent with the universal property $P_2(\Delta, \Delta_0) \approx const./\Delta_0$, at $const. = \chi_0 P_0(0)/3$. Moreover, at $\Delta_{0*} \leq U_0/R_{\max}^3(\ll W)$,

$$P_2(\Delta) \approx \int_{\Delta_{0\min}}^{\Delta_{0\max}} d\Delta_0 P_2(\Delta, \Delta_0) \approx P_0(0)\chi_0\xi, \tag{11.16}$$

at $\chi_0\xi \ll 1$ and $\Delta_{0\,\min} \approx \Delta_{0*}^2/W$. Moreover, the DOSs $P_\nu(\Delta, \Delta_0)$ of collective excitations of low energy $\varepsilon \approx 0(\varepsilon \ll W)$, localized in ν-centre clusters at $\nu > 2$ and characterized by the effective asymmetry (Δ) and tunneling (Δ_0) parameters, seem to be calculable within the perturbation theory in the small parameter $\chi_0\xi \ll 1$, with a similar universal dependence, $P_\nu(\Delta, \Delta_0) \propto P_0(0)(\chi_0\xi)^{\nu-1}/\Delta_0$, for an arbitrary "bare" $P_0(\Delta, \Delta_0) \approx P_0(0)P'(\Delta_0)$.

What is important in the perturbation theory is that, its small parameter $\chi_0\xi$, with $\xi = \ln(R_{\max}/R_{\min})$, logarithmically increases with the size of the region in question. Therefore, strong interaction at $\chi_0\xi \gtrsim 1$ is realisable almost always in macroscopic systems with large enough $R_{\max}$, as directly follows from the occurrence of $1/R^3$ interactions. Indeed, the existence of multi-centre excitations with a probability ~ 1 corresponds to $\chi_0\xi \approx 1$ and, since $\chi_0 \ll 1$, the related region size $R_{\max}(\chi_0\xi \approx 1) = R_*$ is exponentially large,

$$R_*/R_{\min} \approx \exp(1/\chi_0), \quad \text{i.e., } \chi_0 \ln(R_*/R_{\min}) \approx 1. \tag{11.17}$$

Now, the question is whether the total DOS $P_0(\Delta, \Delta_0)$ for the low-energy excitations in the whole variety of multi-centre clusters, with effective energy asymmetry Δ and tunneling amplitude Δ_0, has the universal dependence $P_0(\Delta, \Delta_0) \approx \mathit{const.}/\Delta_0$ for any arbitrary "bare" DOS $P_0(\Delta, \Delta_0) \approx P_0(\Delta)P'(\Delta_0)$. As well as above at $\chi_0\xi \ll 1$, one can start in general, for any $\chi_0\xi$, the analysis of this question with the above Hamiltonian at vanishing $\Delta_{0i} \longrightarrow 0$ and $T = 0$ (or, equivalently, at low enough $T < \varepsilon = [\Delta^2 + \Delta_0^2)$. Let the radius of coherent interaction be $R_1 \gg R_{\min} = (U_0/W)^{1/3}$, and let $P_\nu(\varepsilon, R_1)$ be the DOS of collective excitations on m centres with low energy $\varepsilon < U_0/R_1^3$. If the coherent interaction radius increases to R_2 ($> R_1$), the change $P_\nu(\varepsilon, R_2) - P_\nu(\varepsilon, R_1)$ has to be determined at

$$1 < R_2/R_1 \lesssim R_*/R_{\min} \approx \exp(1/\chi_0). \tag{11.18}$$

One can see from the above results of the perturbation theoretical approach that the probability of a three-centre excitation will be less than that of a two-centre one by an additional factor of $\chi \ln(R_2/R_1)$, and the probability of an ν-centre excitation will be smaller by a factor of $[\chi_0 \ln(R_2/R_1)]^{\nu-2}$

at $\nu > 2$ (so the difference $P_\nu(\varepsilon, R_2) - P_\nu(\varepsilon, R_1)$ can be calculated for two-centre excitations). With Eqs. (11.17) and (11.18), beginning with an arbitrary distribution of parameters and following a renormalisation group analysis, one can find that the excitation spectrum of the renormalised system will be close to that assumed in the STM with the distribution $P(\Delta, \Delta_0) \propto 1/\Delta_0$. That is just the major result obtained in [215], in which the renormalisation group equations are derived. As the coherent interaction radius grows, P_ν decreases, since increase in configurations must satisfy the excitation stability conditions. On the other hand, P_ν can exhibit an increase due to pair interactions of excitations related to small numbers of centres. By considering an excitation of low energy $\varepsilon < U_0/R_2^3$ and following the quoted work, one can obtain the "general renormalisation group" equation

$$\partial\tilde{P}_\nu(\tau)/d\tau = -2\tilde{P}_\nu(\tau)\sum_{\varkappa=1}^{\infty} b_{\varkappa\nu}\tilde{P}_\varkappa(\tau) + \sum_{\varkappa=1}^{\nu-1} b_{\varkappa,\nu-\varkappa}\tilde{P}_\varkappa(\tau)\tilde{P}_{\nu-\varkappa}(\tau), \tag{11.19}$$

where $\tau = \chi\xi(R)$, $\xi(R) = \ln(R/R_{\min})$, $\tilde{P}_\nu(\tau) \equiv \frac{P_\nu(0,\xi(R))}{P_0(0)}$, $b_{\varkappa\mu} = \langle|u_{\varkappa\mu}|\rangle/U_0$. A solution of this nonlinear equation, corresponding to a definite τ (or $\xi(R)$), gives the distribution function for the low $\varepsilon < U_0/R^3$. The solution is found at $R = R_2 = R_{\max}$, where $R_{\max}$ is the limiting radius of the largest spatial region in which the excitations cluster can combine with some other cluster, to form a larger one. For $T = 0$ under consideration, or equivalently at kT less than the cluster related excitation energy ε, $R_{\max}$ effectively is r_ε from $U_0/r_\varepsilon^3 = \varepsilon$, i.e., $R_{\max} \approx (U_0/\varepsilon)^{1/3} = r_\varepsilon$ (for finite $kT > \varepsilon$ see below). Note that actually $R_{\max} \approx R_* \approx R_{\min}\exp(1/\chi_0)$, at $\chi_0\xi \approx 1$. The solution depends on the form of the coefficients $b_{\varkappa\mu}$, i.e., of the average $\langle|u_{\varkappa\mu}|\rangle$ over the distribution $P(u_{\varkappa\mu}) = P(-u_{\varkappa\mu})$ symmetric with respect to $u_{\varkappa\mu} = 0$. The simplest case discussed in [215] is at multiplicative $b_{\varkappa\mu} = b_\varkappa b_\mu$, while $b_{\varkappa\mu} = $ const is a special simplest case. In the latter, with $b_{\varkappa\mu} = 1$, the equation for the total DOS of low-energy excitations, $\tilde{P}_\nu(\tau) = \sum_{\varkappa=1}^{\infty}\tilde{P}_\varkappa(\tau)$, obtained reads:

$$\partial\tilde{P}(\tau)/d\tau = -\tilde{P}^2(\tau), \quad \text{with } \tilde{P}(\tau) = 1/(1+\tau), \tag{11.20}$$

which is the solution with the initial condition $\tilde{P}(0) = 1$. Correspondingly, in this case, $\tilde{P}_\nu(\tau) = (\frac{\tau}{1+\tau})^{\nu-1}\cdot(1+\tau)^{-2}$, while $\tilde{P}_\nu(\tau) = \tau^{-2}$

$\exp[-(\nu - 1)/\tau]$ at $\tau \gg 1$, and the total DOS is $P(\Delta = 0, \xi(R)) \approx 1/\pi U_0 \xi(R)$. Let us take into account here that a multi-centre cluster with a low-energy excitation is characterized by both an asymmetry energy Δ and a coherent tunneling amplitude Δ_0 coupling the excited state to the ground state, with $\Delta_0 \approx W \exp(-\Lambda)$ and Λ the effective quasiclassical tunneling (barrier) parameter. The final expression of the distribution function $P(\Lambda) \approx \sum_\mu P_\mu(\Delta = 0, \Lambda, R = R_{\max})$ for the dimensionless tunneling parameter Λ and low energies ($\varepsilon \approx 0 \approx \Delta$, $R_{\max} \gg R_{\min}$) are also obtained [215] in the simplest cases at $b_{\varkappa\mu} = const.$ and at $b_{\varkappa\mu} = b_\varkappa b_\mu = \sqrt{\varkappa\mu}$. For the most simple case at $b_{\varkappa\mu} = const.$, the distribution function $P(\Delta_0) \approx P[W \exp(-\Lambda)]$ derived indeed has the universal form,

$$P(\Delta_0) \approx const./\Delta_0, \quad const. \equiv P_0(0)/3\chi_0\xi^3_{\max}, \tag{11.21}$$

in the range of low enough energy at $\Delta \approx 0$ and $\Delta_{0*} > \Delta_0 > (\Delta_{0*}/W)^\tau \approx 0$ for $\tau \gg 1$. This expression indeed is universal, being independent of any arbitrary "bare" distribution $P(\Delta_1, \Delta_{01}) = P_0(\Delta_1)P'(\Delta_{01})$. The universal relationship, with a different constant, can also be obtained at $b_{\varkappa\mu} = \sqrt{\varkappa\mu}$. Then, it is reasonably suggested that the earlier postulated in the STM for single-centre TS excitations distribution function $P(\Delta_0) \approx \text{const}/\Delta_0$ is universal (for arbitrary plausible coefficients $b_{\varkappa\mu}$) not only at $\chi_0\xi_{\max} \ll 1$ but also at $\chi_0\xi_{\max} \gtrsim 1$. This result is determined by the dipolar interactions between the single-centre TS excitations independently of the specific form of the distribution function of tunneling amplitude.

The above results concerning the universal form of $P(\Delta_0)$ (as well as of $P(\Delta, \Delta_0)$) for originally uncorrelated Δ and Δ_0) are obtained by assuming $T < \varepsilon$, i.e., actually at $T \approx 0$ for the low excitation energies, $\varepsilon \approx 0$, under consideration. Specific constraints that must be accounted for in a general theory of the universal $P(\Delta_0) \approx const./\Delta_0$ appear at finite $T > 0$, or $kT > \varepsilon$. First of all, the radius of coherent interaction region actually is determined from

$$U_0/R^3_{\max} \approx T, \tag{11.22}$$

since $R_{\max} \approx (U_0/T)^{1/3}$ is less than the upper limit R_c (Eq. (9.22)) of the interaction radius at the low temperatures in question. In fact, at larger distances, the energy of interactions between excitations is less than T and therefore the excitations can be considered as independent ones. Clusters with excitation energies $\varepsilon \sim U_0/R^3$ are involved in the formation of bigger clusters, when $R < R_{\max}$. Then, the energy $\varepsilon > U_0/R^3 > T$, so such clusters can be assumed to occur in the ground state. All features determined by the excitation stability conditions hold here as well. In addition, the maximum radius of the coherent interaction region, determining $\xi_{\max}$ in the constant $\approx P(\Delta_0)\Delta_0$, can be found from

$$\xi_{\max} = \xi_T = \frac{1}{3}\ln(W/T), \quad \text{at } T > \varepsilon = \sqrt{\Delta^2 + \Delta_0^2}, \tag{11.23}$$

or

$$\xi_{\max} = \xi_\varepsilon = \frac{1}{3}\ln(W/\varepsilon), \quad \text{at } T < \varepsilon = \varepsilon. \tag{11.24}$$

Then, multi-centre excitations are expected to be essential at $\chi_0\xi_{\max} \gg 1$, while in many cases the relationship $\chi_0\xi_{\max} \gtrsim 1$ is more than relevant.

11.1.2 *On universality of soft-mode distribution in SMM*

In the SMM approach, the universality of glassy dynamic properties at very low ν and T may appear to follow from the conclusion that two different limit types of basic distribution density $F(\eta, \xi)$ (Eqs. (7.17)–(7.19)) can give rise to the same universal form of the distribution density $P(\Delta, \Delta_0) = P_0/\Delta_0$ (Eq. (8.16)), already in the approximation of independent soft-mode "defects". On the other hand, the SMM of excitations in glasses includes STM as a particular case for very low excitation energies at $0 < \varepsilon < \varepsilon_m = kT_m$ and $T_m \approx 1K$. Then, from the SMM approach viewpoint, the result of ATM may mean that the interactions between soft-mode TS excitations do not violate the available universality of dynamic properties of glasses at very low ε and T, ν. However, the universality problem for dynamic properties becomes different in SMM for moderately low ε and T, ν.

11.2 Moderately low temperatures and frequencies

In SMM, the origin of the boson peak at $\nu \approx \nu_{BP}$, of a large half-width $2\gamma_{BP} \sim \nu_{BP}$, accompanied by a high-frequency sound at $\nu > \nu_{BP}$, are determined by the IRC from a weak inelastic scattering of acoustic phonons by vibrational excitations of soft-modes, to a strong scattering and a resulting strong hybridisation of both types of excitations, around a characteristic crossover frequency identified as the BP frequency, $\nu_{IR}^{(in)} \simeq \nu_{BP}$. As explained in Sec. 10.5, the vibrational instability effect due to elastic interactions between soft-mode vibrational excitations can change quantitative aspects of the SMM, but not its qualitative properties. Therefore, in the SMM, the strong hybridization of phonons with soft-mode vibrational excitations in the Ioffe–Regel crossover for inelastic scattering appears to be the major phenomenon determining the origin of dynamic and thermal anomalies in the glasses under consideration.

As noted above (Sec. 10.2), there is an essential difference between elastic and inelastic scattering of acoustic phonons (and other excitations) and the related Ioffe–Regel crossovers. Elastic scattering does not change the phase of the system wave function, and the appropriate interference effects are anticipated to give rise to the Anderson localisation [129] above the "mobility edge" for acoustic phonons. Elastic scattering of acoustic phonons, with increasing frequency necessarily gives rise not only to the IRC but also to the Anderson localisation of the states at a mobility edge, with frequency considerably higher than IRC one (see, e.g., [202]). On the other hand, taking into account the Heisenberg uncertainty for energy and phase fluctuations [217], one can suggest that inelastic scattering changes the phase and does not necessarily give rise to the localisation. In this sense, the inelastic scattering appears to favor extended excitation states. In fact, the Ioffe–Regel crossover under discussion can be characterised by eigenstates approximated by superpositions of acoustic waves and quasi-local soft-mode states, with coefficients having rather irregular phases. Then, the coherence of the wave phases is violated and the "localising" interference effects related to the hybridised acoustic and soft-mode excitations can become weak. Thus, the eigenstates in the crossover region rather are extended states, propagating waves outside the boson-peak region or non-propagating states inside it. Moreover, the IRC for inelastic scattering can produce propagating waves, high-frequency sound, above the boson peak, while the IRC for elastic scattering cannot. An important question is whether the IRC states can be considered as non-localised ones. Although the rigorous answer in Anderson localisation terms is yet to be given, an argument in favour of a positive answer can be based on an essential difference between elastic and inelastic scattering and related Ioffe–Regel crossovers. Elastic scattering does not change the phase of the system wave function, so the appropriate interference effects are anticipated to give rise to the Anderson localisation above the "mobility edge" for the excitations (acoustic phonons). Therefore, the IRC frequency has to be lower than the "mobility edge" at which the localisation of excitations occurs. In fact, it was explicitly shown in some work, e.g., [202], that acoustic phonons undergo the IRC for elastic scattering from static disorder at a frequency considerably lower than the mobility edge one, the strong scattering transforming the phonons into extended but nonpropagating excitations for intermediate frequencies. On the other hand, taking into account the Heisenberg uncertainty for energy and phase fluctuations [217], one can suggest that inelastic scattering changes the phase of the system wave function and does not necessarily give rise to the localisation. Thus, the inelastic scattering favours extended excitation eigenstates that can be approximated by superpositions of acoustic waves and quasi-local soft-mode states, with coefficients having rather irregular phases, which are either propagating waves outside the IRC region while non-propagating states inside it.

From this viewpoint, a universal feature of glasses, the existence of boson peak, appears to be determined by the occurrence of Ioffe–Regel crossover (IRC) for extended vibrational excitation states of moderately low energies below the mobility (localisation) edge. A non-universal feature is the availability of two distinct essential limit types of the boson peak properties and, in this sense, of glasses: (I) those mainly under discussion, for which two types (or branches) of harmonic vibrational excitation spectrum are separated by a gap and the related IRC for inelastic scattering of acoustic phonons determine both a boson peak (BP) and, above it, a high-frequency sound (HFS) in the inelastic scattering spectra (Ch. 10); (II) glasses for which low energy acoustic phonons only occur, so the Ioffe-Regel crossover at their elastic scattering determines only a BP. Generally speaking, the vibrational spectral intensity $I^{(HV)}(\nu)$ (Eq. (6.8)) may contain both a BP part $I_{BP}^{(HV)}(\nu)$ and a HFS part $I_{HSF}^{(HV)}(\nu)$:

$$I^{(HV)}(\nu) = I_{BP}^{(HV)}(\nu) + I_{HSF}^{(HV)}(\nu). \tag{11.25}$$

The suggestion of the occurrence of two limit types of glasses seems to be consistent with recent data, e.g., for v-SiO_2 as a glass of type I [150] while for v-B_2O_3 as a glass of type II [151] (Chap. 6). The criterion for the existence of glasses of type I appears to be as follows:

$$I^{(HV)}(\nu) \approx I^{(HV)}_{HFS}(\nu) \quad \text{at } \nu_{BP} \simeq \nu^{(in)}_{IR} \quad \text{and}$$
$$\nu_{BP} + \gamma_{BP} < \nu < \nu^{(el)}_{IR} \simeq \nu_{BP}(\ll \nu_D), \tag{11.26}$$

while the criterion for glasses of type II rather is:

$$I^{(HV)}(\nu) \simeq I^{(HV)}_{BP}(\nu) \quad \text{at } \nu \approx \nu_{BP} \simeq \nu^{(el)}_{IR} \ll \nu^{(in)}_{IR}(\ll \nu_D), \tag{11.27}$$

is the criterion for the existence of glasses and vibrational anomalies of type II; here, $2\gamma_{BP}$ is the BP width and $\nu^{(el)}_{IR}$ is the characteristic frequency of IRC for elastic scattering while $\nu^{(in)}_{IR}$ is for inelastic scattering (Eq. (10.8)).

For glasses of type I, the propagating high-frequency sound excitations have to occur for not too high ν, $\nu_{BP} < \nu < \nu_M \equiv p_{\max}\nu_0$, with the crossover frequency $\nu^{(el)}_{IR} \simeq \nu_M$, the highest frequency of elastically scattered high-frequency sound excitations ($p_{\max}$ can be here identified with the experimental value $p^{(exp)}_{\max} \approx 2{\div}3$ for HFS [112]). The hybridisation of the "bare" soft-mode vibrational excitations with the phonons appears to extend the SMM theory to higher ν, $\nu_0 < \nu < \nu_M$, than those, $w < h\nu \lesssim h\nu_{sm}$, of the harmonic vibrational soft-mode excitations, at $h\nu_0 < h\nu_{sm} < h\nu_M$ and $\nu_0 \sim \nu_{sm}$. Moreover, a finite frequency range $\nu_M < \nu < \nu_{loc}$ is expected to occur, in which the excitations are still extended but non-propagating states; ν_{loc} denotes the "mobility edge" for acoustic phonons. A range of higher temperatures T, $T_M < T < T_{loc} \equiv h\nu_{loc}/k_B$ with $T_M \approx 100 \div 150$ K, may then exist, in which another (probably less pronounced) "plateau"-like behavior could be observed in thermal conductivity (Sec. 10.7). For glasses of type I, one can experimentally estimate both $\nu^{(in)}_{IR}$ as the BP frequency ν_{BP} while $\nu^{(el)}_{IR}$ as the upper limit ν_M for the high-frequency sound (since it is still difficult to estimate these characteristic frequencies in terms of other measurable properties, the type of glass can hardly be established at present, independent of inelastic scattering experiments). It follows from above that for glasses of type I, the soft modes and their vibrational excitations, as well as inelastic scattering of acoustic phonons by the excitations are of basic importance for the boson peak in photons (light, X-rays) and neutrons scattering spectra, while elastic scattering of high-frequency sound becomes important only for the higher frequencies sound (above the boson peak).

On the other hand, for glasses of type II, elastic scattering of acoustic phonons is mainly important for the origin of boson peak while soft modes and related inelastic scattering of the phonons are rather unimportant.

As noted above, there are experimental data (see, e.g. [109, 192]) indicating that in amorphous solids, which do not exhibit a BP at moderately low $\nu = \nu_{BP} \sim 1$ THz, no anomalous thermal properties like the specific heat $C \propto T$ and thermal conductivity $\chi \propto T^2$ are observed at very low $T \lesssim 1$ K. This experimental finding can be understood as a pronounced correlation between dynamic and thermal properties of glasses at very low T and ν and those at moderately low T and ν. Since the common view at present is that the properties of glasses are determined at very low T and ν by tunneling-state, non-vibrational excitations, while, by harmonic vibrational excitations at moderately low T and ν, one can conclude, in general, that a pronounced correlation exists between the tunneling and vibrational dynamics in glasses. However, as follows from Chs. 7, 8 and 10, this correlation is naturally predicted at present in the soft-mode model of low-energy excitations in glass dynamics. Hence, one can conclude that the soft-mode model is a relevant mean-field model of the universal dynamic and thermal properties of glasses for both very low and moderately low T and ν. In this connection, the question of whether some amorphous materials like amorphous silicon (a—Si) and arsenic (a—As), which do not seem to exhibit the linear temperature dependence of specific heat at very low temperatures and the boson peak at moderately low frequencies, are true structural glasses hardly seems to be answered positively.

12

OTHER MODELS FOR GLASSES WITH HIGH FREQUENCY SOUND

In the present chapter, other recent theoretical models are briefly described in terms of main statements and results, for interpreting recent experimentally observed main features of inelastic photon and neutron scattering spectra (Ch. 6) [112, 147, 148], at wave-vector shifts q comparable with $q_0(\sim q_D)$ for the first sharp diffraction peak of static structure factor $S(q)$. Then, the statements and results are compared to those of SMM discussed in Chs. 7 and 10, for glasses of type I exhibiting both BP and HFS, as explained in Sec. 11.2. In other words, the most important features of scattering spectra in the glasses include a maximum of an excess non-Debye DOS of harmonic vibrational excitations as a BP and Brillouin peaks in the measured dynamic structure factor, which correspond to both Debye phonons and HFS excitations at $q/q_0 < 0.5$ and $\nu < 0.5\nu_D$ and are characterised by practically temperature independent widths $2\gamma(q) \propto q^{\varkappa}$ at $\varkappa \approx 2$, at least at low enough (but perhaps not too low) $\nu < \nu_{BP}$ and $T \ll T_D$. Atomic dynamics in SMM (Sec. 7.4), for times t in the range $\nu_D^{-1} < t < (\nu_{\min}^{(h)})^{-1} \lesssim 10^{-12}$s, can be interpreted as due to harmonic soft-mode vibrations in the sense that $\nu_{\min}^{(h)} < \nu \ll \nu_D$, where $\nu_{\min}^{(h)}$ is the finite lowest frequency of soft-mode vibrations. Recent molecular dynamics simulations also appear to show that harmonic vibrations of actually soft modes are essential for interpreting the experimental data for a series of glasses.

The basic quantity computed for optically isotropic amorphous systems like glasses is either the dynamic structure factor $S(\mathbf{q}, \nu) = S(q, \nu) = -\pi^{-1}$ Im $\sum_{i,j} \exp{[i(R_i - R_j)]} G_{ij}(2\pi\nu + i\eta)$, with $\eta \rightarrow +0$ and $q = |\mathbf{q}|$ the wave number, or the total vibrational DOS $J(\nu^2) = g(\nu)/2\nu$. Here, $G_{ij}(2\pi\nu + i\eta)$ is the i, j matrix element of retarded Green's function of the harmonic vibrations, in the site representation with R_i, the vector of ith site. Roughly speaking, the dynamic structure factor describes the spectrum of excitations generated by a plane wave with a given q, while the DOS describes the density of all vibrational excitations in the system. In particular, for infinitely narrow Brillouin peaks of Debye phonons described by a delta function $\delta(2\pi\nu - s_0 q)$, one gets $S(q,\nu) = f(q)\delta(2\pi\nu - s_0 q)$ with a weight $f(2\pi\nu/s_0) \propto 1/\nu$, and thus the frequency DOS is $g(\nu) = 2\nu J(\nu^2) = g_D(\nu) \sim \nu^2/\nu_D^3$. As well-known, typically weak interactions of low frequency thermal phonons with each other (inelastic, anharmonic, scattering), with other, non-acoustic, excitations (inelastic, resonant, scattering) and/or with static structure fluctuations (elastic, Rayleigh, scattering) give rise to broadened Brillouin peaks described by a regular pre-limit approximation of $\delta(X)$ at $2\gamma_{ac} \to 0$, e.g., a Lorentzian function $D_1(X) = \pi^{-1}(2\gamma_{ac})[X^2 + (2\gamma_{ac})^2]^{-1}$ with actually small finite width $2\gamma_{ac} \ll \nu$, e.g., anharmonic width $2\gamma_{ac}^{(an)}(\nu, T) \propto \nu^{\varkappa} T^{\lambda}$ at $\varkappa \approx 2$ and $\lambda \approx 2-3$ and/or Rayleigh width $2\gamma_{ac}^{(el)}(\nu, T) \propto \nu^{\varkappa} T^{\lambda}$ at $\varkappa \approx 4$ and $\lambda = 0$ [113]. Unlike such widths, for glasses the typically observed width of a Brillouin peak at moderately low temperatures is $2\gamma_{ac}(\nu, T) \propto \nu^{\varkappa} T^{\lambda}$ at $\varkappa \approx 2$ and $\lambda \approx 0$. The problem of interpreting the origin of the BP and phonon widths is discussed in the present chapter in terms of basic vibrational DOS $J(\nu^2)$ and width $2\gamma_{ac}$ of excitations under discussion or of the dynamic structure factor $S(q, \nu)$ accounting for both the dispersion relation and width of the excitations. It seems worth adding that in terms of $J(\nu^2)$, the BP is identified with the peak of this function while its Debye-like wings correspond to the Brillouin peaks for either low frequency sound below BP or HFS above BP.

Two models [193] and [196] (see also [197]) under discussion actually appear also to be "mean-field" theories based on both computer simulations in specific structure models and analytical calculations in a mean-field approximation. In the models, the major computed quantity is the dynamic structure factor $S(q, \nu)$ and the single excitation approximation actually is applied, $S(q, \nu) = S^{(1)}(q, \nu)$. As seen below, the characteristic variety of peaks in $S(q, \nu)$ contain both primary peaks, e.g., for standard sound frequencies $\nu_{ac} = s_0 q$, and a different, secondary peak that corresponds to large enough $q \approx 0.5q_0 \approx 0.5q_D$ and occurs around a frequency $\nu_{sc} \sim 1$ THZ, practically independent of q, which is considerably lower than the corresponding acoustic frequency $\nu_{ac}(q) \sim 5$ THz. The secondary peak exhibits properties similar to those of the boson peak [112, 147].

12.1 Theoretical mode-coupling model

In the mode-coupling model [193] for description of BP in glasses, exhibiting above HFS, the basic results have been obtained by assuming that the basic version of mode-coupling theory (MCT) of the evolution of atomic dynamics, dealing with simple ergodic liquid systems and varying control parameters, temperature T or mass density ρ, can be extended to low enough T (or high enough ρ) at which a non-ergodic, disordered, glassy structure exists. Actually (Sec. 3.3), MCT is based on a closed system of microscopic classical equations of motions for density correlation functions in a macroscopic system of interacting atoms (molecules) and approximately describes the liquid-glass transition with decreasing T (or increasing density ρ) as a dynamic transition [90]. The approach deals with condensed matter states, of which the (static) structure factor depends smoothly on the control parameters. In fact, recent extensive tests of MCT predictions have shown that the theory describes essential features of structural relaxation in liquids and of the liquid-glass transition. In a simplest version in which rare atomic hopping events are neglected, the equations of motion have been shown to exhibit a singular dynamic transition at a critical value $T = T_c$ (or $\rho = \rho_c$) considerably higher than the standard calorimetric liquid-glass transition temperature T_g (or at corresponding $\rho = \rho_g < \rho_c$), e.g., at $T_c/T_g \simeq 1.2$ for van der Waals systems and T_c/T_g is larger for network systems. In fact, T_g decreases logarithmically with increasing liquid cooling rate and in this sense is ill-defined, whereas the well-defined critical temperature T_c does not depend on the rate; for example, silica T_c is close to 3300 K while the lowest $T_g \simeq 1500$ K at the highest cooling rate. On the other hand, only the extended MCT, accounting for the rare atomic hops, changes the singular transition to a smooth crossover at $T \simeq T_c$, this finding being the major result of MCT obtained from simulations and numerical calculations of analytical formulae. Since the derivation of mode-coupling theoretical relationships was based on canonical averaging procedures while the system was in an arrested non-equilibrium state at $T < T_g$, the applicability of the relationships from a rigorous viewpoint was limited to only $T > T_g$. However, the above mentioned assumption on extending an appropriately modified mode-coupling theory for understanding relevant experimental data in the glassy phase at lower $T < T_g$ seems to be supported by the essential fact that related experimental data do not exhibit anomalies at T around T_g. In [193], the extended MCT was developed, with an appropriate closed system of equations for the glassy phase at $T \ll T_c$. Since the equations are mathematically very complex, most of the works are performed in schematical models in which the system of equations can be truncated for dealing with only a few correlation functions. It was found in some schematic models involving a single correlation function that, as far as the theory can be extended to the glass (non-ergodic) phase at $T \ll T_c$, the density fluctuation spectra in the glass phase exhibit a broad peak that is a superposition of harmonic non-acoustic oscillator spectra. It was also recognized in some works (see References in [193]) that the evolution of such a peak with varying control parameters was similar to what was measured for a BP of some glassy systems, e.g., aqueous LiCl. This conclusion was confirmed in other works in which solutions of schematic models were also applied for describing quantitatively spectra of some glassy materials. Thus the mode-coupling theory, which in the liquid phase usually describes only the long time limit of time correlators of local density $\rho(r)$, can be modified into a theory that describes the glassy phase atomic motions around "arrested" equilibrium positions and related excitations. From the viewpoint applied, the most important for interpreting the origin of the boson peak excitations seems to be that anomalous, non-acoustic, oscillations (AOP) are found in the spectra (see below). Since the equations are mathematically very complex, most of the work is performed in schematical models in which the system of equations can be truncated for dealing with only one or a few correlation functions. It was found in some schematic models that, as far as the theory can be applied to the (non-ergodic) phase at $T \ll T_c$, the density fluctuation spectra in the glass phase exhibit a broad peak that is a superposition of harmonic non-acoustic oscillator spectra. It was also recognized [218] that the evolution of such a peak with varying control parameters was similar to what was measured for a BP in some glassy systems, e.g., in aqueous LiCl, at $T \ll T_c$. This conclusion was confirmed in other works in which solutions of the schematic models were applied for quantitatively describing spectra of glassy materials. Thus, the mode-coupling theory, which in the liquid phase usually describes only the long time limit of time correlators of local density $\rho(r)$, was extended to become a theory of the atomic motions around "arrested" equilibrium positions, as well as of related excitations, in the glassy phase.

The basic quantity characterizing the equilibrium structure of the ergodic (liquid) system discussed in the general mode-coupling theory is the static structure factor $S(\mathbf{q}) = \langle|\rho_{\mathbf{q}}|^2\rangle$ for a given wave vector q, where $\langle\ldots\rangle$ is a canonical average, $\rho_{\mathbf{q}} = N^{-1/2}\sum_\alpha \exp(iqr_\alpha)$ is a Fourier transform of local density $\rho(r)$, $\alpha = 1, 2, \ldots, N$, N is the total number of particles oscillating around random, arrested, "sites" r_α and, for glassy, isotropic, systems, $S(\mathbf{q}) = S(q)$ with $q = |\mathbf{q}|$. The basic variables in atomic dynamics with increasing time t are the density correlators $\phi_q(t) = \langle\rho^*_{\mathbf{q}}(t)\rho_{\mathbf{q}}\rangle/S(q)$, of which a short-time asymptote is a simple function of a characteristic frequency Ω_q, at $\phi_q(t) \simeq [1-(\Omega_q t)^2/2]$. Then, the dispersion law for thermal density fluctuations can be described as $\Omega_q^2 = v^2q^2S^{-1}(q)$, where v is the particle thermal velocity; in particular, in the limit of small q one gets the standard acoustic dispersion relation $\Omega_q \simeq qs_T$, where $s_T = vS^{-1/2}(q)$ is the isothermal sound velocity. The exact equations of motion for the density correlators in the theory is derived to be:

$$\ddot{\phi}_q(t) + \Omega_q^2\phi_q(t) + \Omega_q^2\int_0^t m_q(t-t')\phi_q(t')dt' = 0, \tag{12.1}$$

where $m_q(t)$ is a correlation function of fluctuating forces, i.e., a functional of $\phi_q(t)$ (Sec. 3.3) [10]. The Fourier–Laplace transform $\phi_q(\omega = 2\pi\nu) = i\int_0^\infty dt\phi_q(t)\exp(i\omega t) = \phi_q'(\omega) + i\phi_q''(\omega)$ define the reactive part $\phi_q'(\omega)$ of the correlator and its dissipative part $\phi_q''(\omega)$, which are related by the Kramers–Kronig formulae. Moreover, $\phi_q(\omega) \sim S(q,\nu)/S(q)$ and Eq. (12.1) actually is equivalent to $\phi_q(\omega) = -1/\{\omega - \Omega_q^2[\omega + \Omega_q^2 m_q(\omega)]^{-1}\}$, while the well-known fluctuation-dissipation theorem relates the system's dynamic susceptibility $\chi_q(\omega)$ to $\phi_q(\omega)$, $\chi_q(\omega) = [\omega\phi_q(\omega) + 1]\chi_q^T$, where $\chi_q^T = S^{-1}(q)/(\rho\mu v^2)$ is the isothermal compressibility and μ is the particle mass. As a consequence, Eq. (12.1) is equivalent to the expression (see also

Eqs. (3.21)–(3.23)):

$$\chi_q(\omega)/\chi_q^T = -\Omega_q^2/[\omega^2 - \Omega_q^2 + \Omega_q^2 \omega m_q(\omega)]. \tag{12.2}$$

While in general, the kernel $m_q(t - t')$ can be expressed as a sum of a regular term and a mode-coupling term due to nonlinear interactions (cage effect), the regular term can be neglected as unimportant for the liquid–glass transition of interest. Then, by approximating wave-vector integrals by Riemann sums on an equally spaced wave-vector grid of M "cells", the factor $m_q(t)$, a functional of $\phi_q(t)$, can be approximated as (see Eq. (3.23))

$$m_q(t) \simeq \sum_{k,p=1}^{M} V_{q,kp}\phi_k(t)\phi_p(t). \tag{12.3}$$

Here, $1 \leq q \leq M$, the mode coupling coefficients $V_{q,kp}$ are positive and determined by the equilibrium structure and the variables $\phi_k(\omega)$ describe the dynamic susceptibilities $\chi_q(\omega)$ which, as well as kernels, etc., are M-component vectors, approximating at large $M(\gg 1)$ the continuous liquid systems under discussion. The resulting Eqs. (12.2) and (12.3), in fact, constitute a closed system of nonlinear, "self-consistent", equations that define a unique solution $\phi_q(\omega)$, or the related function $S(q, \nu)$, for all parameters $\Omega_q > 0$ and $V_{q,kp} \geq 0$. The solution has to provide the basic properties of the correlators which are positive definite functions related to non-negative spectra ϕ_q'' and smoothly depend on the system parameters Ω_q and $V_{q,kp}$. In [193], the results of the model under discussion are demonstrated for a modified hard-sphere system (HSS).

In fact, the original HSS model also contains anharmonic interactions between particles (hard spheres), which give rise to anharmonic dynamics. Then, for explaining harmonic vibrational phenomena like BP and HFS, the original HSS model has to be modified by introducing an appropriate cutoff value $r^*(<r)$ for interparticle distances r, in the original mode coupling coefficients, in such a way that harmonic dynamics becomes essential. The applied cutoff has the property that the modified MCT for the HSS model reproduces the result for a Lennard–Jones system and also gives rise to a dynamic crossover at a critical value $T = T_c$ for $\rho = const.$ Then, wave vectors $\mathbf{q}$ and numbers $q \equiv |\mathbf{q}|$ are considered up to a cutoff value $q^*(>q)$ (and they are discretised to $M = 300$ values). Unlike the liquid–glass transition and associated atomic dynamics characterized by very long, macroscopic, times and very low frequencies (e.g., by ν much lower than 1 THz) and by typical frequencies of microscopic vibrational dynamics, of the scale of the Debye frequency, intermediate frequencies $\nu \sim 1$ THz are characteristic of a BP and associated, transient, harmonic dynamics. For the transient dynamics, Eqs. (12.1)–(12.3) are numerically calculated for the modified model, giving rise to the function $\phi_q(\omega)$ that can be discussed by using dynamical (frequency/ time or wave number) windows of about two orders of magnitude. The particle (hard sphere) diameter d is chosen to be as the unit of length, $d = 1$, and the unit of time is chosen such that the thermal velocity of particles is $v_T = 2.5$. Moreover, the packing fraction $\varphi = \pi d^3 \rho/6$ of the particles and its critical value are used instead of the mass density ρ and ρ_c. The results of the calculations are considered for two typical examples with a "large" $q = q_1 \equiv 7.0$ and a "small" $q = q_2 \equiv 3.4$, at $q^* = 40/d \equiv 40 > q_1 > q_2$.

The main results for excitation spectra (actually, for single excitations) associated with the transient dynamics in the glassy phase of modified HSS model, calculated at $\varphi > \varphi_c$ with $\varphi_c = 0.516$, can be briefly described as follows. Two types of characteristic values of φ are considered as examples, $\varphi = 0.60 > \varphi_c$ and $\varphi = \varphi_c(1 + 10^{-n/3}) = 1.1\ \varphi_c$ at $n = 3$. According to MCT [11](see also [90]), at $\varphi > \varphi_c$ each particle is localised in a tight effective cage formed by its neighbours due to existing nonlinear interactions of density fluctuations. In the case at $\varphi = 0.60$, the tight cage is such that the square root of the long time limit of squared displacement, $\delta r = \sqrt{\delta r^2(t \rightarrow \infty)}$, is found to be only 5% of the particle diameter, $\delta r = 0.05 \ll 1$. Then, a particle is shown to jump in its cage with average frequency $\nu \approx v_T \cdot (4\delta r)^{-1}$ and thermal velocity v_T. The estimations can explain qualitatively the existence of a maximum position of $\phi''_q(\nu)$ while no acoustic phonon is found with the linear dispersion law $\nu \approx s_0 q/2\pi$, at a large $q = q_1$. This can be interpreted as follows: while no Brillouin peaks occur with resonances at $\nu_q^{\max} = s_0 q/2\pi$, the above-mentioned AOP excitations peak is present in the spectra. The AOP excitation shape differs qualitatively from a Lorentzian expected for weakly damped harmonic acoustic oscillations: a low-frequency wing of the peak decreases with decreasing ν weaker than its high-frequency wing, so a threshold appears near a characteristic smallest harmonic vibration frequency $\nu_{\min}^{(h\nu)}(\simeq 40/2\pi)$, below which the spectrum is more flat than a Lorentzian one. In the other case at $\varphi = \varphi_c(1 + 10^{-n/3}) = 1.1\varphi_c$ at $n = 3$, the cage is wider, $\delta r = 0.07$, and the oscillation frequency is smaller than those at $\varphi = 0.60$, but also a downward shift of ν_{AOP} occurs. These trends at $q = q_1$ continue if φ further decreases at $\varphi > \varphi_c$, so that the threshold at the vibrational spectrum minimum value for low ν are replaced by a central peak at $\nu < \nu_{\min}^{(h\nu)}$, and the AOP maximum position ν_{AOP} at $\delta r \approx 0.099$ becomes buried under the central peak tail, showing up rather as a shoulder. On the other hand, at a small $q = q_2$ in the glassy phase the spectrum function $\phi_q''(\nu)$ exhibits weakly damped oscillations with almost Lorentzian resonances, the related excitations being qualitatively similar to Brillouin peaks at $\nu = \nu_q^{\max} = s_0 q/2\pi$ for acoustic phonons in crystals. For such a phonon, the frequency decreases with decreasing $\varphi(> \varphi_c)$, corresponding to a glass softening. At the same time, at small $q = q_2$ it is shown that the Brillouin peak positions are located considerably above the AOP maximum position ν_{AOP} and that the phonon resonances do not exhaust the spectrum $\phi_q''(\nu)$, being rather imposed on top of a background. The background appears to exhibit a threshold similar to that at $q = q_1$. It is also found from the calculations for small q that the resonance width exhibits the same wave number dependence expected for sound in an elastic continuum, $2\gamma_{ac}(q) = Dq^2$, but the parameter D does not depend on T. At the small $q(\lesssim 0.1 q_D)$, a single peak of nearly Lorentzian shape exhausts the whole inelastic spectrum $\phi_q''(\nu)$ and the resonance position $\nu_q^{\max}$ is found to follow the acoustic phonon linear dispersion law, $\nu_q^{\max} = s_0 q/2\pi$. On the other hand, the single-peak shape of the phonon and its linear dispersion law are found from the calculations to hold true for all wave numbers q up to about half Debye wave number, $q \lesssim q_D/2$. In addition, the calculated resonance width $2\gamma_{ac}(q)$ is shown to be larger than expected from its behaviour for the small q, $2\gamma_{ac}(q) = D \cdot (2\pi\nu_q^{\max}/s_0)$ when extrapolated to large q. Moreover, it is also found that $\nu_q^{\max} \gtrsim \nu_{AOP}$ at larger $q \gtrsim q_D/4$. For the latter, it is argued that a crossover of Ioffe–Regel type at ν around ν_{AOP} can exist and a hybridisation of high frequency sound (HFS) excitations with AOP ones can become strong. As q increases, a flat spectrum is found to be formed between the low frequency threshold mentioned above and the resonance at $\nu_q^{\max}$. However, for q exceeding $q_D/2$, the spectrum is shown to exhibit two peaks: the lower one can be related to the AOP, whereas the higher can be associated with the continuation of HFS. This two-peak spectrum seems to resemble the spectrum of Eq. (10.4) in SMM, with two branches of frequency levels of which repulsion from each other gives rise to the formation of the inter-branch gap.

Hence, the major result of the MCT model is that the density fluctuation spectra exhibit only Brillouin, or primary, peaks at small $q(< 0.1q_D)$ for Debye phonons, while at large $q \approx q_D/2$ both primary peaks and a secondary AOP exhibit properties similar to those described by experimental data for BP and q^2, i.e., ν^2 dependence $q^2(\propto,\nu^2)$ of the phonon width $\gamma_{ac} \propto \nu^2$. In fact, the properties of an AOP include the following 1. its maximum frequency is considerably lower than the Debye one, corresponding in this sense to soft modes; 2. it is a superposition of harmonic oscillator spectra of various frequencies, of which the distribution has a finite lower value $\nu_{min}^{(h)}$ characteristic of the soft modes; 3. the AOP is broad and asymmetric. Therefore, the theory of AOP can provide a basis for a first principle explanation of a BP in glasses that exhibit a HFS at frequencies than ν_{BP} and, on the other hand, shows similarities with the SMM theory of the BP and HFS. In addition, as the packing fraction φ of the modified HSS model decreases, the MCT model predicts that the frequencies ν_{AOP} and $\nu_{min}^{(h\nu)}$ decrease, tending to zero at the critical value $\varphi = \varphi_c$. In other words, the model predicts that the BP and related vibrational properties of the glass vanish when the temperature reaches the critical value T_c corresponding to the dynamic glass-to-liquid transition, considerably above the temperature T_g of the calorimetric transition.

Another result of simulations in MCT model, which can be in accordance with experimental data in Ch. 6, seems to be that the qualitative behaviour of acoustic phonon width can be approximated by a relationship $2\gamma_{ac}(q = \nu/s_0) \approx Dq^2 \propto \nu^2$, at low enough q ($\ll q_D$), where the parameter D does not substantially depend on T. The independence of acoustic phonon damping of T (or its weak dependence on T) can be interpreted as determined by inelastic scattering of the phonon due to absorption or emission from AOP vibrations related to sound excitations. In this sense, the qualitative meaning of phonon width in MCT model, which appears to be similar to that in SMM, might perhaps support the interpretation of acoustic-frequency-width dependence on the sound frequency in Sec. 10.6. Unfortunately, an analytical derivation of the explicit expression of the width of both low frequency and high frequency sound, which explicitly takes into account the suggested damping mechanism and describes its features, does not yet seem to be presented so far.

12.2 Theoretical random-matrix model

Analytical studies of harmonic vibrations in glasses can be related to the general problem of statistical properties of large random dynamic matrices for the vibrational Hamiltonian and thus are very complex. Usually, perturbation theoretical expansions are applicable for analyzing the harmonic vibrations, if a solvable part of the Hamiltonian can be singled out by some physical and/or mathematical arguments while the remaining part can be considered as a "small" perturbation. This seems to be the case in the soft-mode model (Chs. 7 and 10), in which the solvable part implicitly accounts for some topological features (e.g., soft modes) of glasses and, generally speaking, is not related to properties of a crystalline lattice [219] (see also [196]). As noted in the work under consideration, the simplest perturbation theoretical approximations for analyzing the vibrational Hamiltonian are developed in cases in which the solvable part is in fact associated with some features of a crystalline lattice, while perturbative expansions in powers of disorder parameters are applied and can be partially resummed. Unlike such "on-lattice" models of vibrations in disordered systems, there are also vibrational systems like glasses, where "off-lattice" random matrices are essential, in which a simple separation of a random dynamic matrix into an ordered part and a small disorder related perturbation can hardly be realized. In this respect, alternative theoretical approaches for analytical studies of harmonic vibrational spectra in glasses have recently been developed, best suited for low densities like the liquid phase of the system, with emphasis on the DOS rather than the dynamic structure factor $S(q, \nu)$. A recent efficient approach for investigating dynamics of a glass, i.e., glassy phase of the system, appears to be based on analytical and numerical studies of random matrices [219]. The problem was to find a relevant analytical description of $S(q, \nu)$, of its Brillouin peaks and secondary peak identified with boson peak, as well as of the related vibrational DOS. Let us also add that the origin of BP as an unavoidable dynamic property of glasses was also analysed in a different theoretical "mean-field" model [197], in which the peak is suggested to be related to the occurrence, besides the band of acoustic phonons, also of another band of excitations, "glassons", which rather seem to be of non-acoustic nature, and to the hybridisation of both types of states with each other. At low temperatures, the "glasson" band is found to develop a gap which appears to show up as the BP in the reduced DOS $g(\nu)\nu^{-2}$. In this connection, it seems that there can be a qualitative similarity also between the properties of "glasson"-phonon systems and those of soft-mode-phonon systems.

The random-matrix model was applied for particles harmonically oscillating around their disordered mechanical equilibrium positions at which the forces on particles vanish. For the sake of simplicity, it is assumed that harmonically interacting particles are identical ones with a mass m and move along one direction u, $x_j(t) = x_j^{\text{eq}} + u\varphi_j(t)$, in a scalar model with collinear displacements $\varphi_j(t)$; the equilibrium positions are described by $j = 1, 2, \ldots, N$. Then, the vibrational potential of the system is described as

$$V(\{\varphi_j\}) = \frac{1}{2}\sum_{i,j} f(\mathbf{x}_j^{\text{eq}} - \mathbf{x}_i^{\text{eq}})(\varphi_j - \varphi_i)^2, \qquad (12.4)$$

where $f(x_j^{\text{eq}} - x_i^{\text{eq}}) \equiv f_{ij} = \frac{\partial^2 V}{\partial \mathbf{x}_i \partial \mathbf{x}_j}$ and the units are chosen such that the particle mass $m = 1$ and the characteristic vibrational frequency of the model is also unit. Most important is to analyse the spectrum of the dynamic matrix $\widehat{\mathrm{M}} \equiv \mathrm{M}_{ij} = -f_{ij} + \delta_{ij}\sum_{l=1}^{N} f_{jl}$ for harmonic vibrations under discussion. One can apply here the self-averaging concept (Sec. 10.2) that is shown to be correct for densities of states in disordered systems [204] and is argued to be a correct hypothesis for some other characteristics of the systems, e.g., the dynamic structure factor $S(q, \nu)$. According to the concept, in the thermodynamic limit the spectra obtained for different realisation of the disorder can be identified with those calculated by averaging over all disorder realisations. Since the model is defined by the "spring function" $f(x)$ and the probability distribution of random configuration variables $x \equiv \{x_j^{\text{eq}}\}$, $j = 1, \ldots, N$, the problem under discussion can be solved in a much more simple way than in the original version. Neglecting quantum-mechanical effects for the harmonic vibrations at "high" $h\nu \gtrsim T$ and accounting for only single (1) harmonic vibrational excitations applied in general for excitations of moderately low energies in the present review, the excitation frequency DOS $g(\nu)$ is shown to be formally related to $S(q,\nu) = S^{(1)}(q,\nu)$ in the limit of large $q = |\mathbf{q}| \to \infty$ by the formula

$$g(\nu) = k_B T\,(\nu^2/q^2) S^{(1)}(q \to \infty, \nu), \tag{12.5}$$

at $\nu^2/q^2 = \text{const}$ for acoustic-like excitations, while the dynamic structure factor can be described as follows:

$$\begin{aligned} S(q,\nu) &= S^{(1)}(q,\nu) \\ &= [Tq^2/(2\pi)^3\nu^2]\overline{\sum_n \delta(\nu-\nu_n)|\sum_i e_{n,i}\exp\left(i\mathbf{q}\mathbf{x}_i^{\mathrm{eq}}\right)|^2}. \end{aligned} \tag{12.6}$$

Here, $e_{n,i}$ is the ith component of the nth eigenvector of the dynamic matrix and the eigenfrequencies ν_n are square roots of the respective eigenvalues while the overline means averaging over all random configurations $\{x_j^{\mathrm{eq}}\}$. Since at very high q and acoustic frequencies $\nu \propto q$ the single excitation approximation may not hold, while multi-excitation complexes must be accounted for, the reliability of Eq. (12.5) for describing real glasses is still a problem requiring further studies. As, generally speaking, the configurations are correlated in a real glass (produced, e.g., by quenching a liquid) and accounting rigorously for the correlation is hardly realized at present, it is also argued in the work that a good approximation is to consider the configurations uncorrelated but to apply an effective interaction obtained by substituting an effective spring function $F(x) = f(x)g(x)$ for $f(x)$, where $g(x)$ is a real (e.g., experimental) radial distribution function of the glass in question. At a given effective spring function $F(x)$, the only free parameter of the model is the density ρ. The dynamic structure factor $S(q,\nu)$ can be computed in terms of the system Green's function,

$$G(q,z) = \sum_{jl} \exp\left[i\mathbf{q}(\mathbf{x}_j^{\mathrm{eq}} - \mathbf{x}_l^{\mathrm{eq}})\right](z - \widehat{M})_{jl}^{-1}, \tag{12.7}$$

at $z = \nu^2 + i\eta$ and $\eta \to +0$, while one can see from Eq. (12.5) that

$$S(q,\nu) = S^{(1)}(q,\nu) \approx -k_B T q^2\,\mathrm{Im}\,G(q,z)/2\pi^2\nu \quad \text{at } \eta \to +0. \tag{12.8}$$

In the limit of infinite density, the vibrational spectrum was observed [112] (see also [196]) to coincide with the spectrum $\epsilon(q) = \rho[\widehat{F}(0) - \widehat{F}(q)]$ of an elastic homogeneous medium, as $G_{\rho\to\infty}(q,z) = [z - \epsilon(q)]^{-1}$, where $\widehat{F}(q)$ is the Fourier transform of $F(x)$. A systematic perturbation-theoretical expansion in powers of a "small parameter" ρ^{-1} can be derived in an appropriate region, where ρ is large enough. The aim is to calculate $G(q, z)$ by resumming an infinite subset of terms expected to give the most important contribution to the expansion at appropriate large $\rho = O(z)$ and then by extending the result to smaller values of the density, closer to empirical ones. The terms are described by so-called cactus diagrammes and the result can be presented in a well-known form:

$$G(q, z) = [z - \epsilon(q) - \sum (q, z)]^{-1}, \tag{12.9}$$

where $\nu(q) = \epsilon^{1/2}(q)/2\pi \propto q$ is the dispersion relation for a hydrodynamic sound. At $\rho \to \infty$ the self-energy can be described as $\sum (q, z) = \sum' (q, z) + i \sum'' (q, z) \to 0$, while $S(q, \nu) = S^{(1)}(q, \nu)$ describes the Brillouin peaks for standard acoustic phonons. Then, the real part of self-energy, $\sum' (q, z)$, at a finite ρ renormalises the Brillouin peak frequencies, i.e., their dispersion relation $\nu_{\text{renorm}}(q)$ that remains linear in q, as physically expected. On the other hand, the imaginary part calculated at the renormalised frequency determines the finite width $2\gamma_{\text{ac}}(q)$ by the relationship $\sum'' (q, \nu_{\text{renorm}}(q) + i\eta)_{\eta \to +0} = \nu_{\text{renorm}}(q) = 2\gamma_{\text{ac}}(q)$, at $2\gamma_{\text{ac}}(q) \ll \nu(q)$ for well-defined vibrational excitations. The sum of all cactus diagrams is shown in the theory to be determined by the solution of a self-consistent integral equation:

$$\sum (q, z) = \rho^{-1} \int \frac{d^3\mathbf{q}}{(2\pi)^3} \frac{\rho[\widehat{F}(0) - \widehat{F}(q)]^2}{[z - \epsilon(q) - \sum (q, z)]}, \tag{12.10}$$

presented here for real $3d$ glasses under discussion. As noted in the above references, Eq. (12.10) provides a derivation of the q^2 dependence of the Brillouin peak width in $S^{(1)}(q, \nu)$, which is independent of the force function details for a Debye-like spectrum with $J(\nu^2) \propto \nu^{1/2}$, i.e., $g(\nu) \propto \nu^{3/2}$, at low enough $\nu \ll \nu_D$ or $q \ll q_D$ where $\nu \propto q$ (Eq. (10.7)). Indeed, in the limit of small $q \to 0$, the main contribution $\text{Im} \sum(q, z)$ to the imaginary part of the integral in Eq. (12.10) was shown to be due to large $q' \gg q$. Taking this and Eq. (12.5) into account, one can straightforwardly obtain for a Debye like spectrum, with the basic vibrational DOS $J(\nu^2) \propto \nu^{1/2}$, that $\text{Im} \widetilde{\sum}(q, z) = -(\rho J(\nu^2)/4\pi) \int \frac{d^3q'}{(2\pi)^3} [\widehat{F}(q') - \widehat{F}(q - q')]^2 \propto \nu_{\text{renorm}}(q) q^2$

and thus the Brillouin peak width is found to become as follows:

$$2\gamma_{ac}(q) \propto q^2 T^0 \equiv q^2. \qquad (12.11)$$

In the approach under discussion, the dynamic structure factor, the associated excitation width $2\gamma_{ac}(q)$ and the reduced DOS $g(\nu)/\nu^2$, for single harmonic vibrational excitations have been calculated both by numerically solving the basic Eq. (12.10) resulting from so-called "cactus diagrammes" resummation and by simulations in a box of more than 5×10^5 particles, with the effective spring function described by the Gaussian-like model of $F(x) = (1 - \varsigma x^2)\exp(-x^2/2\sigma^2)$ at $\varsigma = 0$ and $\varsigma = 0.1$ for several values of $\rho\sigma^3 = 1, 0.6, 0.2, 0.1$ (σ is the length unit in the model). An important question in this approach is whether in the Gaussian-like model of spring functions, for the mechanically stable glassy phase with only non-negative vibrational eigenvalues ν^2 occurring at densities ρ larger than the critical value, the basic Eqs. (12.9) and (12.10) are able to give in the low frequency region a peak of the basic vibrational DOS $J(\nu^2) = g(\nu)/2\nu$ (or at least in the reduced DOS) and/or a secondary peak in $S(q,\nu) = S^{(1)}(q,\nu)$, with properties similar to those of a BP.

12.3 Comparison with the soft-mode model

As emphasized in SMM (Ch. 10), the origin of correlated boson peak at $\nu \approx \nu_{BP}$ (with a large width $2\gamma_{BP} \sim \nu_{BP}$) and high-frequency sound (with a narrow width $2\gamma_{BP} \ll \nu_{BP}$) is determined by a Ioffe–Regel crossover from a weak inelastic scattering of acoustic phonons by vibrational soft-mode excitations to a strong scattering and by the resulting strong hybridisation of both types of excitations, around a characteristic crossover frequency $\nu_{IR}^{(in)} \simeq \nu_{BP}$.

Although the relationships between the soft-mode model of harmonic vibrational excitations and two other models of the excitations have to be more specifically investigated, some qualitative similarities and differences, as well as some advancements and problems, of these mathematically different "mean-field" theories, can be summarized here. In fact, each model is able to play its role in understanding the origin of BP and HFS in the glasses so the models can in a sense be complementary to each other.

The SMM advancements can be described as follows.

(I) Non-acoustic low energy excitations, interacting with acoustic ones, in general are available and are identified as anharmonic excitations of soft modes (Chs. 7 and 10), becoming almost harmonic vibrational excitations at moderately low energies while being anharmonic, non-vibrational, tunneling excitations at very low energies (compare also Chs. 8 and 9). The common origin of non-acoustic and acoustic-like soft-mode excitations can naturally explain the experimentally observed correlation [192] of the anomalous dynamic and thermal properties of glasses at very low and at moderately low temperatures and frequencies: the anomalous low-temperature thermal properties are found to be missing in amorphous materials where no boson peak is found.

(II) SMM for harmonic vibrational excitations allows us, independently of the structure model, to assert that the excitations below and above the BP are propagating, i.e., Anderson extended, excitations in glass, whereas the BP excitations are related to strong hybridisation of acoustic phonons with extended, but not propagating, soft-mode excitations in IRC range for inelastic scattering. In particular, the theory appears to show that the harmonic excitations occur only at frequencies ν higher than a threshold at a finite lowest frequency $\nu_{\min}^{(hv)}$, at $\nu_m = T_m/h < \nu_{\min}^{(hv)} \ll \nu_{BP}$ (with $T_m \sim 1\,\mathrm{K}$ and $\nu_{\min}^{(hv)} \approx (0.1-0.2) \cdot \nu_{BP}$). In fact, the BP frequency is located in a range in which propagating excitations are described by an acoustic-like, linear dispersion relation, in accordance with the experimental data noted at the end of Ch. 6.

(III) SMM for harmonic vibrational excitations consistently describes not only the universal dynamic properties of glasses but also the correlated universal thermal properties, like the observed "plateau" and, above it, quasi-linear increase of thermal conductivity, with increasing temperature.

(IV) Since the model of anharmonic soft-mode excitations is a theory that generally is nonlinear in displacements while the theory of harmonic excitations is linear, one can see that the general nonlinear SMM theory gives rise, in particular, to results which are qualitatively similar to those of an intrinsically linear theory (see, e.g., Sec. 12.2).

At the same time, some questions in SMM still remain unanswered:

(1) what are the analytical expressions able to quantitatively describe the IRC for inelastic (almost resonant) scattering of acoustic phonons by soft-mode vibrations (cf. Secs. 10.2 and 10.4);
(2) what are the analytical expressions that can quantitatively describe the intensity and frequency of BP determined by the IRC, as well as the width of vibrational excitations in the HFS region, including the dependence on frequency (or wave number) fitting relevant experimental data (see, e.g., at the end of Chs. 6 and 10).

For the mode-coupling model, advancements can be characterized as follows.

(A) The model originally developed for describing the properties of a liquid near its transition to a glass, and the origin of the transition itself, is extended to glassy phase and also characterises the harmonic vibrational spectra, including the BP and HFS, in the experimentally important region of moderately low frequencies and wave numbers, $0.1 < q/q_D < 0.5$, which shows a finite minimum value and is still characterised by an acoustic-like linear dispersion relation $\nu \propto q$ (cf. above the statement (II) in SMM).
(B) The extended model may explain the formation of an effectively arrested distribution of density fluctuations in the glassy phase below the liquid-glass transition: the arrest is driven by fluctuations with wave numbers $q \approx q_{\max}$ at which the thermal liquid compressibility $\chi_q^T \propto S_q$ is most large at the position $q \approx q_{\max} \approx 2q_D$ of the peak of static structure factor S_q. Since, within the same approach the equations of motion for the density fluctuations are also derived, such a unified treatment of both structure and dynamics of a glass appears to be important, particularly, near the instability at the glass-liquid dynamic transition.
(C) The BP is determined by hybridisation of acoustic phonons with the non-acoustic AOP (anomalous oscillation peak) excitations found in this model, similar to what is noted in the statement (II) for SMM.

The mode-coupling model of BP origin in the glasses [193] appears to be qualitatively consistent with the basic concept of SMM [125] which is associated with a relevant probability distribution of harmonic vibrational soft-mode excitations around the Ioffe–Regel crossover (Eq. (10.8)) for inelastic phonon scattering. A question which remains unanswered in the mode-coupling model, concerns the specific range of the system parameters for which the extension of the model from the supercooled liquid phase to the glassy phase is valid. Taking this and related statements into account, one can assume that the anomalous oscillation peak (AOP) excitations in the mode-coupling model may play a role similar to that of harmonic vibrational soft-mode excitations in SMM. Moreover, the interactions and hybridisation of non-acoustic excitations, either with "mesoscopic wave-length" acoustic phonons in SMM or with "mesoscopic wave-length" density fluctuations in the mode-coupling model, may turn out to produce similar HFS excitations in glass. In addition, the independence of the sound damping mechanism of T (or weak dependence) is interpreted in the mode-coupling model as determined by quasi-resonant absorption/emission of sound excitations by/from AOP related vibrations. This is similar to the interpretation of the phonon width in SMM (Sec. 10.6) as due to quasi-resonant absorption/emission processes. Unfortunately, an analytical derivation of expressions for the width of low and high frequency sound excitations, which explicitly takes into account the suggested damping mechanisms, does not seem to be performed so far in both models.

For the random-matrices model, the basic advancement in the study of BP and HFS excitations appears to be the derived nonlinear integral equation (Eq. (12.10)), where the solution well approximates both the dynamic structure factor $S(q, \nu)$, and the vibrational DOS $g(\nu)$, in the single excitation approximation. This approach (in which the inverse density ρ^{-1} plays a role similar to that of temperature in the MCT model), appears to confirm that the phonon (Brillouin) peaks of $S(q, \nu)$ correspond to propagating sound waves in a glass and that the phonon peak width depends on q, i.e., on $\nu(\propto q)$, as $q^{\varkappa} \propto \nu^{\varkappa}$ at $\varkappa \approx 2$, for small enough $q \ll \pi/a_1$. For larger q, for which the model $S(q, \nu)$ can be transformed to a property similar to the reduced excitation DOS, the Brillouin frequency and width tend to become independent on q, so that the empirical secondary peak in $S(q, \nu)$ exhibits features of the observed BP at "intermediate" q for which the widths and frequency of Brillouin peaks still depend on q. In this model, as well as in SMM and the mode-coupling model, BP seems to be due to hybridisation of sound waves with high frequency non-acoustic modes due to inelastic scattering of both types of excitations from each other. Another result of the model is that the existence of BP can be related to the occurrence of a transition to a mechanical instability of glassy phase at a critical high inverse density ρ_c^{-1}, which is characterised by the appearance of imaginary frequencies of motion modes.

Some questions seem to remain open in the random-matrices model. One question is how the model could unambigously predict BP as a secondary low frequency peak in $g(\nu)/\nu^2$ at realistic glass densities ρ higher than the critical value ρ_c. Another problem is how to derive in this model (as well as in the mode-coupling model) the exponentially decreasing tail of BP at $\nu > \nu_{BP}$ observed in some glasses and is obtained in SMM (Eq. (10.24)) with the tail variation scale $\nu_t \approx 0.5\nu_{BP}$ in agreement with some experimental data noted at the end of Ch. 6. Moreover, both mode-coupling and random-matrices models describe only harmonic vibrations and thus are not related to anharmonic non-vibrational soft-mode excitations of very low energies, associated with atomic tunneling (tunneling states), and therefore can hardly explain the above-mentioned empirical correlation between the linear temperature variation of specific heat due to anharmonic excitations and the occurence of boson peak due to harmonic excitations in glasses.

In all three models under discussion, the excitation states do not seem to be Anderson localised states: rather, they appear to be either well-defined Anderson extended states, outside the BP region, or ill-defined extended states similar to diffusions, inside the BP region, as briefly explained in Sec. 10.2 (the rigorous proof that the excitations are truly well-defined Anderson extended states, outside the BP, does not yet seem to be published so far).

13

RECENT MODELS FOR GLASSES WITH NO HIGH-FREQUENCY SOUND

The soft-mode model of anomalous low-temperature and low-frequency dynamic properties, e.g., boson peak (BP), and thermal ones of glasses, exhibiting the high-frequency sound (HFS) above BP, can also be compared to recent theoretical models for glasses in which HFS excitations do not seem to appear above BP. Generally speaking, in SMM, the acoustic phonon width can be approximated by a sum of contributions of independent lowest-order elastic (el) and inelastic (el) scattering processes, $2\gamma_{\rm ac}(\nu) \approx 2\gamma_{\rm ac}^{(\rm el)}(\nu) + 2\gamma_{\rm ac}^{(\rm in)}(\nu)$, where elastic (el) is the phonon scattering from static disorder and inelastic (*in*) scattering is the scattering from soft-mode excitations. One can conclude from Secs. 10.1, 10.2 and 11.2 that the model of BP for the glasses under discussion rather corresponds to predominant elastic scattering, at $2\gamma_{\rm ac}(\nu) \simeq 2\gamma_{\rm ac}^{(\rm el)}(\nu) \gg 2\gamma_{\rm ac}^{(\rm in)}(\nu)$ and $\nu \approx \nu_{BP} = \nu_{BP}^{(\rm el)} < \nu_{BP}^{(\rm in)}$, where $\nu_{BP}^{(\rm in)}$ corresponds to a "virtual" BP frequency i.e., a possible inelastic Ioffe–Regel crossover frequency $\nu_{IR}^{(\rm in)}(\simeq \nu_{BP}^{(\rm in)})$ is actually not realised at $\nu_{BP}^{(\rm el)} < \nu_{BP}^{(\rm in)}$. Then, the acoustic phonon width in the glasses can also be due to elastic scattering of phonon from static disorder, determined by the Rayleigh frequency dependence $\nu^{\varkappa}$ at $\varkappa \simeq d+1$ (e.g., $\varkappa \simeq 4$ for a bulk glass at $d = 3$), unlike the phonon width in glasses, in which it is proportional to $\nu^{\varkappa}$ at $\varkappa \approx 2$, due to inelastic scattering either from extended soft-mode excitations in SMM or from AOP excitations in mode-coupling theory model.

13.1 Boson peak: Ioffe–Regel crossover at elastic acoustic scattering

One of the earliest models believed to contain the basic physics of BP model in glasses with no high-frequency sound [201, 220] was a harmonic crystalline lattice ($d = 3$) which contained spatially disordered spring-constants and took into account the elastic, Rayleigh scattering, from spatial fluctuations of longitudinal elastic constants in random static field [220, 221, 222] (see references therein). The advantage of such a system is that it can be analytically investigated by applying the well-known coherent-potential approximation (CPA) approach, as well as numerical calculations. In the model, the properties of scalar vibrational modes in a disordered system were compared to those of acoustic plane-wave phonons in the reference crystalline perfect lattice, as well as the role of vector vibrations in disordered lattices was studied in [223] for understanding the effect of disorder on both longitudinal and transverse harmonic vibrational modes. In particular, the analysis was performed for face-centred cubic (fcc) lattices, in contrast to models used in earlier investigations of scalar vibrations. As noted, the problem was considered for only harmonic vibrations, with dynamic matrix elements $D_{i\alpha,j\beta} = \frac{1}{2}k_{ij}(r_{ij})_{\alpha}\cdot (r_{ij})_{\beta}$, where k_{ij} is the spring constant connecting neighbouring atoms i, j and characterised by the equilibrium position vectors r_i and r_j, at $r_{ij} = (r_j - r_i)/|r_j - r_i|$; as usual, the matrix elements obey the sum rule $D_{i\alpha,j\beta} = -\sum_{j\neq i} D_{i\alpha,j\beta}$. The widely applied mean-field CPA was extensively applied for analysing this problem that was reduced to that of an effective crystal in which the atoms are connected by identical springs with an unknown dimensionless, complex spring constants, $\widetilde{k} = k_{ij}/k_0$, with k_0 (= 1 below) the average spring constant in the reference fcc crystal around which $\widetilde{k}$ are distributed in a disordered system. The dimensionless parameter can be found from self-consistent equation corresponding to a requirement that a solution for an effective crystal with one spring constant $\widetilde{k}$ chosen from a random set, averaged over the distribution of $\widetilde{k}$, coincides with the solution for a perfect effective crystal:

$$\{(k_{ij} - \widetilde{k})[1 - (k_{ij} - \widetilde{k})\alpha(\epsilon, \widetilde{k})]^{-1}\}_{av} = 0, \tag{13.1}$$

where $\{\ldots\}_{av}$ means the configuration averaging over the random k_{ij} while $\epsilon = \omega^2 = (2\pi)^2\nu^2$. Following [223] for such disordered systems, the distribution density $\rho(\widetilde{k})$ was supposed to be a rectangular box centred at $\widetilde{k} = 1$, with a half-width Δ. Applying other distributions, e.g., the Gaussian one [221], for a disordered system of classical harmonic oscillators coupled by random spring constants, or $\rho(\widetilde{k}) \propto \widetilde{k}^{-1}$[222], does not change qualitatively the results, at least for the scalar model of acoustic vibrations. The reference crystal properties completely define the function $\alpha(\epsilon, \widetilde{k}) = (2d/z\widetilde{k})\{(\epsilon/\widetilde{k})G^{\rm cryst}(\epsilon/\widetilde{k}) - 1\}$, where $G^{\rm cryst}(y) = \int_0^\infty J^{\rm cryst}(\epsilon)(y-\epsilon)^{-1}$ while $J^{\rm cryst}(\epsilon) \propto g^{\rm cryst}(\nu)/\nu$ is the total vibrational DOS of the crystal, z is the nearest-neighbour coordination number and d is the space dimensionality for vector vibrations ($d = 1$ for scalar vibrations). Another characteristic function is the total vibrational DOS $J^{\rm dis}(\epsilon)$ for the disordered system, which is described as $J^{\rm dis}(\epsilon) = -\pi^{-1}\,{\rm Im}\,\{\widetilde{k}^{-1}G^{\rm cryst}(\epsilon/\widetilde{k})\}$, where $\widetilde{k} = \widetilde{k}' + i\widetilde{k}''$ has to be calculated self-consistently from Eq. (13.1) for the fcc structure with the disorder characterised by the parameter Δ of the box distribution. By omitting calculation details presented in the quoted work, one can see that the disorder gives rise to broadening the acoustic band in the DOS and eventually to the disappearance of van Hove singularities [113] typical of crystals, increasing the dimensionless disorder parameter $\Delta(>0)$. In addition, it has also been found (for scalar acoustic vibrations) that the BP is shifted to much lower frequencies in the case of a distribution $\rho(\widetilde{k})$ weighted to smaller values of the spring constant. It is also suggestive from the calculation results [223, 224] that the BP in such systems is generically related to the lowest-frequency van Hove singularity in the crystalline spectrum. The results have been discussed in connection with the concept of Ioffe–Regel (IRC) crossover for elastic scattering of acoustic phonons (cf. Sec. 10.1), i.e., of the IRC from weak elastic scattering to strong one. Moreover, the trajectory of this IRC with varying disorder Δ has been found, by applying the IRC criterion, which in this case was described as $\nu_{BP}^{-1}(\nu_{IR}^{(\rm el)}) \cdot 2\gamma_{\rm ac}^{(\rm el)}(\nu_{IR}^{(\rm el)}) \approx 1$, with the peak position ν_{BP} comparable to the peak width $2\gamma_{\rm ac}^{(\rm el)}$. The weak-scattering region is found to exist at all frequencies for relatively small values of disorder, $\Delta < \Delta_0 \approx 0.5$, but only at low frequencies for high values of disorder, still less than the critical value $\Delta = \Delta^* = 1.296$ at which negative eigenvalues ν^2 first appear in the CPA results for the disordered lattice becoming mechanically unstable.

It was also established that for most values of spring-constant disorder in the fcc lattice, in the most important low-frequency high-disorder ($\Delta > 0.5$) regime, the BP frequency was close to the IRC frequency, $\nu_{BP}(\nu_{IR}^{(el)}) \approx \nu_{IR}^{(el)}$, at elastic scattering (similar to the frequency for IRC at inelastic scattering of acoustic phonons in Secs. 10.2 and 10.3). Then, it was suggested that the BP at $\nu_{BP} \approx \nu_{IR}^{(el)}$ arouse from general disorder-induced Rayleigh scattering, and related energy-level repelling from each other and states hybridisation in the systems with spring-constant disorder. Moreover, the peak was found to be generically related to the lowest-frequency van Hove singularity in the vibrational DOS of the crystal.

Another feature following from the CPA results for the vibrational DOS concerns the localisation of the vibrational excitations in the disordered system at the highest frequencies near the upper edge of the acoustic band. It is known (see [223] and references therein) that the mean-field CPA is unable to describe strongly localised states, for which the localisation length is comparable with the interatomic spacing. In this connection, the upper frequency limit of the acoustic band of the disordered fcc lattice calculated by the CPA can be considered as a measure of the 'mobility' edge of the vibrational localisation. Then, the IRC for elastic scattering, occurring at the characteristic frequency $\nu_{IR}^{(el)}$ much lower than the localisation ("mobility") edge, can be considered, generally speaking, as not related to the Anderson localisation of acoustic phonons. In other words, the low-frequency IRC for elastic scattering and the BP frequency are well separated from localisation edge frequency at which the transition from extended to localised vibrational states occurs. It also follows from the quoted works that precise results of the related numerical simulations for the extended excitations, acoustic phonons of low frequencies $\nu = \nu(q) \simeq s_0 q \ll \nu_D$, associated with the phonon attenuation due the disorder induced elastic scattering, is found to be proportional to $\nu^{d+1}(q) = \nu^4(q) \propto q^4$, i.e., to be caused by the usual Rayleigh scattering from the static density fluctuations, in bulk glasses ($d = 3$).

In the earliest version of a model for the alternative glasses [201, 220], it was found, in particular, that BP occurs at a frequency ν_{BP} near the frequency of the Ioffe–Regel crossover from weak to strong elastic scattering of acoustic phonons at the characteristic frequency $\nu_{IR}^{(el)}$, i.e., $\nu_{BP} \simeq \nu_{IR}^{(el)}$ (it was shown in the investigations that some crystalline materials also exhibit a non-Debye behaviour of the vibrational DOS due to the flattening of some vibrational dispersion curves, near appropriate van Hove singularities of the related crystal, the cristobalite polymorph of SiO_2 being an example with a peak in the DOS around 2 THz found in inelastic neutron scattering data). The BP positions in such glasses usually correspond to frequencies considerably lower than the van Hove-like peaks in related crystals [223].

13.2 Dynamic and thermal anomalies at elastic acoustic scattering

A different theoretical model of the anomalous harmonic vibrational dynamic properties and related specific heat and thermal conductivity of the glasses, not exhibiting HFS above BP, was suggested to be also based on CPA-like description of a simple cubic crystalline lattice of coupled harmonic oscillators with random scalar spring constants between nearest neighbour particles [221]. Recently, this theoretical model has been essentially generalized and improved by introducing an assumption that static disorder is due to spatial fluctuations of elastic constants, particularly of transverse ones (shear modulus), as described in [225, 226]. An elastic medium is considered in the model, with a given mass density ρ_0, elastic shear modulus G, bulk modulus $K = \lambda + \frac{2}{3}G$ (λ is the longitudinal Lame constant); the local longitudinal and transverse sound velocities are $s_l = [(K + \frac{4}{3}G)/\rho_0]^{1/2} = [(\lambda + 2G)/\rho_0]^{1/2}$ and $s_t = (G/\rho_0)^{1/2}$. It is suggested mainly that the shear modulus G (e.g., at constant elastic longitudinal modulus, $\lambda = \text{const}$) varies randomly in space, where $G(r) = G_0 \cdot [1 + \Delta G(r)]$ and the random $\Delta G(r)$ is a Gaussian fluctuation around G_0 with the "scattering parameter" γ_G describing the "degree of disorder". Applying standard field-theoretical techniques [225], one can derive the related self-consistent Born approximation for the (complex) self-energy function $\sum(\nu) = \sum'(\nu) + i\sum''(\nu)$, resulting in the set of equations:

$$\sum(\nu) = \gamma_G \sum_{q<q_D} [\chi_l(q,\nu) + \chi_t(q,\nu)]. \qquad (13.2)$$

Here γ_G is an effective "scattering parameter" while the longitudinal and transverse dynamic susceptibilities are

$$\chi_l(q,\nu) = q^2 G_L(q, z = \nu + i\eta)_{\eta\to+0}$$
$$= q^2\left\{\left[-(2\pi\nu)^2 + q^2\left[s_{l,0}^2 - 2\sum(\nu)\right]\right]\right\}^{-1} \quad (13.3)$$

and

$$\chi_t(q,\nu) = q^2 G_T(q, z = \nu + i\eta)_{\eta\to+0}$$
$$= \left\{\left[-(2\pi\nu)^2 + q^2\left[s_{t,0}^2 - \sum(\nu)\right]\right]\right\}^{-1}, \quad (13.4)$$

where $\chi_{l,t}(q,\nu)$ are the longitudinal and transverse susceptibilities of the system (the sum over $q < q_D$ gives rise to integration up to the Debye cutoff q_D) while $G_L(q,\nu)$ and $G_T(q,\nu)$ are single-particle Green functions for longitudinal and transverse acoustic phonons. The real part $\sum'(\nu)$ of the self-energy gives rise to renormalisation and dispersion of $s_{l,0}$ and $s_{t,0}$, the sound velocities in a system with no disorder, i.e., to $s_t^2 = s_{t,0}^2 - \sum'(\nu)$ and $s_l^2 = s_{l,0}^2 - 2\sum'(\nu)$, while the imaginary part $\sum''(\nu)$ describes the sound (phonon) attenuation. The applied self-consistent Born approximation is considered as the simplest form of an effective-medium model of the transverse disorder in question, in which the disorder effects give rise to frequency dependent sound velocities and there are only two essential parameters, γ_G and s_l/s_t. Calculations of the DOS $g(\nu)$ for harmonic vibrations in such systems, as well as of the energy diffusivity $D(\nu)$ describing the energy (heat) transport, can be performed by applying the approach of configurationally averaged Green functions $\langle G(r,r';\nu)\rangle$ and their products, based on the equations of motion for frequency dependent displacement vectors in the elastic medium under discussion. The calculations give rise to a mean-field theory for the single-particle Green functions $G_L(q,\nu)$ and $G_T(q,\nu)$ for longitudinal and transverse acoustic phonons and for the related self-energy $\sum(q,z)$, in which

$$g(\nu) = (4/3)\nu \sum_{q<q_D} q^{-2}\,\mathrm{Im}\{\chi_l(q,\nu) + 2\chi_t(q,\nu)\}, \quad (13.5)$$

where the obtained expressions for $\sum(q,\nu)$ and $G_{L,T}(q,\nu)$ generalise the self-consistent Born approximation of earlier version of Green functions not containing the transverse elastic disorder. Then, the system is shown to become unstable at a critical value γ_G^c of the scattering parameter γ_G. It is found that $\gamma_G^c = 1/6$ for longitudinal phonons is considerably less than $\gamma_G^c = 1/2$ for transverse ones, which is interpreted as an indication of a much higher sensitivity of transverse phonons to disorder than that of longitudinal ones. As assumed in [225], a BP occurs at a frequency at which an excitation mean free path becomes comparable to its wavelength. At $\gamma_G < \gamma_G^c$, the reduced DOS $g(\nu)/\nu^2$ describes excess vibrational states (compared to the Debye value) which exhibits a broad peak identified as a BP, of which the height is increased and the frequency is shifted towards lower frequencies as $\gamma_G \to \gamma_G^c$. It is emphasised in the model in question that the BP frequency is determined by the disorder parameter γ_G and therefore is not universal, in a probable contradiction with the result of SMM (Chap. 10) and with the experimental universality of BP frequency scale, $\nu_{BP} \approx 1$ THz.

A main result of the model is that the energy diffusivity of transverse elastic degrees of freedom calculated in the self-consistent Born approximation for transverse phonon scattering can be presented as follows:

$$D(\nu) \propto l_{ph}(\nu)/g(\nu)\nu^{-2} \propto D_0(\nu)/g(\nu)\nu^{-2}, \qquad (13.6)$$

where $l_{ph}(\nu)$ is the scattering mean free path calculated in a standard approximation while the usual kinetic expression for diffusivity is $D_0(\nu) \propto l_{ph}(\nu)$. For calculating the thermal conductivity (see Eq. (6.3) and related comments) with $D(\nu)$ substituted for $D_0(\nu)$, one has to take into account that the well-known divergency of the elastic, Rayleigh, scattering contributions to the phonon mean-free-path $l_{ph}(\nu)$ (and thus to $D(\nu)$), proportional to ν^{-4}, at low $\nu \to 0$, actually has to be cutoff by an appropriate inelastic scattering contribution. Since the latter is assumed in [225] to be due to very-low-energy TLS excitations (see Ch. 9), the results for the thermal conductivity obtained for the case of very low T, with $D(\nu) \to D_{eff}(\nu) = \{(D(\nu))^{-1} + (D_{in}(\nu))^{-1}\}^{-1}$ and $D_{in}(\nu)^{-1} \propto \nu$, are suggested to describe the experimental data in [225–227] better than the simple kinetic formula of Eq. (6.3) at $D(\nu) = D_0(\nu)$ in $D_{eff}(\nu)$. However, this suggestion does not seem to be as well clearly justified for moderately low T for which $D_{in}(\nu)$ is related to inelastic scattering of the phonons from soft-mode harmonic vibrational excitations, rather than to TLS of which the DOS is much lower (see also Sec. 10.7 for the thermal conductivity at moderately low T).

Moreover, the calculations of dynamic structure factor $S(q,\nu)$ of the system in the self-consistent Born approximation can give rise to the expression of width $2\gamma_{ac}^{(el)}(\nu)$ describing the attenuation of acoustic phonon of frequency ν, due to the transverse elastic disorder, for the glasses exhibiting HFS above BP. It also follows from the calculations that the excess (compared to the Debye DOS) vibrational DOS is qualitatively related to the width $2\gamma_{ac}^{(el)}(\nu)$ of phonon (Brillouin peak) in $S(q,\nu)$ for transverse acoustic phonons with $\nu = \nu_t(q) \propto q$ (see Eqs. (10.26), (10.27) and associated comments). In fact, for the transverse contribution (at $\chi_l(q,\nu) \to 0$)

$$\begin{aligned} S(q,\nu) &= \pi^{-1}[1 + n_B(T,\nu)]\,\mathrm{Im}\{2\chi_t(q,\nu)\} \\ &\approx \pi^{-1}[1 + n_B(T,\nu)](q^2/4\pi\nu)\cdot\gamma_{ac}^{(el)}(\nu) \\ &\quad\cdot\{[s_t(\nu)q - 2\pi\nu]^2 + [2\gamma_{ac}^{(el)}(\nu)]^2\}^{-1}, \end{aligned} \tag{13.7}$$

where $2\gamma_{ac}^{(el)}(\nu) = q^2\sum''(\nu)/\pi\nu \approx [4\pi\nu/s_t^2(\nu)]\sum''(\nu)$. From this equation one can determine the width of phonon as $2\gamma_{ac}^{(el)}(\nu) = q^2\sum''(\nu)/\pi\nu \approx [4\pi\nu/s_t^2(\nu)]\sum''(\nu)$. Since the DOS $g(\nu)$ also contains $\sum''(\nu)$, one can predict at $\sum''(\nu) < s_{l,t}^2(\nu)$ the following approximate relation, taking into account the contributions of both longitudinal and transverse phonons:

$$\nu_D[g(\nu) - g_D(\nu)] = f(s_l,s_t)\cdot 2\gamma_{ac}^{(el)}(\nu)\cdot\nu_D^{-1}, \tag{13.8}$$

where $f(s_l,s_t) = \pi^{-2}(s_t/s_l)^2[1 + (s_l/s_t)^4]$ and $n_B(T,\nu) = \{1 + \exp(h\nu/T)\}^{-1}$. Then, the validity of the approximation used in deriving Eq. (13.8) can be verified by comparing the excess DOS with the phonon width multiplied by the factor $f(s_l,s_t)$. One can see from the model [226] that the approximate Eq. (13.8) holds true, at least in the frequency range in which the quantities show approximately a ν^2 dependence, describing another main result of this model: the acoustic phonon attenuation due to elastic scattering from spatial fluctuations of elastic shear modulus and described by the width $2\gamma_{ac}^{(el)}(\nu)$ of phonon in $S(q,\nu)$, at $\nu = \nu(q) \propto q$, is found to be temperature independent and to increase with ν rather as $2\gamma_{ac}^{(el)}(\nu) \propto \nu^{\varkappa} \propto q^{\varkappa}$ at $\varkappa \simeq 2$ (at least at not too low frequencies, higher than the low frequency edge of the excess DOS, e.g., higher than $0.1\nu_{BP}$). In this case, the magnitude of excess DOS is also quantitatively connected with the phonon width, while all parameters (s_t, s_l, ν_D etc.) are macroscopic properties of system, which can be determined independently. Some empirical tests of Eq. (13.6) have been suggested, which were claimed to confirm the theoretical predictions of the model in question.

In the considered model, anomalous dynamic properties, described by Eqs. (13.7) and (13.8) in terms of transverse elastic degrees of freedom affected by static disorder, are characterised by predominant elastic scattering of transverse phonons, whereas the acoustic phonon width $2\gamma_{ac}(\nu) \simeq 2\gamma_{ac}^{(el)}(\nu) \propto \nu^2$ is similar to the phonon width in models for glasses exhibiting HFS above BP, in contrast to the standard Rayleigh scattering of phonons with $2\gamma_{ac}(\nu) \simeq 2\gamma_{ac}^{(R)}(\nu) \propto \nu^{d+1}$ (e.g., at $d+1=4$ for a bulk glass). On the other hand, one can see from Chap. 12 that the mode-coupling and random-matrices models, as well as SMM, describe glasses of type I characterized by Eq. (11.25) and inelastic scattering of phonons, rather than glasses of type II (Eq. (11.26)) with predominant elastic scattering. This contradiction can now be resolved by applying a more recent model [227] which appears to rigorously show that a topologically disordered network of harmonic springs and masses at nodes exhibits only Rayleigh scattering of acoustic phonons with width $2\gamma_{ac}(\nu) \simeq 2\gamma_{ac}^{(R)}(\nu) \propto \nu^{d+1}$ at $d=3$ (at least if the distance dependence of the springs is short-ranged enough with the related Fourier transform of correlation function $K(q)$ finite in the long-wave-length limit $q \to 0$). Then, the conclusion appears to be that the model describes only the above defined glasses of type II with elastic scattering of phonons from spatial fluctuations of longitudinal or/and transverse elastic moduli.

13.3 Boson peak due to spatially random springs constants

Another approach to the boson peak problem for alternative glasses under discussion, which does not seem to be much different from the two models discussed in Secs. 13.1 and 13.2, has recently been suggested in [222]. In this approach, two idealised disordered vibrational models are studied, in which random spring constants $k = k_{ij}$ for bonds between nearest-neighbour particles $(i,j = 1,2,\ldots,N)$ are subject to different distribution densities $P(k)$. Two types of $P(k)$ are discussed in detail. One function is $P(k) \propto k^{-1}$ above a cutoff value $k_{\min}^{-1} > 0$ which can be qualitatively justified by recollecting an earlier thermodynamic free-volume model (Sec. 3.2) of a glass, which considers that an additional, "free", volume is randomly distributed in the glass, compared to that of its crystalline counterpart. Another function $P(k)$ is a binary distribution in which the random spring constants can take only two different values, $k = k_A \equiv k_{\max}$ with probability p_1, $0 \leq p_1 \leq 1$, or $k = k_{\min} \ll k_{\max}$ with probability $p_2 = 1 - p_1$, and the lattice is in fact disordered at $0 < p_1 < 1$. With the binary distribution, the system can be described by a percolation theory [190], in which sites A are connected by the spring constants from clusters and an infinite percolation cluster arises above a critical concentration p_c, e.g., $p_c \approx 0.2488$ in simple cubic lattice. For both cases, the vibrational DOS and the related specific heat were calculated numerically, in a simple cubic lattice N particles of unit masses m_0 at nearest neighbour sites i and j , which are connected by spring constants k chosen from a distribution $P(k)$. The results of numerical calculations of equations of motion of the particles in this model give rise to the vibrational DOS $g(\nu)$ and related specific heat for both types of $P(k)$. The obtained results appear to show that the free-volume model related to $P(k)$ can qualitatively describe the existence of a maximum, of the reduced DOS $g(\nu)/\nu^2$, identified as BP at $\nu = \nu_{BP}$, and of the related hump of specific heat, whereas the binary model hardly can do that, for the alternative glasses considered in this section. A problem that has to be analysed in detail here is whether the vibrational normal modes of the system are Anderson extended states (like acoustic phonons) for all real eigenvalues $\nu_\alpha^2(<\nu_D^2)$, or such a critical eigenvalue ν_{loc}^2 is available that low-frequency modes at $\nu_\alpha^2 < \nu_{\text{loc}}^2 = const.(<\nu_D^2)$ are extended or diffuson-like ones in the BP region (see Sec. 10.2, Eq. (10.8) and comments, and Sec. 11.2), whereas higher frequency excitations at $\nu_{\text{loc}}^2 < \nu_\alpha^2(< \nu_D^2)$ are localised (e.g., when ν_{BP} is close to ν_{loc}). Anyway, the results of this model, as well as of other models considered in Secs. 13.1 and 13.2, must not give rise to the occurrence of high-frequency sound above the BP (unlike the results of the models discussed in Chs. 10 and 12).

13.4 Nakayama model: Boson peak vs. strongly localised modes

A view on the origin of boson peak (BP) in network glasses, which was essentially different from other models discussed above in Chs. 10–13 of different glass types, was presented and discussed by Nakayama and colleagues in a series of papers, including the review [228]. The basic difference of the latter from other above-mentioned models works was that the boson peak was assumed to be determined by well-defined strongly localized low-energy vibrational excitations (modes), whereas by ill-defined non-localised vibrational excitations of low energy in SMM.

In this connection, a simple physical model for network glasses has recently been proposed [229], taking into account the essential features, such as network structure and medium range order, and making use of center-of-mass system for describing the nature of the modes below the Debye cutoff-frequency ν_D. This model consisted of two main chains with constant mass M of molecular units, and those are connected to their nearest neighbours by linear springs with constant strength. The central hypothesis was that in the glasses there should be a certain number of extra vibrational states bounded to locally distorted potentials, which we attach to each main chain with mass M by linear springs of force constant k_i at site i. Moreover, M and k_i were related to characteristic frequencies $\nu_i(= k_i/M)^{1/2}$ of additional vibrational states; the parameters ν_i were random quantities distributed in the range between some $\nu_{\min}$ and $\nu_{\max}$. The distribution of ν_i arouse from additional potentials due to local distortions and strains in glasses (the molecular units connected by spring constant k_i do not necessarily correspond to the true molecular arrangement in glasses). The effective Hamiltonian can be expressed as follows:

$$H = \sum_i \left[\frac{P_i^2}{2M} + \frac{p_i^2}{2M} + \frac{K}{2}(Q_i - Q_{i-1} + a_i)^2 + \frac{k_i}{2}(q_i - Q_i)^2 \right], \tag{13.9}$$

where molecules have mass M, and Q_i and q_i are generalised coordinates describing displacements or changes of angle variables while the corresponding momenta are denoted by P_i and p_i, respectively. Here, capital letters denote quantities for main chains and small letters correspond to additional vibrational states, while a_i represents the term arising from internal strain which, can have the same spatial distribution as the additional vibrational states under condition $\sum a_i = 0$ expressing the balance of forces. The potentials leading to BP are assumed to originate from double-well potentials supposed to be the source of zero-point entropy and are expected to interact with each other via elastic dipole-field, $\propto r^{-3}$. This can be taken into account in the effective Hamiltonian, which can also represent a possible glassy state in a local minimum of the potential map in configuration space. With this Hamiltonian, the dynamic structure factor $S(q, \nu)$ was calculated for a system containing $N = 12{,}000$ molecules of equal masses $M = 1$ and spring constants $K = 1$, under periodic boundary conditions, are taken. The lower and upper limits for the distribution of frequency ν_i^2 are taken such that $\nu_{\mathrm{min}}^2 = 1/(16)^2$ and $\nu_{\mathrm{max}}^2 = 1/4\pi^2$. The force constant $K(= 1)$ should correspond to the largest stretching force constant k_i between molecules, and the force constant k_i should be smaller than K. For instance, the relevant force constant apart from k_i for network-forming structures is the bending one, which is much smaller than K. An important conclusion from calculations of $S(q, \nu)$ is that two bands clearly appear in the calculated spectra as observed in Raman scattering experiments. The lower band (peak), of which the width increases with wave number q, is almost independent of q (non-dispersive), while the higher band depends strongly on q and so contributing modes are "dispersive". In order to clear up the characteristics of modes contributing to these two bands, the density of states (DOS) $G(\nu)$ and localisation lengths $L(\nu)$ have been numerically calculated. The result was that a hump appears in the DOS around $\nu \approx 1$ THz, of which the energy range is the same as that of the lower band in the calculated $S(q, \nu)$. This hump, describing an excess DOS (as compared to the Debye one), can correspond to a hump in elastic neutron scattering parameter $G(\nu)/\nu^{d-1}$, which has been observed in elastic neutron scattering experiments and specific heat data. Moreover, the effective localisation length $L_l(\nu)$ was calculated as a function of ν. The result was that the modes for $\nu_{\mathrm{min}} < \nu < \nu_{\mathrm{max}}$ are strongly localised, while the modes in the lower frequency regime ($\nu < \nu_{\mathrm{min}}$) are weakly localised.

14

ANOMALOUS ELECTRON PROPERTIES OF SEMICONDUCTING GLASSES

Many electron properties of both metallic glasses (see, e.g., [114]) and semiconducting (non-metallic) glasses are anomalous in the sense that they are not characteristic of respective crystals, due to essentially different phenomena. Unlike the anomalies in metallic glasses related to electrons in Anderson extended (non-localized) states, the anomalies in semiconducting glasses are finally due to the occurrence of Anderson localised electron states in the mobility gap of electron energy spectrum, between conduction-band mobility edge, E_c^*, and valence-band one, E_v^*, at the gap width $E_g = E_c^* - E_v^*$. In the present chapter, the electron properties of semiconducting glasses are considered in phenomenological models based on some postulates, whereas in the next chapter the soft-mode model is shown to be relevant for explaining the anomalous electron properties of semiconducting glasses.

14.1 Basic experimental data

The electron properties of semiconducting glasses (SG), which often contain a significant concentration of transition (or rare-earth) atoms, are to large extent determined by the impurity atoms and by structural "defects" corresponding to a local violation of the short-range order in glass (e.g., of local coordination numbers). Experimental data concerning the electron properties and their theoretical analysis are described in many papers and several reviews, in particular, in [79, 190]. It is taken into account in what follows that the data and associated theoretical results are discussed in detail in two books mentioned. Two sets of electron properties of semiconducting (SG), largely chalcogenide glasses, are experimentally observed, which are hardly compatible with each other in the standard single-electron theory of semiconductors (see, e.g., [230]).

Actually, the activation energy W_σ of "interband" electrical conductivity $\sigma_e = \sigma_0 \exp(-W_\sigma/T)$ in both bands and the related position of single-electron Fermi level (per one electron) in the mobility gap, as well as the major photoluminescence peak energy $E_{\rm PL}$, weakly depend on impurity concentration c_i introduced at glass transition, as well as on temperature and external electric field F, at realistic $10^{-6} \lesssim c_i \lesssim 10^{-2} - 10^{-1}, T < T_g$ and $F \lesssim 10^6 V/cm$ at least. The Fermi level is pinned close to the midgap, $E_c^* - \zeta \simeq \zeta - E_v^* \simeq \frac{1}{2}E_g$, actually with $W_\sigma \simeq E_g/2 \sim 1$ eV.

On the other hand, the fundamental optical absorption band width $E_{\rm opt}$ is found to be close to $2W_\sigma \simeq E_g$ and the absorption edge is rather sharp, probably often sharper than in crystalline semiconductors (when subtracting a weak Urbach tail $\sim \exp[-\gamma^*(E_{\rm opt} - \hbar\omega)/\Theta)]$ for $\hbar\omega < E_{\rm opt}$, with $\gamma^* = const. < 1$ and $\Theta = T$ for $T > T_0 \approx \frac{1}{2}\hbar\omega_{\rm D}$, or $\Theta = T_0$, for $T < T_0$).

However, in the standard single-electron theory, the first set of properties corresponds to high electron density of states (DOS) $g(\overline{E})$ at the midgap $E = \overline{E}$ (with $E_c - \overline{E} = \overline{E} - E_v = E_g/2$), which provides Fermi level pinning, whereas the second set indicates a low $g(\overline{E}) \approx 0$. Just in this sense the two sets of properties are incompatible with each other.

Important relations have also been experimentally detected for the photoelectric threshold $E_{\rm PC}$, the photo-induced intragap absorption edge $E_{\rm PA}$ and the photoluminescence excitation threshold $E_{\rm X}({\rm PL})$ in SG. The following simple relations are found:

$$E_g \simeq E_{\rm opt} \simeq 2E_{\rm PL} \simeq 2E_{\rm PA} \simeq E_{\rm PC} \simeq E_{\rm X}({\rm PL}) \simeq 2W_\sigma, \qquad (14.1)$$

(with accuracy $\sim 0.1E_g$), so that the Stokes shift in photoluminescence is large, $\delta_S^{\rm (PL)} = E_{\rm X}({\rm PL}) - E_{\rm PL} \simeq E_g/2 \simeq 1$ eV.

14.2 Negative-*U* centres: Anderson model

The apparent incompatibility of the properties mentioned above was removed by introducing the following fundamental postulate: stationary mobility-gap states (at $T=0$) can only (or mainly) be occupied by singlet ($\sigma=0$) electron pairs with a negative correlation energy,

$$U = E(2) + E(0) - 2E(1) < 0, \quad (14.2)$$

which are called negative-U states (centres) in the Anderson model [231]. Here $E(n)$ stands for a gap energy level occupied by n electrons ($n=0,1,2$) and σ for the spin. The interelectron attraction for $U<0$ is supposed to be related to a distortion of the local atomic configuration which results from the occupation of the gap state, and to the consequent energy gain of the system.

The possibility of realising a negative-U centre is illustrated in the model by applying the simplest model of a small polaron of typical size $\rho_{sp} \simeq a_1 \sim 3$ Å [232, 233, 234], in which the local distortion is described by a single configuration coordinate x. In this model the result is that $E(n) = nE_i - n^2\varepsilon_p + U_c\delta_{n,2}$, with the negative-$U$ centre being a small bipolaron with $U = U_c - 2\varepsilon_p < 0$, if the usual Hubbard repulsion energy U_c [235] is smaller than the sum ε_p of two single-polaron energies (E_i stands for a bare, in the absence of the polaron effect, energy level).

Then, the Hamiltonian of such a simplified gap-state model is as follows:

$$\hat{H} = \hat{H}_e + \hat{H}_{ee} \equiv \sum_i E(\widehat{n}_i), \quad \hat{H}_e = \sum_i E_i\widehat{n}_i \equiv \sum_{i\sigma} E_i\widehat{n}_{i\sigma},$$
$$\hat{H}_{ee} \approx \sum_i U n_{i\uparrow} n_{i\downarrow}, \quad U_i \approx U = const. < 0, \quad \sigma \equiv (\uparrow\downarrow) \equiv \pm\frac{1}{2}, \quad (14.3)$$

where $\widehat{n}_{i\sigma}$ stands for the occupation number operators with eigenvalues $n_{i\sigma}=0$, 1, and i denotes the site. It is supposed in this model that the statistically independent "bare" energy levels E_i and correlation energies U_i in a random lattice (or continuous random network) are characterised by a broad, continuous and significant DOS $\rho(E_i)(\approx 10^{19} - 10^{20}\mathrm{eV}^{-1}\mathrm{cm}^{-3})$ and a narrow distribution $f(U_i)$, practically $f(U_i) \approx \delta(U_i - U)$ with $U_i \simeq U = const. < 0$. The Hamiltonian (14.3) is considered to be relevant only for states (ψ_i, E_i) for which the redistribution of the occupation numbers $n_i(=0,1,2)$ is due to weak intersite transitions and slow enough that the redistribution times τ_i are large, $\tau_i \gg \omega_D^{-1}$.

The gap states, are either occupied by singlet electron pairs (negative-U centres) for energies $E(2) < 2E(1) \simeq 2\zeta$ or empty for $E(2) > 2\zeta$, with ζ the single particle Fermi level. The result is that the ground state of the system is diamagnetic and that simultaneously the actual DOS (per one electron) $g(E)$ is high at the midgap, $g(\overline{E}) \approx 10^{19} - 10^{20}\,\mathrm{cm}^{-3}\mathrm{eV}^{-1}$ with $\rho(E_i)$ suggested (see above), as required for Fermi level pinning at the midgap. The negative-U centres appear to be considered in the model under discussion also as weak covalent bonds, corresponding to energies near the midgap due to a large topological disorder, and the centres with energies close to ζ are to be charged for pinning of the Fermi level, in GS. Then $|U|$ is assumed to be large, $|U| \sim 1$ eV, and comparable to $E_g \sim 1$ eV.

Thermal or optical generation of single-electron (single-hole) metastable excitations is considered to determine σ_e with activation energy $W_\sigma = E_{th} \simeq |U|$ or fundamental absorption band width $E_{\mathrm{opt}} \simeq 2|U|$, the respective gaps in the single-particle excitation spectrum for slow and fast (Frank–Condon) decay processes of negative-U centres. The fast processes are characterised by a polaron-type Stokes shift δ_S, $E_{\mathrm{opt}} = E_{\mathrm{th}} + \delta_s \simeq |U|$. These metastable excitations, with energies in a band of width $\approx U_c$ ($\ll |U|$), are supposed to be responsible for the photo EPR and photo-induced absorption with energy $E_{\mathrm{PA}} \simeq |U|$.

On the other hand, possible two-electron (or two-hole) excitations are not expected to be optically excited and rather contribute to σ_e, though their energy spectrum is gapless. However, the excitations are assumed to contribute to the anomalous properties of SG at very low $T \lesssim 1$ K, i.e. to the TLSs responsible for the anomalies. This contribution can be noticeable for a sufficiently large concentration $c^*_{\mathrm{ex}}(t)$ of two-electron excitations for which the tunnelling transition time is less than the observation time, $\tau_i < t$. Some other particular models of such "electronic" TLSs have been analysed. In [234, 236, 237] the contribution of the "electronic" TLSs is considered to be negligible, since largely $\tau_i > t$ for realistic times $t \lesssim 10^3$s and the very large $|U| \sim 1$ eV supposed in Eq. (14.3). However, this contribution is estimated to be noticeable, for a noticeable $c^*_{\mathrm{ex}}(t)$, in which different (and partly imprecise) approaches to estimate τ_i and $c^*_{\mathrm{ex}}(t)$ and structures of two-particle excitations were presented. The problem still seems far from being solved, since τ_i exponentially depends on the shape of the configuration barrier (the excitation structure), which is difficult to describe accurately enough. In any case, the electronic TLSs appear to be associated with significant atomic displacements for strong coupling of the excitations to the atoms of the random lattice (for $|U| \sim 1$ eV) and in this sense also can be treated as a part of the atomic TLSs in SG.

The concept of negative-U centres and the model of Eq. (14.3) allowed to get an answer, qualitative at least, to some basic questions concerning the experimental data for the anomalous electron properties of SG and has become important in this respect. At the same time, the following problems appeared in the model, which limited its applicability for an interpretation of the basic features of photoluminescence and some other electron properties of GS.

(I) The principal problem concerns the microscopic origin of negative-U centres with a very large $|U| \sim 1$ eV supposed, more specifically of the anomalously strong polaron-like effect actually assumed. The matter is that the usual polaron energy shift $\varepsilon_p \lesssim 0.1 - 0.5$ eV while $U_c \approx 0.1 - 0.5$ eV is estimated from both experimental studies of deep gap centres in crystals and computer calculations. Therefore, $|U| \ll 1$ eV for either $U = U_c - 2\varepsilon_p > 0$ or $U < 0$, so the small-polaron model can be used rather as an illustration, and experimental arguments are decisive in the model (14.3).

(II) It is also important to understand the origin of characteristic nearly discrete energy levels, like $\approx |E^*_{c,v} - \nu E_g/4|$ with $\nu = 1, 2, 3$, in the continuous spectrum of the mobility gap which manifest in some formulae above and correspond to a correlation between $|U|$ and E_g (in fact, $|U| \simeq \frac{1}{2}E_g$).

(III) Finally, it is worth revealing correlations between the negative-U centres and the principal features of the glass structure, which are responsible for the low-temperature properties of SGs.

14.3 Street–Mott and Kastner–Adler–Fritzsche models

An alternative, Street–Mott model of the negative-U centres in SG was proposed by Street and Mott in [238] and developed in [239, 240] and in many other works, in which some answers to the problems (I) and

(III) were presented. The basic idea is that the negative-U centres are related to specific structural "defects", which correspond to nearly discrete energy levels in a largely empty mobility gap. A "defect" is identified with a proper atom having a coordination number different from the normal z, $z'_{\pm} = z \pm 1$. This is considered to apply largely to atoms containing an electron lone-pair, which relatively easily transforms into an extra bond ($z'_{+} = z + 1$) or appears leaving a dangling bond ($z'_{-} = z - 1$), first of all to chalcogen atoms ($z = 2$). An overcoordinated atom is assumed to give rise to a positively charged "coordination defect" D^{+}_{z+1} with charge $e(D^{+}) = |e|$, while an undercoordinated atom gives a negatively charged "defect" D^{-}_{z-1} with $e(D^{-}) = -|e|$. In accordance with the above, D^{-} can be treated as a singlet electron pair ($n = 2$) on a dangling bond, with respect to D^{+} ($n = 0$) as a reference Hubbard state (with an extra bond), while D^{0} ($n = 1$) as a dangling bond containing a single electron.

Thus, the basis of the model is associated with specific chemical properties of proper atoms, rather than with a polaron effect and specific features of the glass structure [cf. the model of Eq. (14.3)]. It was supposed in the model under discussion that transformation of two neutral "coordination defects" D^{0} having a spin $\sigma = \pm\frac{1}{2}$ to a pair of alternatively charged spinless ($\sigma = 0$) "coordination defects" is similar to an exothermic chemical reaction [238, 241]:

$$D^{0} + D^{0} \rightarrow D^{+}_{z+1} + D^{-}_{z-1}. \tag{14.4}$$

The resulting energy gain plays here the role of an effective negative correlation energy of a "valence alternation pair" (VAP) [240]. The latter is a pair of alternatively charged "defects" which do not interact at large enough separation $R > R_g$, with $e^2/k_0 R_g = T_g$ (k_0 is the dielectric constant of the material). A pair of "defects" attracting each other at $R < R_g$ is called an intimate VAP, or IVAP. The properties and effects due to VAP's and IVAP's are different in some respects. VAP's are characterised by the charges $e(D^{+})$ and $e(D^{-}) = -e(D^{+})$, and at a sufficiently high number density N_{VAP} the electron states of the "defect" pairs can determine pinning of the Fermi level at the midgap, if one additionally assumes that $|U| = const. \simeq \frac{1}{2}E_g$, the pinning $\zeta \simeq \overline{E}$ is here determined by the transformations $D^{+} \rightarrow D^{-}$ (or $D^{-} \rightarrow D^{+}$) due to electron transitions between the "defects" D^{+} (D^{-}) and the impurity atoms introduced at the glass transition. The situation differs from that in the model of Eq. (14.3), where $U < 0$ for appropriate individual sites, and the pinning of the electron system Fermi level ζ around the mobility gap middle is associated with the high electron DOS in the continuous energy spectrum in the gap. On the other hand, practically neutral IVAP's, at realistic $R_g \lesssim 10\text{Å}$, possess significant electric dipole moments $\approx(10-1)$ $D(< |e|R_g)$, but generally speaking do not cause Fermi level pinning at the midgap, even with the relation $|U| \simeq \text{const} \simeq \frac{1}{2}E_g$ introduced [242].

In accordance with [243], the law of acting masses for a transformation reaction VAP⇄IVAP at the glass transition rather results in the predominance of either long-lived VAPs, at $Q \equiv N_{\text{IVAP}}(N_{\text{IVAP}} + N_{\text{VAP}})^{-1} \ll 1$, or long-lived IVAPs, at $Q \simeq 1$. The former case appears to apply to SG, with an actual generation energy of VAPs $E_{\text{VAP}} \approx 1\,\text{eV}$ and $N_{\text{VAP}} \sim N_0 \exp(-E_{\text{VAP}}/2T_g) \approx 10^{16}-10^{17}\,\text{cm}^{-3}$ at least, even though $E_{\text{VAP}} > E_{\text{IVAP}}$. In this case both diamagnetism predominates in SG, in accordance with the experimental data, and the Fermi level pinning $\zeta \simeq \overline{E} = E_g/2$ occurs at an impurity concentration $c_i \sim N_{\text{VAP}}a_1^3(\approx 10^{-6}-10^{-5})$. The alternative situation ($Q \simeq 1$) can be characteristic of insulating glasses like a-SiO_2, for which pinning $\zeta \simeq \overline{E}$ does not occur.

In the model under discussion, the ground state ($D^{\pm}$) and metastable excited states ($D^0, D^{\pm}_{\text{ex}}$) of the VAPs (or IVAPs) are the local centres responsible for thermal (slow), optical (fast) and photo-induced effects [79, 238, 239]. A specific hypothesis was introduced by empirical arguments in the Street–Mott model on the positions of nearly discrete (weakly broadened) energy levels of different charge states of the "defects" (D^-, D^0, D^+) in almost empty mobility gap for both slow and fast single-electron transitions. A remarkable symmetry is assumed for both electron (D^-/D^0) and hole (D^0/D^+) occupation levels, with interlevel separations $\approx \nu E_g/4$ with $\nu = 1, 2, 3$, which is characteristic of all the GS in question. The photoluminescence centres are largely identified with excitations like $D^-_{\text{PL}} \equiv \{D^0 + e^-_{\text{ex}}\}$ containing a weakly bound electron e^-_{ex}.

Thermal quenching and fatigue of the luminescence are assumed in the above models to be related to the decay of centres, the weakly bound charge carrier passing to states in the band tail by either thermally activated tunnelling or excitation to the band. In addition, electronic transitions in IVAPs were assumed to give an extra branch of tunnelling TLSs with very low energy $\lesssim$1 K, which can contribute to the anomalous properties of GS at $T \lesssim 1$ K. This contribution seems, however, to be small, insofar as $Q_{IVAP} \ll 1$ and the concentration $c^*_{IVAP}(t)$ of IVAPs with relaxation time $\tau_i \lesssim t$ is small. The coordination defect models appeared to be fruitful for interpreting many experimental data concerning the electronic properties of SG. On the other hand, some problems also appear.

(i) The principal problem of this model concerns the general origin of negative-U centres and the role of polaron-type effects. The matter is that in an essentially quantum-chemical model it seems also necessary to suppose that the large Stokes shift in photoluminescence is associated with an anomalously large polaron energy shift $\varepsilon_p \simeq \frac{1}{4}E_g$. The latter is in fact taken into account in the introduced mobility-gap DOS. Thus, this problem again reduces to the problem of the origin of the anomalously large polaron-type effect in SG, e.g., in the model of Eq. (14.3).

(ii) Another problem concerns the nature of the slowly varying continuous tails of the mobility gap DOS, which seem to be typical in photoconductivity, with the energy spectrum introduced in the model.

(iii) An additional problem concerns the nature of the centres of different processes, like photoluminescence, photo EPR, midgap absorption and photoluminescence fatigue, which hardly reduce to only $D^{\pm}$, D^0 and $D^{\pm}_{PL}$.

Thus the basic problems in the abovementioned models are related to the origin of anomalously large polaron-type effect and basic features of mobility-gap DOS in SG.

14.4 Qualitative analysis of negative-*U* centres

A qualitative theoretical analysis of the abovementioned basic problems has earlier been proposed, which was based on the soft-mode model, not associated with a specific structural model of a glass, and seemed to allow the development of a consistent theory of localised electron states and related phenomena in SG [244, 247]. The theory seems to be able to unify in some aspects two alternative models considered above and to give rise to understanding the origin of anomalously large polaron-like energy shift assumed. The basic idea was that the localised states are related to electron self-trapping in local atomic soft configurations of the glass. The self-trapped (ST) pairs and their excited states in the mobility gap were assumed to be responsible for the anomalous electron properties in a SG (while only single-electron centres and related excitations are important in an insulating glass with a).

The interactions responsible for electron self-trapping in nonmagnetic materials are related to strong coupling of an electron (hole) to neighbouring atoms, which is conventionally characterised by a coupling energy Q_0 of the order of a few eV, $Q_0 \approx 1-3$ eV. Usually, however, these strong interactions occur to a small extent either for a significant average velocity of a (delocalized) electron or for a large local elasticity preventing the distortion of environment (for typical atomic spring constants $k \approx k_0$). Then, the self-trapping energy for an electron (energy gain of the system) W_1 and the related atomic displacement $u_p \equiv a_0 x_p$ are small, $|W_1| \ll \hbar^2/2m_e a_1^2 \sim 1$ eV and $u_p \ll a_0 \approx 1$ Å. The ST electron state in a glass differ in this respect from the well-known polaron states. In fact, the self-trapped states under consideration are formed due to the coupling of an electron in a "bare" (with no self-trapping) mobility-gap state ψ_q to a weakly bound "atom" in a soft configuration in the localisation region of small size $\rho_q \approx a_1$, with the "bare" energy level E_q in the corresponding band tail. The distribution density of random coupling parameter $Q(\rho_q)(>0)$ is supposed to be narrow in the sense that $Q_q(\rho_q \approx a_i) \approx Q_0$ [$\gg Q_q(\rho'_q)$ for $\rho'_q \gg a_1$].

A strong electron-atom coupling is expected to be typical here, because $\rho_q \sim a_1$ and the soft configuration is easily distortable (at $k \ll k^{(0)}$). Then, the atomic displacement $|x|$ is significant, comparable to one, even when a single electron (e.g., with spin $\uparrow$) occupies the state ψ_q. The associated lowering of the electron level $J_q(x) \equiv E_q(x) - E_q(<0)$ is also significant, being comparable to Q_0, whereas the increase of the atomic potential energy is relatively small, $\Delta V(x) \lesssim 0.1\,A \sim Q_0$ at $k \lesssim 0.1\,k_0$. The result is that the total energy of the system also significantly lowers by a value comparable to Q_0, and a ST electron state occurs with significant $|W_1| \lesssim Q_0$ and atomic equilibrium displacement $|x_1| \lesssim 1$. Adding an electron with antiparallel spin ($\downarrow$) gives rise to an increase of the atomic equilibrium displacement $|x_2|(>|x_1|)$ and the ST energy $|W_2|(>|W_1|)$ for a singlet electron pair (with total spin $\sigma=0$), and the largest possible values of $|x_2|$ and $|W_2|$ can be max $|x_2| \sim 1$ and max $|W_2| \sim Q_0$ (see below). In agreement with the general principles of the theory of ST states (see, e.g., [233, 245]), the energy E_n of the ST states can be described by the following formula:

$$\begin{aligned} E_n &= \min_{(x)} E_n(x) \equiv E_n(x_n) = nE_q + W_n + U_c\delta_{n,2} + const., \\ E_n &< nE_q + const., \quad W_n \equiv W_n(x_n) < 0, \quad U_c \equiv U_c(x_2) > 0, \end{aligned} \tag{14.5}$$

where n stands for the state occupation number, e.g., $n=2$ for a singlet electron pair. Moreover, $\min_{(x)} E_n(x)$ denotes minimisation of the adiabatic potential $W_n(x)$ of the system,

$$\begin{aligned} W_n(x) &= E_n(x) - nE_q - \text{const} = V(x) + nJ_q(x), \\ J_q(x) &\equiv E_q(x) - E_q < 0 \text{ for } |x| \neq 0, \end{aligned} \tag{14.6}$$

with respect to x at a fixed n; $U_c(x)$ denotes the interelectron Hubbard repulsion energy. Formulas (14.5) and (14.6) correspond to adiabatic situation for electrons in a state ψ_q, with the inequalities $\hbar\omega_D \ll Q_0 \ll A \simeq \frac{1}{2}k_0 a_0^2 \sim 30\,\text{eV}$ satisfied. It is important in what follows that the adiabatic potential $W_n(x)(n=1,2)$ remains to be a soft-mode potential, with renormalised parameters η and ξ [246]: $\eta \to \eta_n = (\eta + nq^{(2)})[1 + 3\lambda_n^{-1}(\kappa_n^2 + 2\kappa_n)]$, $\xi \to \xi_n = (\xi + nq^{(3)})(1+\kappa_n)$ and $q^{(j)} \equiv A^{-1}\{d^j J_q(x)/dx^j\}_{x=0} (\gtrless 0)$, $|q^{(j)}| \lesssim Q_0 A^{-1} \ll 1$, where κ_n stands for the root of the equation $\kappa^3 + 3\kappa^2 + \lambda_n\kappa + \rho_n = 0$, $\rho_n = 16nq^{(1)}(\xi + nq^{(3)})^{-3}$, and vanishes for $n=0$. In accordance with the above, the correlation energy of a singlet ST electron pair state $U \equiv U(\uparrow\downarrow) = E_2 + E_0 - 2E_1 = W_2 - 2W_1 + U_c$, where $W_n(<0)$ stands for the self-trapping energy ($n=1$ or $n=2$). Thus, an ST pair state is stable with $U<0$ for $U_c < |W_2 - 2W_1|$. Two situations can be distinguished for the relations found by solving the extremum problem in Eq. (14.5), with either a harmonic equilibrium atomic displacement $x_n = x_n^{(H)}$ [for $W_n(x_n) \sim A\eta_n x_n^2$] or an anharmonic one [for $W_n(x) \sim Ax_n^4$]. The resulting expressions can easily be found, and in particular for essential harmonic displacements those are as follows [125]:

$$W_n = W_n^{(H)}(k) \simeq -Q_0^2 n^2/2ka_0^2,$$
$$U = U^{(H)}(k) \simeq U_c - Q_0^2/ka_0^2, \quad \text{for } k_0 \gg k \gtrsim k_n^* \qquad (14.7)$$

and practically continuous spectra of random values $W_n(k)(<0)$ and $U(k)(\lesseqgtr 0)$, as well as of $k(0 \leq k \lesssim k^{(0)})$, are available. In particular, the formulae

$$W_2 \simeq 4W_1 \quad \text{and} \quad n|J_q(x_n)| : V(x_n) = 2 \quad \text{for } n = 1, 2, \qquad (14.8)$$

that are characteristic of conventional polaron states due to the linear electron-phonon coupling in a harmonic lattice, turn out to be relevant also for the ST states under consideration. The anharmonicity, due to nonlinear interactions resulting from expanding $J_n(x)$ in x, leads to deviations from relations (14.7) and (14.8), but the deviations turn out to be relatively small at typical $h\nu_D \ll Q_0 \ll A$.

The general criteria for a stable ST pair state with $U < 0$ can be as follows:

$$U_c < \frac{1}{2}\varepsilon_g(Q_0), \quad k < k^{(l)}, \quad Q_0 A^{-1} > \varphi(\eta, \xi), \qquad (14.9)$$

with $\varphi(\eta, \xi) \approx |\eta - \frac{3}{8}\xi^2|\xi$ for the most important, most numerous single-well harmonic soft-mode potentials. The criteria are satisfied for the majority of soft-mode potentials at typical values $U_c \lesssim 0.5$ eV, $Q_0 \sim 3$ eV and $A \sim 30$ eV, $U_c < \varepsilon_g(Q_0)$. That is why a significant concentration of pair states is expected. The negative-U centres under discussion are here identified with fully occupied stable singlet ST pair states ($n = 2$). Both electron $(2e)_0$ and hole $(2h)_0$ negative-U centres are available, while single-electron (or hole) centres can only be metastable, unlike the conventional ST polaron states for which the situation is usually opposite.

Another most important feature of ST states under discussion is an extremely large value of the pair self-trapping energy, $\max|W_2(k)| = |W_2(0)| \simeq \varepsilon_g(Q_0) \sim Q_0$. Two limit cases can be distinguished:

$$\varepsilon_g(Q_0) \gtrsim E_g \qquad (14.10)$$

or

$$E_g \gg \varepsilon_g(Q_0)(\gg U_c), \tag{14.11}$$

respectively, for a SG with $E_g \approx 1-3$ eV $\lesssim Q_0$ or an insulating glass with $E_g \approx 5\text{--}10\,eV \gg Q_0$.

In the case (14.11), the continuous spectra of values for $|W_2(k)|$ and $|U(k)|$ with $U(k) < 0$ turn out to be limited as follows:

$$|W_2(k)| \lesssim W_{\max} \equiv W_{2,\max} \simeq 4W_{1,\max} \simeq E_g \tag{14.12a}$$

and

$$|U(k)| \lesssim U_{\max} \simeq \frac{1}{2}W_{\max} \simeq \frac{1}{2}E_g \quad \text{for } U(k) < 0. \tag{14.12b}$$

This is a result of the well-known general quantum phenomenon of repulsion of close interacting energy levels [172] for states generated by the conduction and valence bands in the mobility gap. An electron energy level which is lowered at self-trapping, with increasing $|x|$, is repelled from the valence-band mobility edge E_v^* (and similarly for a hole energy level and the conduction-band mobility edge E_c^*). The interactions between gap states, originating from the two bands are related to contributions of random fields in a glass.

Different models for the description and analysis of this phenomenon have recently been developed [247, 248]. The Schrödinger equation for an electron in a two-band disordered system which undergoes self-trapping when the energy level is lowered, was investigated. In this equation,

$$\{E_q(x) - [E_q + J_q(x)]\}\psi_q(x) = \sum_p I_{qp}(x)\psi_p(x), \quad \text{with } |J_q(x)| \lesssim Q_0,$$

$$\sum_p |I_{qp}(x)| \ll Q_0, \quad I_{qp}(0) \equiv I_{qq}(x) \equiv 0 \equiv J_q(0), \tag{14.13}$$

a large part of the interband interaction matrix elements $I_{qp}(x)$, for transitions from a given state ψ_q of one band to states ψ_p of the other band, are finite. Then, Eq. (14.13) characterises a mixing of states of the two bands,

$$\psi_q(x) = c_q(x)\psi_q + \sum_p c_{qp}(x)\psi_p,$$

$$\psi_{q,p}(x=0) \equiv \psi_{q,p}, \quad c_{qp}(0) \equiv c_{qq}(x), \quad \text{with } |c_q(x)|^2 + \sum_p |c_{qp}(x)|^2 = 1. \tag{14.13a}$$

This mixing turns out to be essential for $\Sigma_p |c_{qp}(x)|^2 \sim |c_q(x)|^2$ in the region (x), where $E_q(x)$ approaches the terms $E_p(x)$ close to the mobility edge E^* of the other band. The "bare" DOS $g_0(E_p)$ for $E_p \approx E^*$ is not exponentially small, being in this sense significant, and the terms $E_p(x)$ weakly depend on x, $E_p(x) \simeq E_p \equiv E_p(0) \approx E^*(=E_v^*$ for an electron). This repulsion of the terms slows down the term lowering $J_q(x)$ for large $|x|$, with $|J_q(x)| \approx E_g$, while this lowering is nearly linear, with $|J_q(x)| \propto |x|$, for small $|x|$.

On the other hand, pair self-trapping occurs in all soft configurations, for $0 \leq\leq k < k^{(l)}(<k^{(0)})$, with the following upper limits $W_{\max}$ and $U_{\max}$ for the continuous spectra of $|W_2(k)|$ and $|U(k)|$ with $U<0$:

$$W_{\max} = |W_2(k=0)| \simeq \varepsilon_g(Q_0) \simeq 2U_{\max} \equiv 2|U(k=0)| \ll E_g. \tag{14.14}$$

The states (14.4) predominate for $\varepsilon_g \gg E_g$. However, the level repulsion is unimportant in the case at $U_{\max} \ll E_g/2$. Thus, the properties of the negative-U centres differ in two cases (14.10) and (14.11). In any case, the self-trapping under consideration is characterised by an anomalously large ST energy and equilibrium displacements due to the softness of the configuration.

Note that a triplet ST pair state, with spin $\sigma = \pm 1$, corresponds to an excited state of the singlet pair ground-state with a significant energy (per particle) $\varepsilon^*_{\text{ex}} \sim W_{1,\max}$ in the mobility gap, the difference of the correlation energies $|U(\uparrow\downarrow)| - |U(\uparrow\uparrow)| \sim |U(\uparrow\uparrow)|$ being due to a kind of positive exchange energy. That is similar to the well-known situation for the hydrogen molecule. As usual in such systems, with $E_2(\uparrow\uparrow) = E_2(\uparrow\downarrow) + 2\varepsilon^*_{\text{ex}}$, some effects depending on the state spin ($\sigma = 0$, $\sigma = \pm 1/2$ or $\sigma = \pm 1$) appear in phenomena related to the decay or generation of the pairs (recombination, trapping of charge carriers, photoluminescence, etc.).

Generally speaking, three basic types of negative-U centres can be distinguished, a negative-U centre being a fully occupied ST electron (or hole) pair state in the mobility gap ($n=2$). The three types correspond to different values of the charge of the negative-U centre $e^*_{\mp} = \mp l|e| : l=0$ (neutral centre), $l=1$ or $l=2$; the corresponding "empty" soft configuration ($n=0$) is characterised by the charge $e^*_0 = l_0|e|$ with $l_0 = l - 2sgn(e)$ (and $e<0$ for electrons or $e>0$ for holes). Generally speaking, changes of atomic coordination number do not seem to be necessary for the formation of negative-U centres.

As far as a dangling bond can be related to an atom in a significantly asymmetric double-well potential, the concentration c_{db} of these bonds is expected to be low, in the sense that $c_{db} \ll c_a$, so that $c(D^{\pm}) \ll c_2 \simeq \Sigma_{\pm} c(A^{\pm 2}) + c(A^0)$ in the theory under consideration, with c_2 the total concentration of negative-U centres. The empirical weak dependence of the electric properties of a GS on the concentration c_H of hydrogen atoms, which are able to cancel dangling bonds and related $D^{\pm}$ "defects", up to high $c_H \gtrsim 0.1$, also favours the predominance of $A^{\pm 2}$ and A^0 negative-U centres over $A^{\pm 1}$ centres (compare a-Si:H, where hydrogen atoms cancel dangling bonds and essentially change the properties of amorphous silicon). If also $c(A^0)$ is low enough, say $c(A^0) \ll 10^{-4}$, and $c_2 \approx 10^{-2} - 10^{-3}$ as estimated in Ch. 7, $A^{\pm 2}$ centres appear to be the principal negative-U centres in a GS, in particular in a-Se. This can give a clue for solving problem (iv) of the origin of these centres in a-Se and similar SG.

The $A^{\pm 2}$ negative-U centres, as well as the soft configurations in general, show small atomic spring constants ($k \ll k^{(0)}$), looking as structural "defects" which do not reduce to coordination defects (e.g., $D^{\pm}$). However, these "defects" essentially contribute to the proper structure of a glass.

15

SOFT-MODE MODEL OF LOCALIZED ELECTRON STATES IN THE GLASSES

15.1 General considerations

A negative-U centre in a semiconducting glass, introduced by the above described postulates [79, 231] (see Eqs. (14.2) and (14.3)), can be theoretically described as a singlet electron (or hole) pair self–trapped in a soft atomic configuration that contains most probably a single soft motion mode (Eq. (7.27)) strongly interacting with the pair. In general, the type of self-trapping quasiparticles, like conduction electrons (holes) in a semiconductor, depends on the properties of charge carriers, environment dynamics and their interactions. Whatever the origin of quasiparticle and its interaction with the environment is, in most works on self-trapped (ST) electrons the single-band self-trapping is considered, which occurs if the ST energy $W_{ST} = W_1 \equiv -W(<0)$ is substantially less in magnitude than the mobility (interband) gap width E_g, $|W_1| \ll E_g$. The well-known simplest self-trapping phenomenon is the formation of a polaron of effective radius R_p (large $R_p \gg a_1 \approx 3a_0 (\simeq 3\,\text{Å})$, small $R_p \approx a_1$, or intermediate), determined by self-trapping of a single electron in a bare state ψ_d with polaron energy shift ε_d, due to linear coupling ($-Q_d x a_0 < 0$) of electron to a harmonic vibration mode (x) of a standard spring constant $k = k_0$ [245] (see also, e.g., [233]). Increasing mode displacement gives rise to an increase of the elastic energy ($k_0 x^2/2$) of atomic configuration and a competing electron energy decrease ($\varepsilon_d(x) - \varepsilon_d(0) \simeq -Q_d x a_0$); the competition results in a minimum total energy of the system at an equilibrium mode displacement $x_1 = -Q_d/k$. The related energy decrease (shift) for $n = 0, 1, 2$ electrons is $W_n = -nW$ at $W = Q_d^2/2k_0$ and $x_n = -nQ_d/k_0$, with $W \ll E_g$. Then, a single-polaron state is a ground state, as typically the ST energy magnitude $W \ll E_g$ and the singlet-pair correlation energy $U = W_2 + W_0 - 2W_1 + U_d > 0$ corresponds to an effective Mott–Hubbard electron–electron repulsion energy $U_d (> 0)$ in a bare localised state ψ_d. However, a singlet bipolaron state is a ground state if $U < 0$ (for a singlet pair of effectively attracting electrons). With typical values of the parameters $ka_0^2 \approx k_0 a_0^2 \approx 30$ eV, $Q_d \approx 2 \div 3$ eV, $E_g \sim 1$ eV and a relatively small $U_d \ll E_g$, e.g., $U_d \leq 0.3$ eV, the criterion of single-band ST is expected usually to be satisfied in most crystalline semiconductors.

15.2 Model of negative-*U* centres: Basic relations and approximations

However, in such materials like semiconducting glasses in which important are soft modes ($k \ll k_0$), the ST electron-pair energy shift magnitude $|W_2|$ can become comparable to E_g, and the contribution to electron self-trapping from hybridisation of tail states of the overlapped conduction (c) and valence (v) bands in the mobility gap appears to become essential. The result was a new type of ST, "two-band self-trapping", described in essential details in a general consistent theory [249, 250], which is discussed in what follows (isolated aspects of an inconsistent theory of negative-U centres have been considered earlier in [247, 248]). Apart from single polaron parameters, the generalised theory of two-band ST contains two additional parameters: the gap width E_g and the effective hybridisation energy Δ ($\ll E_g$) accounting for the contribution of hybridised states in the overlapped band tails. The latter contribution enhances the trend to formation of self-trapped singlet electron pair states, which gives rise to a double occupation of nominally empty ($n = 0$) single-electron energy levels as those approach the valence-band mobility edge. This new aspect is added to earlier theories of the self-trapping phenomenon. Since the two-band ST is due to the hybridisation of a bare single-electron site state ψ_d of energy ε_d with states ψ_i of energy ε_i, in both the conduction ($\psi_i^{(c)}$) and the valence ($\psi_i^{(v)}$) bands, the hybridisation matrix elements V_{id} contribute to the true single-electron state $\Psi_{d\sigma}$ in a soft configuration,

$$\Psi_{d\sigma} = C_{dd}^{(\sigma)} \psi_d + \sum_i C_{id}^{(\sigma)} \psi_i \tag{15.1}$$

and thus to the true single-electron energy level $E_{d\sigma}$ (σ denotes the spin projection, $\sigma = \pm 1/2$). The contribution of hybridisation to $\Psi_{d\sigma}$ is described by the sum $\sum_i |C_{id}^{(\sigma)}|^2 = 1 - |C_{id}^{(\sigma)}|^2 \equiv 1 - \gamma_{d\sigma}^2$, so that $\gamma_{d\sigma}^2 \to 0$ for the strongest hybridisation while $\gamma_{d\sigma}^2 \to 1$ alternatively (e.g., with $|C_{id}^{(\sigma)}| \to 0$ at $V_{id} \to 0$). Another related characteristic is the true occupation $\nu_{d\sigma}$ of the bare state ψ_d. Both characteristics vary with the soft-mode (x) variations during the self-trapping, and the hybridisation effects becomes strongest as the bare level ε_d approaches the opposite mobility edge. In this theory, the two-band ST is realised due to large enough soft-mode displacement and competition of soft-mode potential growth with the electron energy decrease.

The goal of the theory of two-band ST is to describe the basic properties of two-band ST, in particular the behaviour of $\gamma_{d\sigma}^2$ and $\nu_{d\sigma}$ with increasing soft-mode displacement to the equilibrium value. This problem is eventually reduced to calculations of the related soft-mode adiabatic potentials Φ_n depending on the soft-mode displacement, as the bare level ε_d is occupied by n electrons. Since for the sake of simplicity, the bare electron state $\psi_d(\varepsilon_d)$ is assumed to be orbitally nondegenerate, two major factors of its nominal occupation by electrons is $n = 0, 1$, or 2. Another problem is to calculate the ST energies W_n and singlet-pair correlation energy U and to find how the hybridisation effects actually favour formation of singlet-pair self-trapped states with $U < 0$, i.e., of negative-U centres, as expected above.

Most important systems exhibiting the two-band ST are semiconducting glasses like chalcogenide glasses, e.g., g-As_2Se_3 and $GeSe_2$, mainly implied in what follows. In these materials the actual mobility-gap states are commonly considered to be singlet electron (hole) pair states with negative correlation energy, i.e., negative-U centres, as assumed earlier in [231, 238] and applied in a series of papers for interpreting a large variety of experimental data (see, e.g., [79]). In the theory presented in what follows, the gap states in semiconducting glasses actually are shown to be negative-U centres with typically large $|U| \simeq E_g/2$, in accordance with the earlier assumptions. This holds in the present theory because $|W_2| > 2|W_1| + U_d$ and $|U| \approx |W_2|/2$ with $|W_2| \simeq E_g$. The latter follows from the hybridisation induced repulsion of the true single-electron energy level off the valence-band, as demonstrated below by direct calculations.

Moreover, it is shown further from calculations that the spectral and thermodynamic properties of the pair self-trapped states are indeed similar to those of negative-U centres mentioned above. In the latter, the apparently conflicting models, in which respectively discrete or continuous features of the gap predominate, may be considered as limiting cases of the present theory.

It follows that two major factors have to be taken here into account, together in the theory for analysing the two-band ST problem:

(i) anharmonicity (or softness) of the substrate soft atomic configuration (x) interacting with conduction electrons (holes);
(ii) hybridisation of the bare state ψ_d with extended states ψ_i of both bands, largely the nonparent band (e.g., the valence band for electrons) in the gap.

Electron self-trapping is mainly considered while hole self-trapping is similar (with trivial substitutions). Then, the Hamiltonian of the model can be described as follows:

$$H(x) = V_{sm}(x) + H_e(x) = V_{sm}(x) + H_e^{(0)} + H_{e,sm}(x), \qquad (15.2)$$

where $V_{sm}(x)$ is the substrate soft-mode potential energy described by Eq. (7.2), $H_e^{(0)}$ is the Haldane–Anderson Hamiltonian of the electrons in the absence of soft-mode motions ($x = 0$), while $H_{e,sm}(x)$ characterises the electron-soft-mode interaction as described in the soft-mode model (Sec. 7.4). The electron Hamiltonian, introduced in a different context in [251] and described as follows:

$$\begin{aligned} H_e^{(0)} &= H_e(x=0) \\ &= \sum_{i,\sigma} \varepsilon_i n_{i\sigma} + \sum_{\sigma} \varepsilon_d^{(0)} n_{d\sigma} + (1/2)U_d \sum_{\sigma \neq \sigma'} n_{d\sigma} n_{d\sigma'} \\ &\quad + \sum_{i,\sigma} \{V_{id}^{(0)} a_{i\sigma}^{+} d_{\sigma} + \text{h.c.}\}, \end{aligned} \qquad (15.3)$$

takes into account the hybridisation ($V_{id}^{(0)}$) of a bare local state [$\psi_d^{(0)}, \varepsilon_d^{(0)} = \varepsilon_d(x = 0)$], in a reference atomic configuration ($x = 0$), with band states

(ψ_i, ε_i), where $n_{j\sigma} = a_j^+ a_j$, at $j = i$ or $j = d$, and $\{a_{j\sigma}^+, a_{j\sigma}\}$ are the electron operators. Coulomb interactions of electrons in band states with electrons in the local state are not accounted explicitly, being assumed small, in the sense that $U_d \ll E_g$ (as argued above and can be seen in more detail in [249, 250]). The bare local state $(\psi_d^{(0)})$ is considered as a state with the reference energy level $(\varepsilon_d^{(0)})$ which belongs mainly to the gap or to the conduction band, not too high above its mobility edge E_c. Finally, the electron-soft-mode interaction contains contributions, both diagonal $[\Delta\varepsilon_d(x)]$ and nondiagonal $[\Delta V_{id}(x)]$ in the electron states,

$$H_{es}(x) = \Delta\varepsilon_d(x) \sum_{\sigma} n_{d\sigma} + \sum_{i,\sigma} \{\Delta V_{id}(x) a_{i\sigma}^+ d_\sigma + h.c.\}, \qquad (15.4)$$

where $\Delta\varepsilon_d(x) \equiv \varepsilon_d(x) - \varepsilon_d^{(0)}$ and $\Delta V_{id}(x) = V_{id}(x) - V_{id}(0)$. With the Hamiltonian $H(x)$, an essential problem is to calculate the correlation energy U of a self-trapped singlet electron pair state in a soft atomic configuration with its parameters (η, ξ) (Eq. (7.27)) and to find out, under which conditions U is negative for the systems in which the hybridisation of states is decisive.

15.3 Adiabatic potentials and electron energy

The correlation energy can be expressed in terms of adiabatic potentials $\Phi_n(x)$ of configuration distorted at self-trapping in the bare state, for different nominal electron occupations $n = 0, 1$ or 2:

$$\begin{aligned} U &\equiv U(\eta, \xi) \\ &= \Phi_2(x_2) + \Phi_0(x_0) - 2\Phi_1(x_1) = W_2 + W_0 - 2W_1, \end{aligned} \qquad (15.5)$$

while the self-trapping energies

$$W_n \equiv W_n(\eta, \xi) = \Phi_n(x_n) - \Phi_n(x_0), \qquad (15.6)$$

are negative by definition of the self-trapping. The potential extrema (minima, maxima), including the equilibrium self-trapping displacements $x_n = x_n(\eta, \xi)$ along the soft mode x, are found as usual from the equation

$$d\Phi_n(x)/dx = 0 \quad \text{at } x = x_n(\eta, \xi). \tag{15.7}$$

Equations (15.5)–(15.7) take into account the adiabaticity of the self-trapping electron motion, with respect to atomic motion in the slow soft mode x, as the related parameter is small, $\varepsilon_{\text{exc}} \cdot |E_{d\sigma}(x) - \varepsilon_d^{(0)}|^{-1} \approx \varepsilon_{\text{exc}} \cdot E_g^{-1} \ll 1$, for actual large equilibrium displacements $|x| \approx |x_n| \sim 1$ and low energies $\varepsilon_{\text{exc}} \ll h\nu_D$ for the soft-mode excitations [249, 250]. However, since V_{id} depends on x, the true occupation of the bare level is also a function of x, so the adiabaticity of the soft-mode motion with respect to variations of the electron state occupation often assumed for single-band ST does not hold precisely.

The Green's function approach is applied and the total electron energy of the system with Hamiltonian $H(x) = V_s(x) + H_e(x)$ (Eq. (15.2)) is calculated in the Hartree–Fock (HF) mean-field approximation. Then, the total electron Hamiltonian is approximated in a standard way as

$$H_e(x) \equiv H_e[\varepsilon_d(x), V_{id}(x)] \simeq H_{\text{HF}}(x) \equiv \widetilde{H}_e[\varepsilon_{d\sigma}^{\text{eff}}, V_{id}(x)] - \delta E, \tag{15.8}$$

where $\widetilde{H}_e[\varepsilon, V_{id}] \equiv H_e[\varepsilon, V_{id}] - \frac{1}{2}U_d \sum_{\sigma\neq\sigma'} n_{d\sigma}n_{d\sigma'}$, $\delta E \equiv \frac{1}{2}U_d \sum_{\sigma\neq\sigma'} \nu_{d\sigma}\nu_{d\sigma'}$, $\varepsilon_{d\sigma}^{\text{eff}} = \varepsilon_{d\sigma}^{\text{eff}}(x) = \varepsilon_d(x) + U_d\nu_{d,-\sigma}(x)$ the effective single-particle level substituted for the bare energy level ε_d, and $H_{\text{HF}}(x)$ is bilinear in electron operators while $\nu_{d\sigma}$ is the true occupation of the bare state with spin projection $\sigma = 1/2$. This approximation is assumed to be relevant, since in the materials, Hubbard energy $U_d \ll E_g$ [252] while the resulting pair correlation energy is negative and much larger in magnitude, $|U| \sim E_g$, for most pairs under discussion.

In the Green's function approach the electron energy contribution to $\Phi_n(x)$ and bare state occupation are described by the following formulae [249, 252]:

$$\begin{aligned}
\Delta\Phi_n(x) &= E_e^{(n)}(x) \\
&\equiv \Phi_n(x) - V_s(x) \\
&= 4\pi \sum_\sigma \int_{(R_n)} d\nu \ \text{Im}\, G_{dd}^{(\sigma)}(\nu) \cdot \left[1 - d\sum_d{}' (\nu)/d\nu\right] - \delta E
\end{aligned} \tag{15.9}$$

and

$$\nu_d(x) = \sum_\sigma \nu_{d\sigma}(x),\ \nu_{d\sigma}(x) \equiv \langle n_{d\sigma}(x)\rangle = 2\int_{(R_n)} d\nu\, \mathrm{Im}\, \mathrm{G}_{dd}^{(\sigma)}(\nu). \tag{15.10}$$

Here (R_n) stands for the range of frequencies ν of the nominally occupied states only, one has to apply the formula $Tr\mathrm{G}^\sigma(\nu) = \frac{d}{2\pi d\nu}\ln\{2\pi\nu - \varepsilon_{d\sigma}^{\mathrm{eff}} - \sum_d'\}^{-1}, \sum_d' \equiv \mathrm{Re}\sum_d$, and the diagonal matrix element of the Green's function operator for the Hamiltonian is

$$\mathrm{G}_{dd}^{\sigma}(\nu) = \left[2\pi\nu - \varepsilon_{(d\sigma)}^{\mathrm{eff}} - \sum_d{}'\right]^{-1}, \tag{15.11}$$

whereas the respective self-energy

$$\sum_d \equiv \sum_d(\nu, x) = \int d\varepsilon \Delta(\varepsilon, \nu; x)(2\pi\nu - \varepsilon)^{-1}, \tag{15.12}$$

is determined by the effective hybridisation energy

$$\Delta(\varepsilon, \nu; x) = g_0(\varepsilon)|V(\varepsilon, 2\pi\nu; x)|^2, \tag{15.13}$$

where $N^{1/2}V_{id}(x) \equiv V(\varepsilon, \nu; x), N$ is the total number of atoms, and $g_0(\varepsilon)$ is the bare electron density of states (DOS). Moreover, an important characteristic of two-band ST is the degree of hybridisation of the bare local state with the band states:

$$\begin{aligned} 0 \leq \gamma_{d\sigma}^2(x) &= 2\int d\nu\, \mathrm{Im}\, \mathrm{G}_{dd}^{(\sigma)}(\nu) \\ &= \left[1 - d\sum_d{}'/d\nu\right]^{-1} \leq 1, \quad \text{at } h\nu = E_{d\sigma}(x). \end{aligned} \tag{15.14}$$

The degree of hybridisation $\gamma_{d\sigma}^2(x)$ is the probability to find the bare state electron in the cloud of the true state $\Psi_{d\sigma}$ [Eq. (15.1)] with energy $E_{d\sigma}$ in the gap. In particular, $\gamma_{d\sigma}^2(x) \simeq 0$ corresponds to nearly complete hybridisation, at large $|x| \sim 1$ and $\gamma_{d\sigma}^2(x) \simeq 1$, to negligible hybridisation at small $|x|$. The true, hybridisation renormalised single-electron energy level

$E_{d\sigma} \equiv E_{d\sigma}(x)$ is the pole of $G_{dd}^{(\sigma)}(\nu; x)$ in the (mobility) gap and can essentially differ from the effective level $\varepsilon_{d\sigma}^{\text{eff}}$ for large $|x| \sim 1$, at

$$E_{d\sigma} - \varepsilon_{d\sigma}^{\text{eff}}(x) - \sum_d{}' (E_{d\sigma}) = 0, \qquad (15.15)$$

while $\sum_d''(2\pi\nu) \equiv \text{Im} \sum(2\pi\nu) = 0$ at $E_v \equiv 0 < 2\pi\nu = E_{d\sigma} < E_c \equiv E_g$. The right-hand side of Eq. (15.14) follows from Eqs. (15.11) and (15.15).

Then the problem of two-band ST characteristics and hybridisation effects is reduced to calculations of the total electron energy $E_{\text{el}}^{(n)}(x)$, the effective hybridisation energy $E_{d\sigma}(x)$, as well as of $\nu_d(x), \gamma_{d\sigma}^2(x)$, and of the equilibrium ST displacements $x = x_n(\eta, \xi)$ along the soft mode x. The basic parameters include, besides $\{\varepsilon_d^{(0)}/E_g, U_d/E_g, \Delta(\varepsilon, \nu; x = 0)\}$ for a reference ($x = 0$) soft-mode, also the soft-mode (x) related characteristics $\Delta(\varepsilon, \nu; x)/E_g$ and $Q_d/E_g, A/E_g$), with the following approximation for hybridisation energies:

$$\Delta(\varepsilon, \nu; x) \simeq \Delta_v(x)\theta(\varepsilon - E_v')\theta(E_v - \varepsilon) + \Delta_c(x)\theta(\varepsilon - E_c)\theta(E_c' - \varepsilon). \qquad (15.16)$$

Here $\theta(z) \equiv \{1 \text{ at } z > 0; 0 \text{ at } z < 0\}$ and $\Delta_{v,c}(x)$ are, generally speaking, x-dependent characteristics of extended states of the valence and conduction bands, with $E_v' - E_v \equiv D_v, E_c' + E_c \equiv D_c$ and $D_{c,v}$ the band widths. This approximation for the gross features of the energy spectrum implies that the dependence of Δ on ε does not affect the qualitative features of two-band ST and related estimations. Typical values of hybridisation energies can be estimated as follows:

$$\Delta_{v,c}(x) \approx N|\Delta V_{id}(x)|^2 g_0(E_{v,c}) \approx V_1{}^2/D_{v,c} \approx 0.01\text{–}0.1 \text{ eV} \ll E_g \approx 1\text{–}3 \text{ eV}, \qquad (15.17)$$

where $|V_{id}| \equiv V_1 N^{-1/2} \approx 0.3\text{–}0.5$ eV, and $g_0(E_{v,c}) \sim 1/D_{v,c}, 5 \lesssim D_{v,c} \lesssim 10$ eV.

As noted, hybridisation of states influences the total electron energy $E_{\text{el}}^{(n)}$ of the system in question. The hybridisation of the bare state (ψ_d) is taken into account not only with the nonparent band states but also, under self-trapping ($x \neq 0$), with the parent band states, the efficiency of hybridisation depending on the position of the bare energy level $\varepsilon_d(x)$ in the gap. If the level is far from the band edges $E_{c,v}$, $\Delta \ll |\varepsilon_d(x) - E_{c,v}| \ll E_g$, the interactions of the bare state with the band states are not essential and the true occupation ν_d of the self-trapping state within the mobility gap is close to its nominal occupation n. However, the situation changes drastically as the bare level approaches a band edge or penetrates a band. If the energy level $\varepsilon_d(x)$ is located within the conduction band $[\varepsilon_d(x) > E_c]$, the hybridisation of the state ψ_d with unoccupied conduction-band states gives rise to such an electron (or electron pair) state, which mainly consists of the conduction-band states, rather than that of the state ψ_d. Therefore, as follows from Eq. (15.10), the true occupation of the state ψ_d becomes close to zero, $\nu_d \simeq 0$, independently of its nominal occupation n, actually due to the well-known quantum-mechanical dispersion of the wave packet in the conduction band. A similar situation takes place when the bare level $\varepsilon_d(x)$ appears within the valence band $[\varepsilon_d(x) > E_v]$. The difference is that, hybridisation leads to almost complete occupation of the state ψ_d by two electrons coming from the valence-band states, so that $\nu_d \simeq 2$, independently of the nominal state occupation n. Essential changes $\Delta\nu_d \equiv \nu_d - n$ in the state occupation occur within an energy range around the appropriate band edge, of which the width is of the order of magnitude of the effective hybridisation (or, in some sense, interaction) energy Δ. Since actually $\Delta/E_g \ll 1$ and $U_d/E_g \ll 1$, simple approximate relation can be derived for the electron energies $E_{\text{el}}^{(n)}(x)$ of the system, taking into account the obvious expressions characteristic of the hybridisation-free case:

$$E_{\text{el}}^{(n)}(x) = n\varepsilon_d(x) + U_d\delta_{n,2}. \tag{15.18}$$

In this connection let us assume, for the sake of simplicity, that the hybridisation generates abrupt changes of the state occupation as the energy level $\varepsilon_d(x)$ crosses a band edge with increasing soft-mode displacement. As just noted, if the energy level $\varepsilon_d(x)$ is located in the conduction band $[\varepsilon_d(x) > E_c]$ and the state ψ_d is singly occupied ($n = 1$), the true occupation of the state becomes close to zero, $\nu_d \simeq 0$, and the electron occupies the band bottom, $E_{\text{el}}^{(1)}(x) \simeq E_c$. Then the true and nominal occupations nearly coincide ($\nu_d \simeq n = 1$), if $E_v < \varepsilon_d(x) < E_c$, and the relation (15.18) at $n = 1$ is applicable. On the other hand, if $\varepsilon_d(x)$ is located within the valence band, the true occupation $\nu_d \simeq 2$, so the same relation may also be applied at $n = 2$. The approximate expressions for $E_{\rm el}^{(n)}(x)$ follows:

$$E_{\rm el}^{(n=0)}(x) \equiv E_{\rm el}^{(0)}(x) \approx 2\varepsilon_d(x) + U_d \quad \text{at } \varepsilon_d(x) < 0;$$
$$0 \quad \text{at } \varepsilon_d(x) > E_v \equiv 0; \tag{15.19}$$
$$E_{\rm el}^{(1)}(x) \approx E_g \quad \text{at } \varepsilon_d(x) > E_c \equiv E_g; \quad \varepsilon_d(x) \quad \text{at } 0 < \varepsilon_d(x) < E_g;$$
$$2\varepsilon_d(x) + U_d \quad \text{at } \varepsilon_d(x) < 0 \tag{15.20a}$$

and

$$E_{\rm el}^{(2)}(x) \approx 2E_g \quad \text{at } \varepsilon_d(x) > E_g;$$
$$2\varepsilon_d(x) + U_d \quad \text{at } \varepsilon_d(x) < E_g. \tag{15.20b}$$

In accordance with the note above, it is taken into account that $E_{\rm el}^{(n)}(x) \approx nE_g$, if $\varepsilon_d(x) > E_g \gg U_d$. The expressions (15.19)–(15.21) give a simple qualitative picture of the electron energy changes under self-trapping. This is supported and quantified by the results of numerical calculations, discussed in what follows for actual positive soft-mode displacements which correspond to $\varepsilon_d(x)$ approaching and penetrating the valence band, i.e., to $\varepsilon_d(x) < E_g$. Although the hybridisation energy Δ is finite and therefore the related changes in the state occupation are continuous, the latter indeed are rather sharp for actual Δ/E_g, the sharper the smaller will be Δ/E_g. Note also that the conduction-band states only slightly contribute to the electron energy $E_{\rm el}^{(n)}(x)$, the true occupation $\nu_{d\sigma}(x)$ and the degree of hybridisation of the bare state $\gamma_{d\sigma}^2(x)$, and thus to the self-energy $\sum_d (\nu = \nu_{d\sigma})$, as long as $E_{d\sigma}(x)$ approaches E_v and penetrates the valence band at large displacements, $1 \gtrsim x \geq x_g = E_g/Q_d$. In this actual case implied in what follows, the expressions for $E_{\rm el}^{(n)}(x)$ and $\gamma_{d\sigma}^2(x)$ can be simplified, with

$$E_{d\sigma}(x) \simeq \varepsilon_{d\sigma}^{\rm eff}(x) \quad \text{and} \quad \gamma_{d\sigma}^2(x) \simeq \left[1 + \frac{\Delta}{E_{d\sigma}(x)}\right]^{-1}. \tag{15.21}$$

The full expressions for $E_{\rm el}^{(n)}(x)$, as well as for other characteristics $E_{d\sigma}, \nu_{d\sigma}, \gamma_{d\sigma}^2$ and x_n, the true soft-mode displacement in ST, which are used in numerical calculations, are derived in the approximation (15.16) and (15.17), and in more detail the above results are obtained and discussed in [249, 250].

15.4 Basic features of self-trapped states and negative-U centres

Equations (15.7), (15.10) and (15.15) have been numerically solved and the basic characteristics $E_{\mathrm{el}}^{(n)}(x)$, $\nu_{d\sigma}$ and $\gamma_{d\sigma}^2$ have been calculated for self-trapped states at equilibrium soft-mode displacements $x_n(\eta, \xi)$ for different nominal occupations $n = 0, 1, 2$ which exhibits in a most pronounced way the basic features of the self-trapped states and the related atomic dynamics. Hybridisation effects, giving rise to these features, actually are strong just for the substrate soft configurations (Eq. (7.27)) of which anharmonicity limits the equilibrium displacement to realistic $|x_n(\eta, \xi)| \leq 1$. Characteristic values $(\eta_n^{**}, \xi_n^{**})$ of the soft-mode parameters (η, ξ), which mark essential changes in the bare state occupation ν_d and related atomic adiabatic potentials (atomic dynamics), are also calculated. However, analytical approximations for estimating $(\eta_n^{**}, \xi_n^{**})$ can be obtained only for quasiharmonic soft configurations belonging to the range of (η, ξ), actually at $1 \gg \eta \gg \eta_L$ (since important contributions come from $\xi \simeq \overline{\xi}$, the mean value, which is small, $\overline{\xi}^2 \ll 1$, practically being chosen as $\overline{\xi} \simeq 0$).

It is worth noting that two-band self-trapping is realised as long as bare electron energy level $\varepsilon_d(x) \simeq \varepsilon_d(x) - Q_d x$ moves toward and penetrates the valence, nonparent, band with increasing soft-mode displacement $x(> 0)$, and strong hybridisation of the electron state, with the valence-band states occurs at large enough positive equilibrium displacements $x_n(\eta, \xi) \sim 1$. The soft-mode asymmetry (ξ) in Eq. (7.5) does favour $x > 0$ and $x_n(\eta, \xi) > 0$ at $\xi < 0$ while it does not at $\xi > 0$. Therefore a finite subrange of $(\eta, \xi > 0)$ may exist, in which $\overline{x}_n(\eta, \xi) < 0$ and $\varepsilon_d(x)$ moves toward (and may penetrate) the conduction, parent, band and hybridisation, with these band states becoming strong. Self-trapping in such soft configurations does not contribute to the two-band self-trapping under discussion and is a special type of single-band self-trapping which does not favour formation of negative-U centres. Numerical calculations of $\overline{x}_n(\eta, \xi)$ show that such a subrange A_2 of (η, ξ), although finite, is small compared to the main subrange A_1 of (η, ξ), which corresponds to two-band self-trapping with large $\bar{x}_n(\eta, \xi) \geq x_g = E_g/Q_d$ and $\varepsilon_d(x_n) < 0 < \varepsilon_d(0) \sim E_g$. As shown in the next section, only the contribution of the main subrange A_1 of (η, ξ) to the electron density of states (DOS) and thermal equilibrium statistics of the system under consideration is important. In this connection in what follows we focus on the two-band self-trapping and related soft modes in the main subrange A_1 of their parameters (η, ξ). In fact, as seen in Sec. 7.1, only a small part of this subrange A_1 is important in thermal equilibrium properties, and corresponds to quasiharmonic single-well soft-mode potentials (Eq. (7.5)), at $1 \gg \eta \gg \eta_L$. However, the whole subrange A_1, including anharmonic non-single-well soft-mode potentials, and the related spectrum of self-trapping energies $W_n(\eta, \xi)$, pair correlation energies $U(\eta, \xi)$, and atomic adiabatic potentials $\Phi_n(x; \eta, \xi)$ are considered in what follows, since those may be important for nonequilibrium properties of the system. An example is a correlation between photoinduced electron effects, associated with generation of negative-U center excitations (photoluminescence, photoconductivity, etc.), and changes in atomic structure (photostructural changes), in atomic soft-mode potentials and low-energy dynamics, as well as in the associated low-temperature properties, of glassy semiconductors, which will be briefly discussed below in Sec. 16.2.

Three most important effects can be revealed, which are due hybridisation of states in the gap and are characteristic of the two-band self-trapping and the related spectrum $W_n(\eta, \xi)$, $U(\eta, \xi)$, and $\Phi_n(x; \eta, \xi)$ under consideration, as seen from, and explained in detail in [249].

(i) As the electron energy level $\varepsilon_{d\sigma}^{\text{eff}} \equiv \varepsilon_{d\sigma}^{\text{eff}}(x) = \varepsilon_d(x) + U_d \nu_{d,-\sigma}(x)$ approaches under ST the valence-band mobility edge $E_v \equiv 0$ (Eqs. (15.19)–(15.21)), an increasing hybridisation of states gives rise to repulsion of the true level $E_{d\sigma}(x)$ off the edge. The hybridisation and its characteristic energy Δ are negligible as the effective energy level $\varepsilon_{d\sigma}^{\text{eff}}(x)$ is far from E_v, at

$$E_{d\sigma}(x) \simeq \varepsilon_{d\sigma}^{\text{eff}}(x) \gg \Delta \quad \text{and} \quad \nu_{d\sigma}(x) = 1 - \gamma_{d\sigma}^2(x) \ll 1, \qquad (15.22)$$

whereas it becomes decisive as $\varepsilon_{d\sigma}^{\text{eff}}(x)$ approaches E_v and penetrates the band, $\varepsilon_{d\sigma}^{\text{eff}}(x) \leq E_v$. In the latter case, $E_{d\sigma}(x)$ stops near E_v,

$$E_v \equiv 0 < E_{d\sigma}(x) < E_v + \Delta \equiv \Delta \quad \text{and} \quad \nu_{d\sigma}(x) \simeq 1. \qquad (15.23)$$

In fact, the effective electron-mode coupling energy $Q_{d\sigma}^{(\text{eff})}$ can be defined, for large displacements and $E_{d\sigma}(x)$ close to E_v at least, as follows from Eqs. (15.14) and (15.15),

$$Q_{d\sigma}^{(\text{eff})}(x) \equiv dE_{d\sigma}(x)/dx \simeq \gamma_{d\sigma}^2(x)Q_d. \tag{15.24}$$

One can see from above that $E_{d\sigma}(x)$ approaches the edge E_v mainly in an exponential way,

$$E_{d\sigma}(x) \approx E_v \exp[\varepsilon_{d\sigma}^{\text{eff}}(x)/\Delta], \tag{15.25}$$

for $|\varepsilon_{d\sigma}^{\text{eff}}(x)| = -\varepsilon_{d\sigma}^{\text{eff}}(x) \gg \Delta$ that $Q_{d\sigma}^{(\text{eff})}(x) \ll Q_d$ as $\gamma_{d\sigma}^2(x) = 1 - \nu_{d\sigma}(x) \ll 1$ for large $|x|$. The essential decrease of $Q_{d\sigma}^{(\text{eff})}(x)$ can be considered as the ultimate cause of true energy level repulsion off E_v.

(ii) The hybridisation of states is responsible also for an increase of a bare state occupation from a nominal occupation $n = 0$ to a finite true occupation $\nu_{d\sigma} > 0$ due to the states hybridisation. This corresponds to a change in an equilibrium ST displacement from a nominal electron displacement $x_0 \simeq 0$ to a finite $x_0 \neq 0$, as follows from Eq. (15.6). A single displacement $x_0 = x_{01} \simeq 0$ only occurs for large enough $\eta > \eta_0^* = \eta_{n=0}^*$, at weak hybridisation with $\nu_{d\sigma} \ll 1$. However, an additional large displacement $|x_{02}| \sim x_g = E_g/Q_d \leq 1$ appears for smaller $\eta < \eta_0^*$, which corresponds to a metastable minimum and, for still smaller η, to an absolute minimum. Thereby, the effective electron level $\varepsilon_{d\sigma}^{\text{eff}}(x)$ reaches E_v and penetrates the valence band, the true level $E_{d\sigma}(x)$ is repelled off E_v, and nearly complete hybridisation occurs, at $\gamma_{d\sigma}^2(x) \ll 1$. Then x_{02} should be close to the equilibrium displacement x_2 for a nominally doubly occupied state ($n = 2$) in a hybridisation-free case, so $|x_{02}| \simeq |x_2| \leq 1$ should be true for soft enough configurations, at $\eta_0^* \leq Q_d/A \sim 0.1$ obtained in harmonic approximation. Note also that the ST of a nominally free site ($n = 0$) is entirely determined by hybridisation. The latter originally is weak ($\nu_{d\sigma}(x) \simeq \Delta/E_g \ll 1$ for $|x| \ll 1$) but finite, and then increases with growing $|x|$ up to $x_{02} \approx x_2$, due to increasing contributions from extended valence-band states. The resulting self-trapped pair state is still localised but its effective size ρ increases, as compared to the bare state size $\rho_0 \approx (1 \div 2)a_1$, typically with $a_1 \sim 2$ Å. For realistic not too small true energy levels $E_{d\sigma}(x_{02})$, $1 < \Delta/E_{d\sigma}(x_{02}) \leq 10^2$, it was obtained that $1 < (\rho/\rho_0) \approx [1 + \Delta/E_{d\sigma}(x_{02})] \leq 5$.

The related effects of true single-electron energy level $E_{d\sigma}(x)$ repelled off the valence-band edge E_v and of the increase of bare state occupation may be interpreted as follows. The true occupation $\nu_{d\sigma}$ with increasing $|x|$ changes from $n_\sigma = 0$ to $\nu_{d\sigma} = 1$ due to contributions of valence-band electrons which flow into the region of bare state, strongly interacting with the soft mode. Thereby, in accordance with behaviour of the total electron energy, the bare level efficiently drops and penetrates the valence band, effectively as $\varepsilon_d(0) - \nu_d Q_d x$ at $\nu_d \to 2$. However, the contributions of bare state to the true, hybridisation–renormalised state $\Psi_{d\sigma}$ are characterised by $\gamma_{d\sigma}^2(x) \to 0$ for large displacements $|x| \to x_{02}$. This rather corresponds to the fact that the true energy level, as a solution of the effective Schrödinger equation for Hamiltonian of Eq. (7.27), behaves in such a way as if it is unaffected by the true occupation of bare level, and may be characterised by the nominal occupation which is zero in the case in question.

(iii) A competition between an increase of the substrate atomic potential energy and an electron energy gain due to growing hybridisation, with increasing $|x|$, can generate for sufficiently soft substrate configurations additional anharmonic features of the adiabatic potential

$$\Phi_0(x) = V_{sm}(x) - U_d[\nu_{d\sigma}(x)]^2 - 2[E_{d\sigma}(x) - \varepsilon_{d\sigma}^{\mathrm{eff}}(x)] \qquad (15.26a)$$

and thus of the related local atomic dynamics. A signature of the potential anharmonicity is softness and/or appearance of an extra potential minimum as a metastable one for an excited self-trapped state at $\eta < \eta_0^*$, and then as a lowest-energy minimum for a ground state at smaller $\eta < \eta_0^{**} < \eta_0^* \ll 1$. The value $\eta_0^{**} \simeq 0.01$ is found to be a representative value of the parameter. An approximate expression of $\eta_0^{**} = \eta_{0h}^{**}$ for a harmonic substrate follows from

$$\begin{aligned}\Phi_0(x_{01} \simeq 0) &\simeq 2\Delta \ln (D_v/E_g)\\ &\simeq \Phi_0(x_{02} \simeq x_2) \simeq 2E_g - Q_d^2/A\eta_{0h}^{**} + U_d, \qquad (15.26b)\end{aligned}$$

so $\eta_{0h}^{**} \simeq Q_d^2/A(2E_g + U_d) \approx 0.05$ is not far from η_0^{**}.

One can conclude that the lowest-energy minimum position of the adiabatic potential depends on the value of η/η_0^{**}, unlike the universal single minimum at $x_0 = 0$ for single-band self-trapping. For $\eta < \eta_0^{**}$, the electron ground state related to the lowest-energy potential minimum at $x_{02} \simeq x_2$ is the self-trapped state for a singlet electron pair $[\nu_d(x_{02}) \simeq 2]$, so ST rather requires overcoming the interval barrier. Since η_0^{**} actually is rather small, related extra anharmonic features really occur for soft enough configurations in a semiconducting glass. A similar picture is typical for the adiabatic potential $\Phi_1(x)$, of a nominally singly occupied bare state at $n = 1$.

The situation is drastically different for $\Phi_2(x)$ of a nominally doubly occupied bare state ($n = 2$) that is weakly sensitive to hybridisation as it is filled originally. Therefore,

$$\Phi_2(x) \simeq V_{sm}(x) + U_d + 2\varepsilon_d \quad \text{and}$$
$$\Phi_2(x_2) \simeq 2\varepsilon_d^{(0)} + U_d + W_2, \qquad (15.27a)$$

as if the influence of the valence band vanishes. Then the adiabatic potential has the same structure as the soft-mode one $V_{at}(x)$. For instance, it would be harmonic, with the singlet pair ST energy

$$W_2(x_2) \simeq -Q_d^2/A\eta a_0^2 \quad \text{at } x_2 \simeq Q_d/A\eta, \quad \text{at } V_{sm} \simeq A\eta x^2. \qquad (15.27b)$$

It is worth noting that the lowest energy minimum position for adiabatic potential at two-band ST depends on the value of $\eta < \eta_0^{**}$, unlike the universal single minimum at $x_0 = 0$ for a single-band ST.

The hybridisation-induced changes in adiabatic potentials at $n = 0$ and $n = 1$, as well as in soft-mode displacement x_n, for the lowest energy minima give rise to significant deviations of the total electron energies $E_{el}^{(n)}(x)$, ST energies $W_n = \Phi_n(x_n) - \Phi_n(x_0)$ and the pair correlation energy $U = \Phi_2(x_2) + \Phi_0(x_0) - 2\Phi_1(x_1) = W_2 + W_0 - 2W_1$ from standard formulae:

$$W_0 = 0, \quad W_1 = W_2/4 = -W = -Q_d^2/2k^{(0)}a_0^2,$$
$$E_{el}^{(n)}(x) = n(\varepsilon_d^{(0)} - Q_d x) + \delta_{n,2}U_d \quad \text{and} \quad U \equiv U_h = -2W + U_d, \qquad (15.28)$$

for single-band self-trapping in a harmonic soft-mode potential. The deviations of the characteristics for two-band ST from those for single-band ST become essential for small enough $\eta \leq \eta_1^{**}$ as the nominally singly occupied electron ground state ($n = 1$) is related to the extra minimum at $x_1 = x_{12} \simeq x_2$ for $\Phi_1(x)$ and actually doubly occupied [$\nu_d(x_{12}) \simeq 2$].

Not entering into details of the two-band ST theory within soft-mode model [249], one can emphasize that the hybridisation of states in two-band ST, generally speaking, changes the relationships between $E_{d\sigma}\,(x_n)$ and $W_n(\eta, \xi)$, as well as the behaviour of $U(\eta, \xi)$), as compared to those for single-band ST. The variations are minor, for not soft enough modes, at $1 \gg \eta \gtrsim \eta_1^{**}(> \eta_0^{**})$, for which the potentials are quasiharmonic; η_1^{**} is a parameter bigger (but not necessarily in scale) than η_0^{**}. However, the variations become considerable for such softer modes for which $\eta < \eta_1^{**}$, and even larger for $\eta < \eta_0^{**}(< \eta_1^{**})$, for which the soft mode potentials are anharmonic and may become non-single-well ones (actually, double-well potentials). Then, in the subrange of (η, ξ), at $1 \gg \eta^2 > (\eta_1^{**})^2$, in which the negative correlation energy $U(\eta, \xi)$ is close to its lowest value at $U \approx U_m \approx -E_g/2$, an electron-hole symmetry (assumed in earlier theories, e.g., [124, 125, 248]) can be proven in the sense that

$$U_m^{(e)} \simeq -E_g/2 \simeq U_m^{(h)}, \tag{15.29a}$$

with accuracy to small corrections $\lesssim \Delta/E_g \ll 1$. This symmetry is important as the subrange of (η, ξ), in which Eq. (15.29a) holds, determines the principal spectral and thermal equilibrium properties of mobility gap states in semiconducting glasses. Note that the pair self-trapping energies also exhibit such a symmetry,

$$W_{2e} \simeq -E_g \simeq W_{2h}, \tag{15.29b}$$

although $1/4 \leq |W_1/W_2| < 1/3$, rather than $W_2 = 4W_1$ for single-band self-trapping in a harmonic lattice.

15.5 Density of states and thermal equilibrium properties

For the semiconducting glasses under discussion for which the electron subsystem actually is not excited, the principal characteristic of the "mobility" gap states, their density of states (DOS) per particle

$$g(E) = g^{(e)}(E) + g^{(h)}(E), \tag{15.30}$$

additively contains electron (e) and hole (h) contributions of both single-particle states of positive correlation energy U, $g_1^{e,h}(E)$, and singlet-pair states of negative U, $g_2^{e,h}(E)$. The respective expressions are

$$g^{e,h}(E) = g_1^{(e,h)}(E) + g_2^{(e,h)}(E), \tag{15.31}$$

$$g_1^{e,h}(E) = \int_{-\infty}^{\eta^*} d\eta \int_{-\infty}^{\infty} d\xi \int d\varepsilon_d^{(0)} g_0^{e,h}(\varepsilon_d^{(0)}) F(\eta,\xi)\delta[E - E_1(\varepsilon_d^{(0)};\eta,\xi)] \tag{15.32a}$$

and

$$g_2^{e,h}(E) = \int_{-\infty}^{\eta^*} d\eta \int_{-\infty}^{\infty} d\xi \int d\varepsilon_d^{(0)} g_0^{e,h}(\varepsilon_d^{(0)}) F(\eta,\xi)\delta[E - E_2(\varepsilon_d^{(0)};\eta,\xi)/2], \tag{15.32b}$$

where the total energies of states nominally occupied by electrons (holes) can be described as follows:

$$E_0 = 0,\ E_1(\varepsilon_d^{(0)};\eta,\xi) = \varepsilon_d^{(0)} + W_1(\eta,\xi) - W_0(\eta,\xi), \tag{15.33a}$$

$$E_2(\varepsilon_d^{(0)};\eta,\xi) = 2\varepsilon_d^{(0)} + W_2(\eta,\xi) - W_0(\eta,\xi) + U_d, \tag{15.33b}$$

with

$$\int_{-\infty}^{\infty} d\eta \int_{-\infty}^{\infty} d\xi F(\eta,\xi) = 1, \quad \int d\varepsilon_d^{(0)} g_0^{e,h}(\varepsilon_d^{(0)}) = 1. \tag{15.34}$$

The DOS $g(E)$ determines both the total concentration of self-trapped states and the related thermal-equilibrium properties, in particular the concentration c_2 of the occupied negative-U pair states (negative-U centres) and the position of the chemical potential ζ, or the Fermi level ζ_0 (at low $T \to 0$), in the gap. Then, the concentration of positive-U ($n = 1$) and negative-U ($n = 2$) centres is

$$c_n = c_n^{(e)} + c_n^{(h)}, \tag{15.35}$$

$$c_n^{(e,h)} = \int_{-\infty}^{\eta^*} d\eta \int_{-\infty}^{\infty} d\xi \int d\varepsilon_d^{(0)} g_0^{e,h}(\varepsilon_d^{(0)}) F(\eta,\xi)\phi_n(\varepsilon_d^{(0)};\eta,\xi), \tag{15.36}$$

where the Gibbs occupation factor is

$$\phi_n(\varepsilon_d^{(0)};\eta,\xi) = Z^{-1} g_n \exp\left\{T^{-1}\left[n\zeta - E_n\left(\varepsilon_d^{(0)};\eta,\xi\right)\right]\right\},$$
$$Z = \sum_{n=0}^{2} g_n \exp\left\{T^{-1}\left[n\zeta - E_n\left(\varepsilon_d^{(0)};\eta,\xi\right)\right]\right\}, \quad g_0 = g_2 = 1, \quad g_1 = 2. \tag{15.37}$$

In what follows, low T are only implied, so temperature dependent effects are not considered in detail, and some approximations are introduced, which simplify the calculations concerning in particular the electroneutrality equation

$$c_2^{(e)}(\zeta) = c_2^{(h)}(\zeta), \tag{15.38}$$

for the chemical potential $\zeta \equiv \zeta(T)$ of an intrinsic semiconducting glass (the approximations can be justified by results of calculations, as described in more detail in [249]).

As follows from general arguments, the electron-hole symmetry in Eq. (5.29) corresponds to the Fermi level ζ_0 close to the midgap $\overline{E} = E_g/2$,

$$\zeta_0 \equiv \zeta(T=0) \gg T(\to 0) \quad \text{and}$$
$$\zeta \simeq \zeta_0 = E_g/2 + \Delta\zeta \simeq E_g/2 \gg |\Delta\zeta|, \tag{15.39}$$

whereas it should be also taken into account that

$$g(\zeta) \simeq g_2(\zeta_0) \gg g_1(\zeta_0). \tag{15.40}$$

Then, the most important range of $(\varepsilon_d^{(0)};\eta,\xi)$ corresponds to equation $E_{2e}(\varepsilon_d^{(0)};\eta,\xi) = 2\varepsilon_{de}^{(0)} + W_{2e}(\eta,\xi) - W_{0e}(\eta,\xi) + U_d^{(e)} \simeq 2\zeta_0 \simeq E_g$ so that $\zeta_0 - E_{1e} = \zeta_0 - \varepsilon_{de}^{(0)} - (W_{1e} - W_{0e}) \simeq \frac{1}{2}(W_{2e} - W_{0e} + U_d^{(e)}) - (W_{1e} - W_{0e}) <$

0 for actual $|W_{2e} - W_{0e}| > 2|W_{1e} - W_{0e}| + U_d^{(e)}$. Therefore,

$$Z \simeq 1 + \exp\left[\frac{2\zeta - E_2}{T}\right] \gg 2\exp\left[\frac{\zeta - E_1}{T}\right], \tag{15.41}$$

for low temperatures in question. It is worth noting that the subrange of (η, ξ) in Eq. (15.41) rather corresponds to the inequality $\Phi_0(x_0 \simeq 0) \approx \Delta \ll |U_m| \simeq E_g/2$, not much different from the situation for negative-U centres in the earlier theoretical model of negative-U centres, applying the relations of single-band ST.

In approximations (15.38) and (15.41), the total low-temperature concentration of negative-U and positive-U centres in the gap is

$$c_{tot} = c_1 + c_2 \simeq c_2 = c_2^{(e)} + c_2^{(h)} \gg c_1^{(e)} + c_1^{(h)} \tag{15.42}$$

and c_2 can be expressed in terms of $g_2(E)$ and $\phi_2(\varepsilon_d^{(0)}; \eta, \xi)$. Here

$$c_2^{(e)} = \int d\varepsilon_d g_0(\varepsilon_d^{(0)})\psi(\varepsilon_d^{(0)}) = \int_{E_v}^{E_c} dE g_{2e}(E)\overline{\phi}_{2e}(E),$$

$$\psi(\varepsilon_d^{(0)}) = \int_{-\infty}^{\eta^*} d\eta \int_{-\infty}^{\infty} d\xi F(\eta, \xi)\phi_{2e}(\varepsilon_d^{(0)}; \eta, \xi) \tag{15.43}$$

and

$$\begin{aligned}\overline{\phi}_{2e}(E) &\equiv \overline{\phi}_{2e}(E_2/2) = 1 - \overline{\phi}_{2h}(E)\\ &\equiv \phi_{2e}(\varepsilon_d^{(0)}; \eta, \xi) \simeq \left\{1 + \exp\left[-\frac{2(\zeta_0 - E)}{T}\right]\right\}^{-1}\end{aligned} \tag{15.44}$$

at $E = E_2(\varepsilon_d^{(0)}; \eta, \xi)$. Moreover,

$$\overline{\phi}_{2e}(E) = 1 - \overline{\phi}_{2h}(E) \to 0 \quad \text{for } E > \zeta \quad \text{while } \overline{\phi}_{2e}(E) \to 1 \quad \text{for } E < \zeta \quad \text{at } T \to 0. \tag{15.45}$$

Numerical calculations of Eqs. (15.38) and (15.42)–(15.45) may justify the relations (15.39)–(15.41). The calculations can be carried out by finding $\varepsilon_d^{(0)} = f(E, \eta, \xi)$ from the definition of E, $2E = E_2(\varepsilon_d^{(0)}; \eta, \xi)$, so that $g_2(E) = \int_{-\infty}^{\eta^*} d\eta \int_{-\infty}^{\infty} d\xi F(\eta, \xi)Q(\eta, \xi; E)$ at $g_0(\varepsilon_d^{(0)}) \equiv Q(\eta, \xi; E)$.

The bare DOS $g_0(\varepsilon_d^{(0)})$ of two-band system and the distribution density $F(\eta, \xi)$ are approximated by often applied Gaussian-like functions

$$g_0^{(e)}(\varepsilon_d^{(0)}) = g_{0e}^* \left\{ \exp\left[-\left(\frac{E_c - \varepsilon_d^{(0)}}{w_t^{(e)}} \right)^2 \right] \right.$$

$$\left. \times \theta(E_c - \varepsilon_d^{(0)}) + \alpha_e \theta(\varepsilon_d^{(0)} - E_c) \right\}, \qquad (15.46)$$

$$F(\eta, \xi) = F_0(\eta) \exp\left[-\left(\frac{\overline{\eta} - \eta}{\Delta\eta} \right)^2 - \left(\frac{\overline{\xi} - \xi}{\Delta\xi} \right)^2 \right], \qquad (15.47)$$

where $g_{2e,h}^* \equiv g_0^{(e,h)}(E_{c,v})$, $\alpha_{e,h} = const.$, $\overline{\eta} \approx 1$, $\overline{\xi}^2 \ll 1$; $\Delta\eta$ and $\Delta\xi$ stand for the respective distribution widths, and $F_0(\eta) = F_0 = \text{const}$ or $F_0(\eta) = |\eta|\Phi_0$, $\Phi_0 = \text{const}$ for two limiting types of $F(\eta, \xi)$ at $|\eta| \ll 1$, in Eqs. (7.17) or (7.18) of the soft-mode model (Ch. 7). The constant F_0 or Φ_0, as well as ratio α_e/α_h, can be estimated by taking into account the normalisation Eq. (15.34), actually at $w_t^{(e,h)}/E_g \ll 1$, e.g., $w_t^{(e,h)}/E_g \approx 0.1$. The variation scale $\delta\eta$ of $F(\eta, \xi)$ in η for $|\eta| \ll 1$ appears to be $\delta\eta \simeq (\Delta\eta)^2/2\overline{\eta} \approx 0.1 \sim (\Delta\xi)^2$ for typical soft-mode concentrations $c_d \sim 10^{-2}$ in glasses (Ch. 7). The parameters used in calculations include both the parameters in Eqs. (7.17)–(7.19) and those in Eqs. (15.43)–(15.47). Then, the results of numerical calculations for $g(E)$, $c_2(\zeta)$, and ζ can be found. At least in order-of-magnitude estimates, the DOS $g(E)$ around the midgap and the related thermal equilibrium properties are not expected to strongly change at typical variations of the parameters in Eqs. (7.17)–(7.19) and in Eqs. (15.43)–(15.47) and of the type of $F_0(\eta)$. Moreover, the results are weakly sensitive to the type of $F_0(\eta)$, as quasiharmonic soft modes with not very small $|\eta| \gg \eta_L \sim 10^{-2}$ are found to mostly contribute to the properties under discussion while the difference in $F_0(\eta)$ is important for very small $|\eta| \leq \eta_L$.

More in detail the results of calculations are described and discussed in [249, 250]. Let us here only note that Eqs. (15.29), essential for the above-mentioned calculations. Particularly, with Eqs. (15.28) and

(15.39)–(15.42), the relations (15.29a) and (15.29b) still hold so the discrete levels

$$E_j = E_c - j|W_1| \quad \text{or} \quad E_j = E_v + j|W_1|, \tag{15.48}$$

with $j = 1, 2, 3$ and $E_g/4 \le |W_1| < E_g/3$, $|W_1| - E_g/4 \ll E_g$, are characteristic of the mobility gap spectrum and coexist with the continuum spectrum DOS $g(E) \simeq g_2(E)$ around $E = \overline{E} \equiv E_g/2$. In other words, the coexistence of continuum features and discrete levels, which is found in experiments to characterise negative-U centers in mobility gap of semiconducting glasses (see, e.g., [124, 125]) can also be established in a rather natural way, in the theory under discussion.

15.6 Concluding remarks

A consistent theory of two-band self-trapping of electron (holes) in a semiconducting glass is presented above (in Ch. 15) in which negative-U centres in a mobility (interband) gap, between conduction and valence bands, are formed as singlet electron (hole) pairs self-trapped in substrate atomic soft-mode configurations. The latter appear to be characteristic of the glasses (and possibly of some types of defects in crystalline lattices) and exhibit anharmonicity in single soft motion modes. Both softness of motion mode and hybridisation of bare electron (hole) state with extended states of a nonparent band (e.g., of valence band states for an electron) are essential for formation of self-trapped states and negative-U centres. The hybridisation significantly changes the substrate soft configurations introducing extra anharmonicities in related atomic dynamics, i.e., giving rise to an additional softness and adding or subtracting a minimum in adiabatic potentials. The resulting singlet electron pair correlation energy U becomes negative and, for most negative-U centres, comparable in magnitude to the gap width, $|U| \approx E_g/2$. The dependence of $U(\eta)$ on the configuration softness (η) is nonmonotonic and exhibits a deep minimum with large magnitude $|U_m| \simeq E_g/2$ characteristic of negative-U centres. For the latter, the DOS is rather high, $g_2(E) \lesssim 10^{-2}$–10^{-3} and determines the position of the low-temperature Fermi level ζ near the midgap $\overline{E} \equiv E_g/2$. The related low-temperature concentration of negative-U centres is relatively high, $c_2 \lesssim 10^{-3}$–10^{-4}, corresponding to a low electron spin resonance (ESR) signal and to Fermi level pinning observed in glassy semiconductors. The high susceptibility of soft configurations and related negative-U centres to external fields like hydrostatic pressure in semiconducting glasses gives rise to significant effects, which are predicted to be considerable even at not very high pressures $p \approx 10^4$–10^5 bar and still finite gap width, as discussed in Ch. 16.

The earlier models of negative-U centres in semiconducting glasses are characterised either by continuum gap spectrum with high DOS $g_2(\overline{E}) \sim 10^{-2}$ and $c_2 \sim 10^{-3}$ or by quasidiscrete levels $\approx E_c - \nu E_g/4$ at $\nu = 1, 2, 3$ in the gap with much lower $c_2 \lesssim 10^{-5}$. The theory under consideration, as well as the earlier theory of negative-U centres in soft configurations, establishes the coexistence of both continuum features of the gap and discrete levels $E_\nu \approx E_c - \nu\varepsilon_0$ at $E_g/4 \leq \varepsilon_0 < E_g/3$ in the gap, as observed in thermal equilibrium and nonequilibrium phenomena in semiconducting glasses. In this connection, the earlier models of negative-U centres in semiconducting glasses may be considered as simplified limiting cases of negative-U centres at very low DOS $g_2(E)(\rightarrow 0)$ and concentration $c_2(\rightarrow 0)$.

16

ADDITIONAL MANIFESTATIONS OF SOFT MODES IN GLASSES

Additional manifestations of atomic-motion soft modes in glasses may be available. One of those is represented in what follows in a preliminary form, by applying both qualitative suggestions and scale estimations. Let us confine ourselves by taking into account only the vast majority of soft-mode potentials being single-well harmonic potentials, of which the concentration $c_{sm}^{(1)} \simeq c_{sm}$, the total soft-mode concentration.

16.1 Negative-*U* centres model of photostructural changes in semiconducting glasses

As observed in experiments (see, e.g., review [253]), light of frequency $\nu \approx \nu_0 = E_{\text{opt}}^0/h$ corresponding to original optical gap width E_{opt}^0, produces long-lived "photostructural changes", i.e., observable changes in medium-range order structures and related change in properties of semiconducting glasses (SG), particularly of chalcogenide glasses, with the mobility-gap width $E_g \approx (1-2)$ eV and $E_{\text{opt}}^0 \simeq E_g$. Related substantial changes for temperatures far below the empirical (calorimetric) glass transition temperature T_g are produced in a variety of macroscopic physical and chemical properties of these glasses. A rather surprising effect is the "photodarkening" (PD) that is a gap-light induced decrease ΔE_{opt} of the original optical gap width E_{opt} by up to around 10%. Interesting anisotropic effects (e.g., dichroism, birefringence) associated with the photostructural changes have also been revealed recently in the experiments.

The nature of the photostructural changes was one of challenging problems in the physics of semiconducting glasses since they were discovered more than three decades ago. Some models, involving assumptions concerning microscopic mechanisms, have been proposed to account in general for photo-induced metastable defects in semiconductors and insulators [234, 253, 254], in which the metastability was due to generation of electronic excitations with finite life-time. Such phenomenological models were based on an assumption that adiabatic potentials of local atomic configurations in essential motion modes were different in their ground and excited (e.g., by gap-light) electron states, and on plausible suggestions that agreed with known features of the phenomenon. In fact, even for crystalline materials such models have been put on a proper theoretical footing only for defect generation in several kinds of materials, e.g., alkali halides and crystalline SiO_2, whereas in general the important atomic modes, as well as the parameters of local configurations, actually were not well-defined in earlier theories. For example, such a model was postulated [254] to be characterised by a two-branch adiabatic potential in the configuration space, in which the lower branch was a double-well potential for a ground electron state while the upper branch was a single-well potential for an excited electron state, separated by a splitting energy $\Delta \gg h\nu_v$ (ν_v is a typical vibration frequency). The process of creation and destruction of metastable defects was supposed to be due to two basic effects in the configuration space: (I) light-induced Franck–Condon (FC) effect, corresponding to "vertical" transitions between two (lower and upper) branches of the adiabatic potential at the same configuration $x = x_{\min}$ (vertical transitions); (II) thermal-activated transitions between double-well potential minima in the lower potential branch. As well known (see, e.g., [113, 172]), vertical transitions between two branches of adiabatic potential are determined usually by a small nonadiabaticity of electron-lattice interactions, of which the typical scale is small, $\sim (m/M)^{\varkappa} \ll 1$ at $\varkappa \sim 1$, where m or M is typical mass scale of electron or atom in atomic lattice. A standard kinetic equation for the metastable defects concentration c_d was analysed and realistic features of its dependence on ν and T were discussed in the phenomenological models. Experimental studies (e.g., [253, 255]) of the photodarkening parameter ΔE_{opt}, assumed to be proportional to c_d, have shown qualitative agreement of observed features with the expected ones in phenomenological models.

However, in spite of the stimulating role of such models for experimental studies of photostructural changes and their interpretation, a number of basic questions seemed to remain unanswered in the models: (1) Why photostructural changes occurs mainly in semiconducting glasses? (2) What are the atomic motion modes involved in these changes, and what is the mechanism that gives rise to the changes as a result of the occurrence of electronic excitations produced by gap light? (3) What is the criterion for the existence of postulated adiabatic potentials, and why is the criterion realised in such glasses? (4) What are the analytical expressions for transition probabilities in kinetics of the photostructural changes?

The questions (1)–(4), as well as some others mentioned below, can be answered at least qualitatively and scalewise, in the theoretical model under consideration, in which the basic charge carriers interacting with gap light are negative-U centres described in Ch. 15. In other words, occupied localised states in the mobility gap of a glass with typical $E_g \approx 1-2$ eV, are in fact expected to be negative-U centres. The theoretical results can be applied for interpreting a variety of observed data for anomalous electron properties of the glasses, e.g., the absence of substantial paramagnetism, pinning of the Fermi level near the mobility gap middle, anomalous photo-induced effects and ac conductivity (see, e.g., [79]). Let us remind that the first well-known self-trapping effect for a single electron was a polaron of large-size ($R_p \gg a_1 \sim 2$ Å) formed due to a strong interaction of a single conduction electron with phonons in a crystalline semiconductor [245]. In this system, an electron self-consistently changes the neighbourhood and strongly interacts with the resulting environment, so that the total energy decreases due to appearance of self-trapping energy, $W_1 < 0$, which is small in the sense that its magnitude is $|W_1| \ll E_g$, with only a single (e.g., conduction) band involved in the process. On the contrary, a singlet electron pair self-trapping is due to much stronger pair-soft-mode interactions, involving both conduction (parent) and valence (non-parent) bands, and the pair self-trapping energy can analytically be estimated in scale as (see Eq. (15.29))

$$W_2 \approx -2Q^2/k_{\text{eff}} \approx -E_g \quad \text{at } k_{\text{eff}} \simeq k_g = Q^2/E_g \ll k_0,$$
$$\text{so } |W_2| \approx E_g \gg |W_1| \approx Q^2/k_0. \qquad (16.1)$$

Here Q is a typical scale of electron-vibration coupling energy, e.g., $Q \approx 3\,\text{eV}$, and the inequality $|W_2| \gg |W_1|$ holds because of small spring constants of soft modes, $k_{\text{eff}} \ll k_0$. This scale estimation also can follow from direct calculations of the model Hamiltonian in Eq. (15.2). The hybridisation of the electron pair state with valence band states becomes particularly strong at the valence mobility edge E_v and creates a renormalised single-electron state $\Psi_{0\sigma}$ of a size $R_0 \gtrsim a_1$, with an energy level $E_{0\sigma} \gtrsim E_v$. Indeed, for small soft-mode spring constants, $0 < k \lesssim 0.1k_0$ (Eq.(7.5)), the self-trapping energy magnitude might exceed E_g but then the pair energy level repels off the valence band edge states due to the hybridisation, so the characteristic soft-mode spring constant $k_{\text{eff}} \approx k_g$.

The basic idea of the theory under consideration [256, 257] is that the photo-structural changes are metastable "defects" in the original glass medium-range structures, generated around gap-light excited negative-U centres. Actually, the excited state of a negative-U centre can be formed due to absorption by the ground-state centre of a gap light photon and is accompanied by a soft-mode displacement Δx_{exc}. On the other hand, the formation of a ground state centre is provided by the strong self-trapping of a singlet electron (or hole) pair, associated with a soft-mode displacement Δx_{ST}, its typical magnitude is relatively large at $|\Delta x_{\text{ST}}| \sim 1$ (the soft-mode displacements x, as well as the mean atomic separation $a_1 \approx 2$–3, are measured in atomic length unit $a_0 = 1\,\text{Å}$). From above one can suggest that in scale $|\Delta x_{\text{exc}}| \sim |\Delta x_{\text{ST}}| \sim 1$ at $\Delta x_{\text{exc}} \approx -\Delta x_{\text{ST}}$, so both typical displacements are comparable in magnitude. Moreover, the excited electron states of a negative-U centre are suggested to be essential for non-equilibrium phenomena, including the photostructural changes, in the glasses.

Let us emphasise that the soft-mode potentials under consideration are most probably single-well harmonic potentials with small spring constants, as those are characterised by the most large concentration $c_{sm}^{(1)}$, at $c_{sm} \simeq c_{sm}^{(1)}$. Hence, soft modes with anharmonic potentials, in particular with double-well potentials, are neglected in what follows.

As the localised system under consideration is a singlet electron pair (with spatial coordinates r_1, r_2) strongly interacting with a soft mode x, the negative-U centre occurs either in ground state, with wave function $\Psi_2^{(0)}(r_1, r_2)$, or in excited state, with wave function $\Psi_2^{(1)}(r_1, r_2)$.

The appropriate system Hamiltonian $H_{2e,sm}$, can be written in the following form:

$$H_{2e,sm}(\mathbf{r}_1, \mathbf{r}_2; x) = H_{2e}(\mathbf{r}_1, \mathbf{r}_2) + H_{sm}(x) + H_{\text{int}}(\mathbf{r}_1, \mathbf{r}_2; x), \qquad (16.2)$$

where $H_{2e}(r_1, r_2) = T_{2e} + V_{ee}$ is a standard Hamiltonian of "fast" subsystem (singlet pair of electrons of effective mass m), including the kinetic energy term T_{2e} and electron–electron repulsion energy V_{ee}, while $H_{sm}(x) = T_{sm} + V_{sm}(x)$ is a Hamiltonian of "slow" soft-mode subsystem, with a harmonic potential energy for most soft modes, $V_{sm}(x) \simeq kx^2/2$ at $0 < k \lesssim 0.1\,k_0$, and a kinetic energy T_{sm} describing a relatively small non-adiabatic perturbation to the applied adiabatic approximation, at a typical small parameter $(m/M_{sm})^{1/2} \approx 10^{-2}$–$10^{-3}$. The interaction H_{int} between an electron pair and a finite soft mode displacement x in a basically covalent glass in question can be considered as a short range one and approximated by an additive form like

$$H_{\text{int}}(\mathbf{r}_1, \mathbf{r}_2; x) \simeq \sum_{j=1,2} \delta\Phi(\mathbf{r}_j; x), \quad \text{where } \delta\Phi(\mathbf{r}; x) \equiv \delta\Phi(r, \Theta; x)$$
$$\simeq (D_1(\Theta)x + 0.5 D_2(\Theta)x^2) \exp(-r/L_0). \qquad (16.3)$$

Here $\mathbf{r} \equiv (r, \Theta, \varphi)$, whereas $D_1 \approx C_0 L_0^{-1}$ and $D_2 \approx C_0 L_0^{-2}$ in general give rise to extra contribution to soft-mode spring constant, $k \equiv k_{sm} \longrightarrow k + \Delta k$. This appears to be a simplest possible form of $H_{\text{int}}(r_1, r_2; x)$ explored below, which gives rise to a small split of the crossing ground- and excited-state adiabatic potentials from each other (another limit case with angle dependent $\delta\Phi(r, \Theta; x)$, which can also correspond to a rather large splitting, is considered elsewhere). As usual, it is assumed that the interaction length is of short range order scale, $L_0 \sim a_1 \approx (2$–$3)$ Å, while the energy C_0 is of scale of a typical atomic elasticity energy measured below in an energy unit $\varepsilon_0 = 1$ eV, so $|C_0| \sim 10$. As noted earlier [231, 238, 249], a negative-U centre in ground state can play the role of an extra, weak, covalent bond so that $\Delta k \lesssim 0.1\,k_0$. In accordance with SMM, the distribution of random C_0 and L_0 is suggested to be so narrow that their average values can be used instead.

As seen below, the resulting adiabatic potential $E_2(x)$ for the soft mode, indeed has a structure that consists of two (quasi) parabolic curves crossing each other at a certain $x = x_c$, at which generally speaking exhibits a splitting into two renormalised curves, so that a lower branch is a double-well potential branch, $E_2^{(0)}(x)$, either with deeper potential well, or with a deeper double-well potential, related to ground state wave function $\Psi_2^{(0)}(r_1, r_2)$ of negative-U centre, and another, shallower, single-well branch, $E_2^{(1)}(x)$, associated with a gap-light generated excited state wave function $\Psi_2^{(1)}(r_1, r_2)$ of the centre (the potential is qualitatively similar to a two-branch adiabatic potential postulated in [254] for interpreting empirical data and briefly described above). In order to describe the adiabatic potentials $E_2^{(\alpha)}(x)$ $(\alpha = 0, 1)$, it is necessary to specify the wave functions $\Psi_2^{(\alpha)}(r_1, r_2)$, which are approximated in a usual way [172]:

$$\Psi_2^{(\alpha)}(\mathbf{r}_1, \mathbf{r}_2) \approx (1/2)^{1/2}[\phi_1^{(\beta)}(\mathbf{r}_1)\phi_1^{(\delta)}(\mathbf{r}_2) + \phi_1^{(\beta)}(\mathbf{r}_2)\phi_1^{(\delta)}(\mathbf{r}_1)]. \quad (16.4)$$

Here, $\phi_1^{(0)}(r) = (\pi\rho_0^3)^{-1/2}\exp(-r/\rho_0)$ is a single-electron ground-state wave function of size ρ_0, while $\phi_1^{(\beta,\delta)}(r)$, at $\{\beta,\ \delta\} = \{0 \text{ and/or } 1\}$ is a single-electron excited-state, localised (of finite size $\rho_1 < \infty$) or nonlocalized (of size $\rho_1 \longrightarrow \infty$) wave functions $\phi_1^{(\beta,\delta)}(r)$. For example, a single-electron excited state can be a singly-ionized state (near the mobility edge) like orthogonalised plane wave, or excited bound state, of size $\rho_1 \longrightarrow \infty$. With such choices, the pair wave function $\Psi_2^{(1)}(r_1, r_2)$ can become orthogonal to $\Psi_2^{(0)}(r_1, r_2)$, and the theoretical approach can become a qualitative and scalewise one.

In the present approach, the adiabatic potential $E_2(x)$ related to a "defect" appears to consist of two "branches" (at $\alpha = 0$ and $\alpha = 1$) and to be described in a standard way as follows [256, 257]:

$$\begin{aligned} E_2(x) &\equiv \{E_2^{(\alpha)}(x)\} \\ &= \iint (d\mathbf{r}_1)(d\mathbf{r}_2)\Psi_2^{(\alpha)}(\mathbf{r}_1, \mathbf{r}_2)H_{2e,sm}(\mathbf{r}_1, \mathbf{r}_2; x)\Psi_2^{(\alpha)}(\mathbf{r}_1, \mathbf{r}_2) \\ &= T_2^{(\alpha)} + 2Q^{(\alpha)}x + (1/2)\,k^{(\alpha)}x^2 \\ &= W_2^{(\alpha)} + (1/2)\,k^{(\alpha)}(x - x_2^{(\alpha)})^2. \end{aligned} \quad (16.5)$$

Here the functions $E_2^{(\alpha)}(x)$ for the ground ($\alpha = 0$) and excited ($\alpha = 1$) electron states describe two potential branches with renormalised soft-mode spring constants $k^{(\alpha)}$, energy minima $W_2^{(\alpha)} = E_2^{(\alpha)}(x_2^{(\alpha)})$, and coordinates $x_2^{(\alpha)}$, which are functions of the model parameters C_0 and L_0. Moreover, in Eq. (16.5), $Q^{(\alpha)}$ is an effective electron–soft-mode interaction coefficient, $x_2^{(\alpha)} = -2Q^{(\alpha)}/k^{(\alpha)}$ are the soft-mode coordinates for two energy minima, $W_2^{(\alpha)} \simeq T_2^{(\alpha)} - 2(Q^{(\alpha)})^2/k^{(\alpha)}$, whereas, in terms of $\Lambda_\alpha = 1 + \rho_\alpha/2L_0$, the kinetic energy of electron pair is $T_2^{(\alpha)} = (2-\alpha)\, E_0\, (\Lambda_\alpha - 1)^{-2}$, where $E_0 = (\hbar^2/8mL_0^2)$ and m is an electron effective mass. In addition, $Q^{(\alpha)} = D_1/\Lambda_\alpha^3$ and $Q^{(1)} \simeq Q^{(0)}/2$, while the changes $\Delta k^{(\alpha)}$ in soft-mode spring constants are given by $k^{(\alpha)} = k + \Delta k^{(\alpha)}$, with $\Delta k^{(0)} = k^{(0)} - k \simeq D_2/\Lambda_0^3$ and $\Delta k^{(1)} \simeq \Delta k^{(0)}/2$. As follows from Eq. (16.1), most soft modes contributing to the formation of negative-U centres are characterised by $k^{(0)} \approx k_g = 2(Q^{(0)})^2 E_g^{-1}[1 + 2E_0(\Lambda_0 - 1)^{-2}E_g^{-1}]^{-1}$ while $\Delta k^{(0)}$ apparently is less than k_g for realistic values of the parameters. For the sake of simplicity, it is reasonable to concentrate on the case (16.3) at $D_2 \simeq 0$, so then all spring constants actually become equal, $k^{(\alpha)} \simeq k \simeq k_g$.

Thus, the calculations give rise to an approximate formula (16.5) for the adiabatic potential $E_2(x)$, with its two branches which cross each other at a soft-mode displacement $x = x_c$, and eventually can give rise to a small energy splitting Δ of two branches. An effective frequency of transitions between the two branches at a crossing point is $P_0 = w_{LZ}$, the probability (per sec) of Landau–Zener, semi-classical, transition [172] due to small nonadiabatic perturbation of the potential; in such a simplest case, a typical estimation appears to be $w_{LZ}/\nu_{sm} \lesssim 10^{-2}$ for realistic values of model parameters and the energy-level small splitting, $\Delta \approx hw_{LZ} \ll h\nu_{sm}$, can be neglected. In this sense, the adiabatic potential corresponds to two single-well curves, crossing each other at $x = x_c$ in the configuration space, in which the upper branch corresponds to excited states of negative-U centres, whereas the lower branch to ground states of the centres.

The calculations also show that the value of Λ_0 (i.e., of ground-state size ρ_0) may be obtained by minimising the ground-state potential energy $E_2^{(0)}(x_2^{(0)})$ with respect to ρ_0, and by solving the resulting equations

$$E_0/(\Lambda_0-1)^3 \approx 3D_1^2/(k_g\Lambda_0^7) \quad \text{at } \gamma\Lambda_0(\Lambda_0-1)^3 \approx E_0, \tag{16.6}$$

where $\gamma = \hbar^2/3mL_0^2E_g \lesssim 0.1$ for realistic values of electron effective mass $m \approx m_0$, the free electron mass. For such γ, one can obtain a scalewise estimation $\rho_0(\gamma)/2L_0 \approx \gamma^{1/3}(1+\gamma^{1/3}/3)$ (as also seems to be confirmed by numerical calculations). Equation (16.6) and an extra energy minimum requirement, $E_0/(\Lambda_0-1)^4 > 7D_1^2/(k_g\Lambda_0^8)$, appear to give rise to inequalities $1 < \Lambda_0 < 1.75$, i.e., $0 < \rho_0 < 1.5\ L_0$, which can be realised at the above approximation of $\rho_0(\gamma)$. Finally, the criterion for the existence of metastable "defects" is that the soft-mode displacement at the above-mentioned crossing point x_c of two adiabatic potential branches occurs between the coordinates $x_2^{(\alpha)}$ of the ground-state and excited-state minima. Taking into account Eq. (16.6) and, by definition, $E_2^{(0)}(x_c) = E_2^{(1)}(x_c)$, one can find after simple algebra that

$$x_c \approx 3D_1(1-\Lambda_0)/(k_g\Lambda_0^4). \tag{16.7}$$

Then, the criterion of a "defect" state occurrence appears to be that Λ_0 should exist approximately between 1 and 3. Since actually for the model in question the inequalities $1 < \Lambda_0 < 1.75$ hold true, the criterion can indeed be realised for SGs. In accordance with Eqs. (16.1) and (16.4)–(16.6) at $E_g = 2D_1^2/(k_g\Lambda_0^6)$, the energy $E_2^{(\alpha)}(x_c)$ at the crossing point $x = x_c$ exceeds the minimum of excited state energy by the activation energy

$$\begin{aligned} E_{1c} &= k_g(x_c - x_2^{(1)})^2/2 \\ &\approx [2D_1^2/(2k_g\Lambda_0^6)](1-\kappa/\Lambda_0)^2 \equiv E_g\zeta^2, \end{aligned} \tag{16.8}$$

where $\kappa \approx 1.5$ and $\zeta^2 \sim 0.1$ and so $E_{1c} \sim 0.1E_g \ll E_g$. However, the activation energy E_{1c} can be close to its experimental scale that rather is defined as $0.1 < E_{1c}^{(\mathrm{exp})} \lesssim 1$ eV so the scale of ζ^2 may be estimated by inequalities $0.1 < \zeta^2 \lesssim 1$. Actually, in Eq. (16.8), the values of κ and of ζ^2 should not be understood too literally in the scalewise (not quantitative) analysis. Therefore, if the difference between the scale estimations $\zeta^2 \sim 0.1$ and $\zeta^2 \lesssim 1$ can be assumed negligible, the activation energy E_{1c} may be close to empirical scale, $0.1 < E_{1c} \lesssim 1$ eV (unfortunately, more precise estimations of the parameters are not yet available).

Let us now introduce, for the photon flux F giving rise to photostructural changes, the cross-section $\sigma^{(r)}_{1\to 0} = P^{(r)}_{1\to 0}/F$ of a radiative transition, from one branch of two-branch adiabatic potential (Eq. (16.5)) and a single-photon state $|1\rangle$ to another branch and a zero-photon state $|0\rangle$, due to photon emission (see e.g., [217]). That is actually the cross-section $\sigma^{(FC)}_{1\to 0}$ for a well-known "vertical", Frank–Condon (FC) transition at photon emission ($1 \to 0$), from a vibrational state around in a potential well (1) to a state in another potential well (0),

$$\sigma^{(r)}_{1\to 0} \equiv 1/FP^{(r)}_{1\to 0} \simeq \sigma^{(FC)}_{1\to 0}. \tag{16.9}$$

The FC cross-section $\sigma^{(FC)}_{1\to 0}$ for negative-U centres can explicitly be calculated by applying to adiabatic potentials for negative-U centres the well-known Lax formula (see, e.g., [257]) for a radiative, e.g., photon-emission, transition which is essentially associated with multi-phonon processes characteristic of transitions among self-trapped states. In this case (see Eq. (16.1)), the FC cross-section is as follows:

$$\sigma^{(FC)}_{1\to 0}(\nu, T) \approx \sigma_0 \tanh\left(h\nu_{sm}/2T\right) \exp\left[-E^{(1)}_a(\nu)/\theta_T\right], \tag{16.10}$$

where $E^{(1)}_a(\nu) = (E^{(1)}_m - h\nu)^2/E_g$ plays the role of a characteristic activation energy at "high" $T > h\nu_{sm}/2$, and M_{tr} is the relevant electric-dipole transition matrix element in the expression $\sigma_0 \approx 8\pi^2\nu|M_{tr}|^2(2M_{sm})^{1/2}/c_m\varepsilon(\nu)hE_g^{1/2} \sim 10^{-14}\,\text{cm}^2$ ($\approx L^2_{sm}$), c_m is the light velocity and $\varepsilon(\nu)$ is the dielectric parameter of semiconducting glass. Moreover, in Eq. (16.10), $\theta_T = (h\nu_{sm}/2)\coth(\hbar\nu_{sm}/2T)$ and a typical soft mode frequency is $\nu_{sm} \sim 10^{12}$/s. One can rather simply find that $E^{(1)}_m = \gamma_1 E_g/4$ and $E^{(0)}_m = \gamma_0 E_g < E_g$ at $3/4 \lesssim \gamma_0 \lesssim 5/6 < \gamma_1 \lesssim 4/3$ (one can also see that there is no contradiction with the fact that the observed maximum of light excitation spectrum occurs in the so-called Urbach absorption tail [79] at $\nu \approx \nu_{\max} < E_{opt}/h$). Then, the FC cross-section changes with T as $T^{-1}\exp[-E^{(\alpha)}_a(\nu)/T]$ for $T > h\nu_{sm}/2 \approx 25$ K while it almost does not change at lower $T < h\nu_{sm}/2$. In addition, taking into account Eq. (16.1), one can indeed see that the effective "activation energy" $E^{(1)}_a(\nu)$ is simply related to E_g (formulae for photon-absorption transition, $0 \to 1$, are similar, with trivial change).

On the other hand, the "non-radiative" probability (per sec) of relaxation transition, can be approximated as

$$P_{1c0}^{(nr)} \simeq P_0 \exp(-E_{1c}/T), \tag{16.11}$$

which increases with T by a thermal-activation law with activation energy $E_{1c} = E_g\zeta^2 < E_g$, as far as the finite time of vibrational relaxation is small enough, $\tau_{\rm ph} \ll P_0^{-1}$ (the estimation of "non-radiative" probability $P_{0c1}^{(nr)}$ is similar). Indeed, in accordance with the comment to Eq. (16.6) and with Eqs. (16.7)–(16.8), the effective frequency of relaxation transition actually is the product of $P_0 = w_{LZ}$, the probability (per sec) of Landau–Zener transition at the crossing point $x = x_c$ (Fig. 18), by the thermal activation probability to the crossing point.

In the theoretical model under discussion (Eq. (16.5)), in which [256, 257] photostructural changes are formed around metastable "defects" identified as gap-light excited negative-U centres, a qualitative and scalewise analysis can be presented on their kinetics in terms of probabilities of transitions between adiabatic potential curves (and potential wells $E_2^{(\alpha)}(x)$, $\alpha = 0, 1$). The kinetic equations, describing the time (t) evolution of concentrations $c_0(t)$ of ground-state negative-U centres and $c_1(t) \equiv c_d(t)$ of the "defects", can be presented as follows:

$$\begin{aligned} \frac{dc_d(t)}{dt} &= \left[P_{0\to1}^{(r)}c_0(t) - P_{1\to0}^{(nr)}c_d(t)\right] \quad \text{and} \\ \frac{dc_0(t)}{dt} &= \left[P_{1\to0}^{(r)}c_0(t) - P_{0\to1}^{(nr)}c_d(t)\right], \end{aligned} \tag{16.12}$$

for a simple case of a gap-light flux $F(t)$ switched in time as a step function, $F(t) = \{F_0,\ \text{at}\ t > 0; 0,\ \text{at}\ t < 0\}$, and of a given concentration ($c_0 = const. > 0$ at $t \le 0$) (at typical $c_0(0) \sim 10^{-3}$ and $c_1(t \le 0) = 0$). The transition probabilities (per sec) $P_{1\to0}^{(r)}$, $P_{0\to1}^{(r)}$ for gap-light induced, radiative, transitions (between two branches of the adiabatic potential) and $P_{0c1}^{(nr)}$ and $P_{1c0}^{(nr)}$ for thermal, non-radiative, relaxation transitions (via thermal activation state at $x = x_c$). A useful approximation of Eq. (16.12) is to reduce both balance equations to a single equation for the concentration

$c_1(t) \equiv c_d(t)$ of excited-state negative-U centres,

$$dc_d(t)/dt \simeq [P^{(r)}_{0\to 1}c_0(t) - P^{(r)}_{1\to 0}c_d(t)] - P^{(nr)}_{0\to 1}c_d(t), \qquad (16.13)$$

by assuming a conservation of total concentration under consideration, $c_{tot}(t) = c_0(t) + c_d(t) = C_0 = const.$, and the characteristic relation $c_0(t) \simeq C_0 - c_d(t) \gg c_d(t)$ for essential, not too large times.

Actually, Eq. (16.13) is derived in the soft-mode model for describing kinetic processes in a system of negative-U centres in ground (0) states and gap-light excited states in adiabatic-potential wells (Eq. (16.5)) of the centres in which the excited states are metastable "defects". This equation is similar to the balance equation suggested earlier in [254] for one of the functions, say $c_d(t) \equiv c_1(t)$:

$$dc_d(t)/dt \simeq F(\sigma_0 c_0 - \sigma_1 c_d) - c_d/\tau. \qquad (16.14)$$

Here, F_0 is the photon flux that produces the metastable defects, σ_0 is a cross-section of photon absorption followed by Franck–Condon transition from ground-state negative-U centres to excited-state, metastable "defect" ones while σ_1 is a cross-section of an inverse transition from "defect" states to ground states of the centres, while τ is the lifetime of metastable state due to a thermal transition above an energy barrier. It was assumed in this approximation that the inverse lifetime depends on temperature by an activation law, $\tau_d^{-1} = \nu_0 e^{-\Delta_d/T}$, where the depth of metastable state potential well, $\Delta_d \gg T$ (e.g., at $\Delta_d \sim 1$ eV), and ν_0 is a typical vibration frequency in a potential well. It was also suggested that cross-sections σ_0 and σ_1 depend on temperature and photon energy in a rather standard way [254]:

$$\sigma_j(\nu) = \sigma_j^{(0)} \frac{h\nu}{T} e^{-E_j(\nu)/T}, \quad j = 0, 1, \qquad (16.15)$$

where $E_j(\nu)$ is a related transition energy, $\sigma_j^{(0)} = |4\pi x_{12}|^2 e^2/\hbar c n(\nu)$, $n(\nu)$ is the refractive index at frequency ν and ex_{12} is the electric-dipole matrix element. For a light flux switched in time as a step function, $F(t) = \{0,$ at $t < 0;\ F_0 = const.$ at $t > 0\}$, the standard solution of Eq. (16.13) and (16.14), is found to be as follows:

$$\Lambda(t) \equiv c_d/(c_d + c_0) = c_d/C_0 \simeq \Lambda_\infty + (\Lambda_0 - \Lambda_\infty)e^{-\theta t}, \qquad (16.16)$$

where $\Lambda_0 = \Lambda(t=0)$, $\Lambda_\infty = F_0(\sigma_0+\sigma_1)+1/\tau$ is the steady-state value of metastable defects population and $\theta = F_0(\sigma_0+\sigma_1)+1/\tau$. If the flux is large, $F_0(\sigma_0+\sigma_1) \gg 1/\tau$, then $\Lambda_\infty \simeq \sigma_1 \cdot (\sigma_0+\sigma_1)^{-1}$. After the gap light is switched off at $t > 0$, the metastable defects population decreases exponentially, $\Lambda(t) \simeq \Lambda_\infty \cdot \exp(-t/\tau)$, with increasing time. This describes an experimentally well-known effect of thermal bleaching with the gap-light decrement $1/\tau \propto \exp(-const./T)$. Moreover, the T-dependence of Λ_∞ is found in this approach to be described by the following expression:

$$\Lambda_\infty = \left\{1 + (\sigma_1^{(0)}/\sigma_0^{(0)}) \cdot \exp[(E_0(\nu) - E_1(\nu))/T]\right\}^{-1}. \qquad (16.17)$$

This means that Λ_∞ either decrease, if $E_1(\nu) - E_0(\nu) > 0$, or increase, if $E_1(\nu) - E_0(\nu) < 0$, with increasing temperature. The dependence of Λ_∞ on photon energy $h\nu$ is determined by Eq. (16.17) and its character also depends on the ratio $E_1(\nu)/E_0(\nu)$. The above described results hold at some conditions (in particular, at $w_{LZ} \ll \nu_{sm}$) at which Eqs. (16.6)–(16.13) can be applied for calculations of the basic kinetic properties under discussion. It appears that the kinetic properties obtained in the soft-mode model of negative-U centres [256, 257] include, in particular, the properties found in the phenomenological kinetic model [254]. Moreover, the experimental test [255] of this model (see, e.g., Eqs. (16.16) and (16.17)) seems to agree with both models, which hold at some conditions, where Eqs. (16.6)–(16.13) can be applied for calculations of the kinetic characteristics.

As noted above, the described results hold at typical negligibly small w_{LZ}/ν_{sm}, at which Eqs. (16.6)–(16.11) can be applied for calculations of basic characteristics of the kinetic processes in question. Then, the above-mentioned questions (1)–(4) appearing in the phenomenological model can be answered in the soft-mode model as follows: (1) the photostructural changes occur just in semiconducting glasses in which the basic charge carriers are negative-U centres; (2) atomic-motion modes involved in the changes are soft modes (Chs. 7–10), and the mechanism providing the changes is due to the occurrence of electronic excitations of negative-U centres, produced by gap light; (3) the criterion for the existence of required adiabatic potentials is described above and indeed realised in the glasses; (4) the analytical expressions for transition probabilities in the kinetics of photostructural changes, as well as the relations between energies characteristic of the changes, are justified and discussed above. There are also other questions which can be answered in the soft-mode model of glasses (see below), in particular, why the PD parameter $\Delta E_{\rm opt}$ typically decreases with increasing both temperature T, at $T_m \approx 1\,{\rm K} \ll T < T_D$ and with increasing hydrostatic pressure p in the range of moderately high one, at $10^6 > p \gtrsim 10^5$ bar.

Let us now rather naturally suggest, within the negative-U centre model based on Eqs. (16.5)–(16.14), that the photodarkening parameter $\Delta E_{\rm opt}$ is a regular function of steady-state metastable "defect" concentration c_∞. Since the latter by its definition is small, $c_\infty \ll 1$, one can conclude that $\Delta E_{\rm opt} \propto c_\infty$. Then, one can readily find from Eqs. (16.5)–(16.13) that, in general, both radiative, e.g., photon emission, processes and non-radiative, relaxation, processes contribute to the parameter $\Delta E_{\rm opt}$ in terms of simple analytical functions at $t > 0$:

$$\Delta E_{\rm opt}/h \propto c_\infty \propto F_0\sigma^{\rm (FC)}_{1\to 0}[F_0(\sigma^{\rm (FC)}_{0\to 1} + \sigma^{\rm (FC)}_{1\to 0}) + P_0 \exp{(-E_{1c}/T)}]^{-1}, \tag{16.18}$$

(accounting for FC absorption, radiative, processes within the same approach is similar, with trivial change).

16.2 Gap-light frequency dependence

As seen from Eqs. (16.18), the dependence of $\Delta E_{\rm opt}(\propto c_\infty)$ on ν is related to the frequency dependence of FC cross-section $\sigma^{\rm (FC)}_{1\to 0}$ for gap-light emission by a "defect" being an excited state of a negative-U centre. In this case, the cross-section $\sigma^{\rm (FC)}_{1\to 0}$ for the negative-U centre (Eq. (16.10)) is an expression that is essentially different from a standard expression for a single-band strong-coupling polaron [245] by accounting for the contribution of self-trapped states of a singlet electron pair, with eigenstates determined by hybridisation of states in both conduction band and valence band, and for associated multi-phonon processes, as described in Ch. 15, particularly in Eq. (16.10). In the case under discussion, the established expression in Eq. (16.10) is essentially different from that applied in earlier phenomenological models [254] mostly because a FC cross-section

$\sigma_{1\to0}^{(FC)}(\nu, T = \text{const})$ for photon emission exhibits a strong maximum at $h\nu = E_m^{(0)}$ (for photon absorption, a FC cross-section $\sigma_{0\to1}^{(FC)}(\nu, T = \text{const})$ exhibits a strong maximum at $h\nu = E_m^{(0)}$; in what follows, for the sake of simplicity, we mainly analyse the dependencies of PD parameter ΔE_{opt} on ν and T for cases of sufficiently narrow distributions in which the maximum can be approximated by average values). Then, Eq. (16.10) gives rise to

$$c_\infty \propto \tanh(T_{sm}/T) \cdot \exp[-(E_m^{(1)} - h\nu)^2/\Theta_T)]. \tag{16.19}$$

Hence,

$$dc_\infty/d\nu \propto c_\infty \cdot (E_m^{(1)} - h\nu)/E_g\Theta_T \tag{16.20}$$

so that both at $T < T_{sm}$ and $T \geq T_{sm}$

$$dc_\infty/d\nu \lesseqgtr 0 \quad \text{at } E_m^{(1)} - h\nu \lesseqgtr 0, \tag{16.21}$$

with $dc_\infty/d\nu = 0$ for a "resonance" frequency $\nu = \nu_r = E_m^{(1)}/h(\lesssim E_g/h)$. Thus the steady-state concentration c_∞ of metastable "defects" and the related PD parameter ΔE_{opt} exhibit here a monotonous behaviour with increasing ν and a maximum at $h\nu = h\nu_r = E_m^{(1)}$ in the absorption band, close to the optical-gap edge. The predicted ν-dependence is qualitatively similar to an experimental feature found in elemental SG (S, Se), with a maximum at $\nu = \nu_r$. The non-monotonous behaviour of f_∞ does not seem to directly follow from earlier phenomenological models [253, 254], in which the applied absorption FC cross-sections $\sigma_{\alpha t}^{(r)}(\nu, T)$ do not contain the characteristic factor $\exp[-E_m(\nu)/\Theta_T]$ related to singlet electron pair self-trapping in negative-U centres. The deviation of Eqs. (16.19)–(16.21) from the empirical ν-dependence [253] for SG being compounds As_2S_3 or As_2Se_3 and exhibiting after the described increase of c_∞ a change to a plateau-like behaviour or a weaker increase, with increasing ν, might be explained by assuming that the compounds are characterised by larger disorder induced spatial fluctuations in negative-U centres' energy levels and thus by substantially broader probability distributions for characteristic energies $E_a^{(m)}$ than those for the elemental SG.

16.3 Temperature dependence

The dependence of $\Delta E_{\rm opt}$ on T (Eq. (16.18)) is substantially related to temperature dependence of, e.g., a FC cross-section $\sigma^{(FC)}_{1\to 0}(\nu, T)$ for gap-light photon emission by an excited-state negative-U centre transforming into a ground-state centre. It follows from Eq. (16.18) that, for low $T < T_{sm}$, $\sigma^{(FC)}_{1\to 0}(\nu, T)$ is almost independent of T, while that is not the case for high $T > T_{sm}$. Then, for asymptotically low T, the theory predicts that

$$\Delta E_{\rm opt} \propto c_\infty \propto \sigma^{(FC)}_{1\to 0}\{\sigma^{(FC)}_{0\to 1} + \sigma^{(FC)}_{1\to 0} + P_0 \exp{(E_{1c}/T)}\}^{-1}, \qquad (16.22)$$

may strongly decrease with decreasing T at realistic, e.g., empirical, values $E_{1c} \sim 1$ eV. On the other hand, for asymptotically high T, a prediction rather can be that

$$\Delta E_{\rm opt} \propto c_\infty \propto \exp{[-E^{(1)}_a(\nu)]/T)]} \cdot \{\exp{[-E^{(0)}_a(\nu)]/T)]} + \exp{[-E^{(1)}_a(\nu)]/T)]} + P_0 \exp{(E_{1c}/T)}\}^{-1}, \qquad (16.23)$$

which may also decrease with increasing T for activation energies $E_0(\nu) = (E^{(1)}_m - h\nu)^2/E_g$ at a typical $E^{(1)}_m \approx \gamma_0 E_g (\gamma_0 \approx 1)$ with empirical energy E_{1c} of the scale of 1 eV and gap-light frequency $\nu \approx E_{\rm opt}/h \simeq E_g/h$, as far as it may be $dc_\infty/dT \propto \{[(E_a(\nu) - E_{1c})/T] - 1\} < 0$. The predicted decrease of $\Delta E_{\rm opt}$ with increasing T seems to be in a qualitative agreement with an empirical finding at ν around $\nu_M \lesssim E^0_{\rm opt}/h$.

It follows from Eqs. (16.10), (16.19), (16.22) and (16.23) that the behaviour of $\Delta E_{\rm opt}$ with increasing T may clearly differ from the dependence predicted in the earlier phenomenological model [254, 255], first of all because of the difference in the expressions for absorption cross-sections.

17

SUMMARY, CONCLUSIONS AND PROBLEMS

The present book contains a review of structure and dynamics of glasses, their formation due to liquid-to-glass transition and related relaxation processes (Chs. 1–4), but mainly it describes anomalous, universal, dynamic and thermal properties of glasses at low temperatures and/or respective frequencies (Chs. 5–16). The basic results of soft-mode model (SMM) of anomalous properties of glasses at temperatures and frequencies lower than their Debye values, are mostly presented and discussed. It is shown that SMM can explain (mostly qualitatively and scalewise) experimentally observed phenomena and processes and probably predict other ones.

A summary of the book can be described as follows

(i) For very (but not too) low temperatures, $0.1 < T \lesssim T_m \approx 1\,\text{K}$, and for respective frequencies ν, $0.1 < \nu \lesssim \nu_m = T_m/h$, experimental data appear to show that the specific heat $C(T) \propto T^{1+\alpha}$ and thermal conductivity $\kappa(T) \propto T^{2-\rho}$ at finite, small α and ρ (e.g., $0.1 \lesssim \alpha,\ \rho \lesssim 0.3$), while the Standard Tunneling Model gives rise to the specific heat $C \propto T$ and thermal conductivity $\kappa \propto T^2$ at $\alpha = 0 = \rho$. The formulae of SMM appear to be able to qualitatively explain the deviation of α and ρ from $\alpha = 0 = \rho$ and experimentally found more complex dependence of the specific heat on measurement time.

(ii) The experimental temperature dependence of acoustic and electromagnetic wave attenuation $\alpha(\nu = \mathit{const.}, T)$, exhibiting a broad

Debye-like maximum around $T \sim 10\,\text{K}$, can be explained (Sec. 8.2), and a crossover can be predicted at a characteristic $T = T^*$, from thermal TS excitations of very low energy $\varepsilon \lesssim h\nu_m$ at $T < T^*$ to classical, overbarrier excitations at intermediate temperatures, $T^* \lesssim T \lesssim \varkappa w$ with $1 < \varkappa \lesssim 2$.

(iii) SMM can explain why the Tunneling Model interprets qualitatively and scalewise the experimental data for dynamic and thermal properties of glasses only at very low frequencies, $\nu \lesssim \nu_m \approx 3 \cdot 10^{-2}\,\text{THz}$, and temperatures, $T \lesssim T_m \approx 1\,\text{K}$, but not at all low $T \ll T_D$ and $\nu \ll \nu_D\,\text{THz}$. The explanation is that SMM reduces to Tunneling Model only for predominant soft-mode TS excitations of very low energy, $\varepsilon \lesssim \varepsilon_m = h\nu_m = T_m$, as described in Secs. 8.1 and 9.1, whereas for moderately low energies (Ch. 10), $3\varepsilon_m \lesssim \varepsilon \lesssim \varepsilon_M = T_M$ at $T_M \sim 10^2$ K, predominant are basically different soft-mode harmonic vibrational excitations that scatter acoustic phonons.

(iv) SMM explains for the glasses the existence of boson peak and high-frequency sound due to the occurrence of Ioffe–Regel crossover for acoustic phonons inelastically scattered by soft-mode harmonic vibrational excitations.

Several major conclusions from the SMM of low-energy excitations in glasses are also presented

One conclusion is that two types of glasses can be revealed: glasses of type I which exhibit a universal boson peak at a moderately low frequency $\nu = \nu_{BP}$ and a high-frequency sound at higher frequencies, $\nu_{BP} < \nu < \nu_M$, up to not too high frequency $\nu_M \sim 10^2\,\text{K} < \nu_D$ (e.g., at $\nu_D \approx 300$–$500\,\text{K}$), and glasses of type II which do not exhibit a high-frequency sound above the peak.

Another conclusion is that the excess, non-Debye, excitations of low energies $\varepsilon \ll h\nu_D$ in glasses are soft-mode excitations (Ch. 7) that are essentially different at very low energies, $\varepsilon \lesssim \varepsilon_m \sim 1\,\text{K}$, and at moderately low energies, $3w \sim \varepsilon_m \lesssim \varepsilon \lesssim \varepsilon_M \sim 10^2\,\text{K}$. One of two fundamental quantum-mechanical simple ($1D$) systems [172], a strongly anharmonic two-level system or a harmonic oscillator, actually characterises, respectively the non-vibrational soft-mode excitations at very low energies $\varepsilon \ll w$ or harmonic vibrational excitations at moderately low energies in the range $\varepsilon_m \lesssim \varepsilon \lesssim \varepsilon_M$. The latter are harmonic vibrational excitations of random frequency subject to a rather broad distribution, whereas the former are non-vibrational excitations that are related to strongly anharmonic two-level systems (more specifically, to atomic tunneling states) of random energy also subject to a relatively broad distribution (Chs. 7, 8 and 10). In this connection, a *correlation* exists in the soft-mode model between dynamic and thermal properties at very low temperatures/frequencies and the properties at moderately low ones in glasses, because of the common origin of "excess" excitations. This conclusion can naturally explain the following empirical *correlation* [192]: the universal thermal properties of glasses at very low temperatures are found to be missing in amorphous materials where no boson peak in moderately-low energy vibrational dynamics is found. Since at present such a correlation is found only in the soft-mode model, the latter is suggested to be a relevant mean-field theory for general qualitative and scalewise description of the universal dynamic and thermal properties of glass (at least of type *I*) at both very low and moderately low temperatures, $T \ll T_D$, and frequencies, $\nu \ll \nu_D$. One can also suggest from a comparison of SMM (Chs. 7–10 and 11) with other models of glassy anomalous properties (Chs. 12 and 13) that the existence of atomic soft modes and their excitations is probably unavoidable in low-energy atomic dynamics of glasses. In this connection, the glasses under discussion apparently can also be defined as amorphous solids that exhibit the universal dynamic and thermal properties. Then, a question whether amorphous materials like amorphous silicon, which do not seem to exhibit a pronounced linear temperature dependence of specific heat at very low temperatures and a boson peak at moderately low frequencies, are genuine structural glasses does not seem to be answered positively.

As noted in Ch. 10, generally speaking, soft-mode excitations of low energies $\varepsilon \ll h\nu_D$ do not appear to be Anderson localised states around the Ioffe–Regel crossover energy $h\nu_{IR}^{(in)}$ for inelastic scattering, at $h\nu_{IR}^{(in)} < h\nu_{IR}^{(el)} < \varepsilon_{loc}$ (Eq. (10.8)). Moreover, this conclusion appears to be relevant even around the crossover energy $h\nu_{IR}^{(el)}$ for elastic scattering, at

$\varepsilon_{\rm loc} > h\nu_{\rm IR}^{(\rm in)} > h\nu_{\rm IR}^{(\rm el)}$. The states by definition becomes localised ones only at energy ε higher than the localisation edge, called also "mobility edge" $\varepsilon_{\rm loc}$, i.e., at $\varepsilon > \varepsilon_{\rm loc}$ (see e.g., [79]). In this sense, the soft-mode model is essentially different from models of low-energy atomic dynamics, in which low-energy excitations are identified as Anderson localised states, as is the case for the model [228].

Since SMM of low-energy atomic dynamics of glasses is a mean-field theory, some problems concerning the model basis and applications do not yet seem to have rigorous solutions

The problems are related to: 1. Elaboration of non-perturbative theory of Ioffe–Regel crossover for inelastic scattering of acoustic phonons from soft-mode excitations (Ch. 10); 2. Analysis of the behaviour of basic distribution density $F(\eta, \xi)$ for random soft-mode parameters (Sec. 7.2); 3. Analysis in more detail of vibrational excitations with moderately low energy (Ch. 10), including well-defined Anderson non-localised states below BP, ill-defined diffuson-like (extended) states around BP and well-defined HFS states above BP, and probably also of Anderson well-defined localised states above HFS (Sec. 11.2); 4. Analysis in more detail of universal and non-universal features of soft-mode dynamics at moderately low energies (Sec. 11.2).

Acknowledgments

Discussions with W. Goetze, V. Halpern and D. Osheroff, concerning some aspects of glass physics considered in the present book, are gratefully acknowledged.

Appendix A

CONVOLUTION OF SOFT-MODE VIBRATIONAL DOS AND TRANSFORMATION KERNEL IN THE DOS OF BP AND HFS EXCITATIONS

In this appendix, approximate analytical expressions for the total vibrational DOS $J(\varepsilon = \nu^2)$, describing the spectrum of vibrational eigenvalues $\varepsilon = \varepsilon(q) = \nu^2(q)$ with the dispersion law $\nu(q)$ of Eq. (6.7), are derived by applying a procedure developed in [194] both for independent soft modes, with the soft-mode vibrational DOS $J_\alpha^{(\mathrm{HV})}(\varepsilon') = C_\alpha \nu_0^{-2}(\varepsilon'/\nu_0^2)^{(2\alpha-1)/2}$ at $\alpha = 1$ or 2 (Eq. (10.5)), and for interacting soft modes, e.g., with $J_2^{(\mathrm{HV})}(\varepsilon') = J^*(\varepsilon')$ (Eq. (10.22)).

The total vibrational DOS can be calculated by applying its general expression (Eq. (10.9)) to the spectrum Eq. (10.7) (see, e.g., [204]):

$$J(\varepsilon \equiv \nu^2) = \int_{\varepsilon_1'}^{\varepsilon_2'} d\varepsilon' J^{(\mathrm{HV})}(\varepsilon') I(\varepsilon; \varepsilon'), \qquad \text{(A.1)}$$

where the transformation kernel, from soft-mode vibrational eigenvalues ε' to eigenvalues ε resulting from soft-mode–acoustic interaction, is

$$I(\varepsilon; \varepsilon') = (\pi/2) \cdot \Sigma_{j=1,2} \int_{\Omega(q)} dq q_D^{-3}\, q^2 \cdot D[\varepsilon - \epsilon_j(q; \varepsilon'); \kappa^2]. \qquad \text{(A.2)}$$

Here, the dispersion relation $\varepsilon \equiv \epsilon(q;\varepsilon')$ is independent of the wave vector direction, and $\Omega(q)$ is the range of allowed wave-number values, e.g., at $-q_D/2 \leq q \leq q_D/2 = \pi/a_1$, while the integration limits in Eq. (A.1) are $\varepsilon_1' = (3w/h)^2 \approx 0.1\nu_0^2$ and $\varepsilon_2' = \nu_0^2 (\ll \nu_D^2)$ (Eq. (7.30)). For well-defined vibrational excitations sufficiently below or above the spectral pseudogap of Eq. (10.7), e.g., for Debye phonons with $\epsilon(q) = \nu^2(q) = s_0^2 q^2$, the function $D(X;\kappa^2)$ has to be by definition an appropriate "pre-limit" function for the δ-function at $\kappa^2 \equiv \kappa_1^2 = 2\gamma/\nu \ll 1$, i.e., $\delta(X) = \lim D(X;\kappa_1^2)$ at $\kappa_1^2 \to +0$ [113]. The small width Γ of the excitation eigenvalue $\epsilon = \nu^2$ can be approximated from the complex frequency $\overline{\nu} = \nu - i\gamma$ with its width $\gamma \ll \nu$, $\overline{\nu}^2 \simeq \nu^2 - 2i\gamma\nu$, so $\Gamma \simeq 2\gamma\nu \equiv \kappa_1^2\nu^2 \ll \nu^2$. Then, the function can be approximated by an often used Lorentzian function $D_1[X]$, for typical $\nu \sim \nu_0$:

$$D_1(X;\kappa_1^2) = \pi^{-1}\{\Gamma/[X^2 + \Gamma^2]\} \simeq \pi^{-1}\{\kappa_1^2\nu_0^2/[X^2 + \kappa_1^4\nu_0^4]\}, \qquad (A.3)$$

at $X = \varepsilon - \epsilon$ and small $\kappa_1^2 \ll 1$. By taking into account comments below Eq. (10.10), one can approximate the kernel $I(\varepsilon;\varepsilon') = I_1(\varepsilon;\varepsilon')$, which accounts for the contribution of IRC and gives rise to Eq. (10.11) (instead of Eq. (10.10)):

$$I_1(\varepsilon;\varepsilon') = I_{(ac)}(\varepsilon;\varepsilon') + I_{(\mathrm{IRC})}(\varepsilon;\varepsilon'), \qquad (A.4)$$

where

$$I_{(ac)}(\varepsilon;\varepsilon') \simeq (\pi/2q_D^3)\cdot\Sigma_{j=1,2}\int q^2dqD[\varepsilon - \epsilon_j(q;\varepsilon');\kappa_1^2 \ll 1]$$
$$\times[\Theta(\varepsilon_l - \epsilon_j(q;\varepsilon')) + \Theta(\epsilon_j(q;\varepsilon') - \varepsilon_u)], \qquad (A.5)$$

and

$$I_{(\mathrm{IRC})}(\varepsilon;\varepsilon') \simeq (\pi/2q_D^3)\cdot\Sigma_{j=1,2}\int q^2dqD[\varepsilon - \epsilon_j(q;\varepsilon');\kappa_2^2 \approx 1]$$
$$\times[\Theta(\epsilon_j(q;\varepsilon') - \varepsilon_l) + \Theta(\varepsilon_u - \epsilon_j(q;\varepsilon'))], \qquad (A.6)$$

where $\Theta(X) = 1$ at $X > 0$ and $\Theta(X) = 0$ at $x < 0$. The function $I_{(\mathrm{IRC})}(\varepsilon;\varepsilon')$ and $D[\varepsilon - \varepsilon_j(q;\varepsilon');\kappa_2^2 \approx 1]$ describe the region around and in the pseudogap, containing ill-defined excitations with large excitation widths, whereas $I_{(ac)}(\varepsilon;\varepsilon')$ and $D[\varepsilon - \varepsilon_j(q;\varepsilon');\kappa_1^2 \ll 1]$ characterise two regions of acoustic-like well-defined excitations, Debye excitations and HFS ones, below and above the pseudogap respectively; the results of numerical calculations of $I_{(ac)}(\varepsilon;\varepsilon')$ are practically the same as to those obtained with the δ-function $\delta(X)$ substituted for $D[X;\kappa_1^2 \ll 1]$.

By introducing new, dimensionless, variables and parameters,

$$u \equiv \varepsilon/\nu_0^2 \equiv \nu^2/\nu_0^2, \quad t \equiv \varepsilon'/\nu_0^2, \quad \delta \equiv \Delta/\nu_0^2 \quad \text{and} \quad z \equiv \epsilon/\nu_0^2. \tag{A.7}$$

and by using, in particular, $D(X;\kappa_2^2) = D_1(X;\kappa_2^2)$, the resulting total vibrational DOS can be described by the following expression:

$$J(\varepsilon) = 0.5(\nu_0/\pi\nu_D^3)\cdot\Sigma_{j=1,2}\int_{\varepsilon_1'}^{\varepsilon_2'} d\varepsilon' J^{(HV)}(\varepsilon')I_1(\varepsilon;\varepsilon')\int_{\Omega(\epsilon_j,\varepsilon')} Q(\epsilon_j;\varepsilon')d\epsilon_j, \tag{A.8}$$

where $Q(\epsilon;\varepsilon') = q^2(\epsilon;\varepsilon')(\partial q(\epsilon,\varepsilon')/\partial\epsilon) = (1/2s_0^3)[\epsilon(\epsilon-\varepsilon'-\Delta)/(\epsilon-\varepsilon')]^{1/2}\cdot[1+\varepsilon\Delta/(\epsilon-\varepsilon')^2]$ and $q(\epsilon;\varepsilon')$ is the appropriate solution of the equation $\epsilon(q;\varepsilon') = \epsilon$, while $\Omega(\epsilon,\varepsilon')$ denotes the variation range of such ϵ and ε' that $Q(\epsilon;\varepsilon') \geq 0$. Moreover, $J(\varepsilon)$ can straightforwardly be reduced to the dimensionless function $I(u)$:

$$J(\varepsilon) = 0.5\cdot C\cdot(\nu_0/\nu_D^3)(\nu_0/\nu_D)^{2\alpha+1}\cdot I(u), \tag{A.9}$$

where

$$I(u) = \int_{\Omega(z,t)} dt \int dz\sqrt{z}g(z,t,\delta)\cdot\kappa^2[(z-u)^2+\kappa^4]^{-1}. \tag{A.10}$$

It is taken into account in the derivation of Eq. (A.9) that $\int d^3\mathbf{q}\{\ldots\} = 4\pi\int q^2dq\{\ldots\} = 4\pi\int_{\Omega(\epsilon)} K(\epsilon)d\epsilon\{\ldots\}$, where

$$K(\epsilon) = (\nu_0/2s_0^3)\sqrt{z}\cdot p(z,t;\delta), \tag{A.11}$$

with

$$p(z,t;\delta) = (z-t-\delta)^{1/2}(z-t)^{-1/2}\cdot[1+t\delta\cdot(z-t)^{-2}] \tag{A.12}$$

and the integration range $\Omega(\epsilon)$ for the system two-valued continuous spectrum $\epsilon(q,\varepsilon')$ in Eq. (10.7) can readily be transformed in Eq. (A. 10) to the two-valent integration range $\Omega(z,t)$.

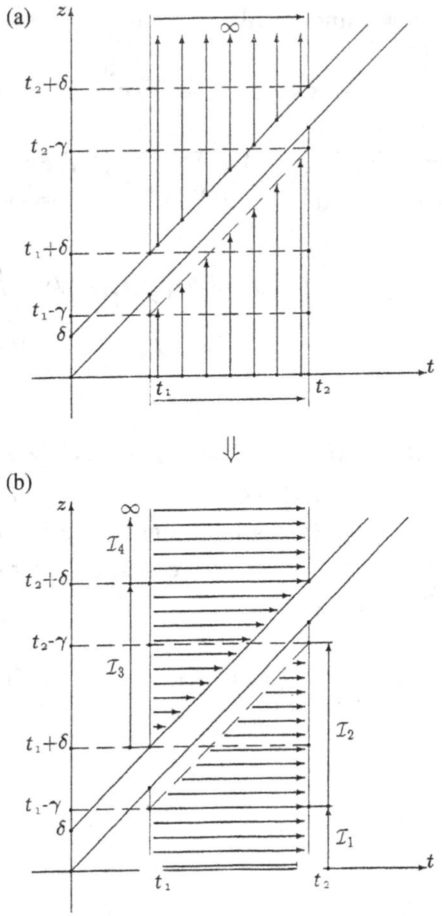

Figure 16: (a) Initial integration range $\Omega(z,t)$ in the double integral of Eq. (A.9), defined in Eqs. (A.12) and (A.13). (b) Integration range $\Omega(z,t)$ in the double integral of Eq. (A.9), after changing the sequence of integrations.

The latter includes the range $t_1 \leq t \leq t_2$ and is shown in Fig. 16(a), consisting of four two-dimensional strips:

$$t_1 \leq t \leq t_2, \quad z_{\min} \leq z \leq t_1 - \varsigma, \quad t_2 + \delta \leq z \leq z_{\max}; \tag{A.13}$$

$$t_1 - \varsigma \leq z \leq t_2 - \varsigma, \quad z + \varsigma \leq t \leq t_2; \quad t_1 + \delta \leq z \leq t_2 + \delta,$$
$$t_1 \leq t \leq z - \delta; \tag{A.14}$$

where $t_i \equiv \varepsilon_i'/v_0^2$ $(i = 1, 2)$ and ε_i' is defined in Eqs. (10.9). Then, after changing the sequence of integrations in the double integrals of Eq. (A.10) in accordance with Fig. 16(b), the integral $I(u)$ can explicitly be described as follows:

$$I(u) = I(u; \delta, \kappa) = \sum_{\varrho=1,2,3,4} I_\varrho(u; \delta, \kappa), \tag{A.15}$$

where

$$I_1(u; \delta, \kappa) = \int_0^{t_1-\varsigma} \kappa^2\sqrt{z}[(z-u)^2 + \kappa^4]^{-1} dz \int_{t_1}^{t_2} p(z, t; \delta)dt;$$

$$I_2(u; \delta, \kappa) = \int_{t_1-\varsigma}^{t_2-\varsigma} \kappa^2\sqrt{z}[(z-u)^2 + \kappa^4]^{-1} dz \int_{z+\varsigma}^{t_2} p(z, t; \delta)dt; \tag{A.16}$$

$$I_3(u; \delta, \kappa) = \int_{t_1+\delta}^{t_2+\delta} \kappa^2\sqrt{z}[(z-u)^2 + \kappa^4]^{-1} dz \int_{t_1}^{z-\delta} p(z, t; \delta)dt;$$

$$I_4(u; \delta, \kappa) = \int_{t_2+\delta}^{\infty} \kappa^2\sqrt{z}[(z-u)^2 + \kappa^4]^{-1} dz \int_{t_1}^{t_2} p(z, t; \delta). \tag{A.17}$$

Note that at $\Delta = 0$ the function $K(\epsilon)$ in Eq. (A.10) is the well-known Debye function, $K(\epsilon) = K_D(\epsilon) = (2\pi)s_0^{-3} \cdot \sqrt{\epsilon}$, as expected.

REFERENCES

[1] P. W. Anderson, in: *Ill-Condensed Matter*, ed. R. Balian (North-Holland, Amsterdam, 1979).

[2] R. Zallen, *The Physics of Amorphous Systems* (Wiley, New York, 1983).

[3] J. Phillips, *Solid State Physics*, Vol. 37, eds. H. Ehrenreich and D. Turnbull (Academic Press, New York, 1982).

[4] J. Jaeckle, *Rep. Progr. Phys.* **49**, 171 (1986).

[5] S. A. Brawer, *Relaxation in Viscous Liquids and Glasses* (American Ceramic Society, New York, 1983).

[6] F. Yonezawa and T. Ninomiya (eds.), *Topological Disorder in Condensed Matter* (Springer, Berlin, 1983).

[7] S. R. Elliott, *Physics of Amorphous Materials* (Longman & Wiley, New York, 1990).

[8] F. Sommer, *Rapidly Quenched Metals*, Vol. 1, eds. S. Steeb and Warlimont (North-Holland, Amsterdam, 1985).

[9] F. Luborski, (ed.), *Amorphous Metallic Alloys* (Butterworths, London, 1983).

[10] J. P. Hansen and I. R. McDonald, *Theory of Simple Liquids*, 2nd edn., (Academic Press, London, 1986).

[11] W. Goetze, in *Liquids, Freezing and Glass Transition*, eds. J. P. Hansen, and D. Levesque, J. Zinn-Justin (North Holland, 1991); W. Goetze and L. Sjoegren, *Rep. Prog. Phys.* **55**, 241 (1992).

[12] W. Goetze, *Complex Dynamics of Glass-Forming Liquids: A Mode-Coupling Theory* (Oxford University Press, 2009).

[13] K. Kawasaki, *Phys. Rev.* **150**, 291 (1966); *Ann. Phys.* (NY) **61**, 1 (1970).

[14] H. Vogel, *Z. Phys.* **22**, 645 (1922).

[15] G. Fulcher, *J. Am. Chem. Soc.* **8**, 339 (1925).

[16] G. Berry and T. Fox, *Adv. Polym. Sci.* **5**, 261 (1968).

[17] C. Angell and W. Sichina, *Ann. NY Acad. Sci.* **279**, 53 (1976).

[18] J. C. Maxwell, *Philos. Mag.* **27**, 299 (1864).

[19] J. Jaeckle, *Physica B* **127**, 79 (1984).

[20] G. Johary, *Ann. NY Acad. Sci.* **279**, 117 (1976).
[21] S. F. Edwards, *Polym. Sci.* **17**, 933 (1976).
[22] J. *Jaeckle*, *Philos. Mag. B* **44**, 533 (1981).
[23] R. G. Palmer, *Adv. Phys.* **31**, 669 (1982).
[24] M. Goldstein, *J. Chem. Phys.* **77**, 667 (1973).
[25] P. Gupta and C. Moynihan, *J. Chem. Phys.* **65**, 4136 (1976).
[26] R. Kohlrausch, *Ann. Phys. (Lpz)* **12,** 393 (1847).
[27] G. Williams and D. Watts, *Trans. Farad. Soc.* **66,** 80 (1970).
[28] A. K. Jonscher, *Dielectric Relaxation in Solids* (Chelsea Dielectric Press, London, 1983).
[29] J. D. Ferry, *Viscoelastic Properties of Polymers*, 3rd edn. (Wiley, New York, 1980).
[30] R. Zwanzig, *Phys. Rev. Lett.* **54**, 364 (1985).
[31] L. D. Landau and E. M. Lifshitz, *Fluid Mechanics* (Pergamon, Oxford, 1959).
[32] Phillips, A. Barlow and J. Lumb, *Proc. R. Soc. London, Ser. A* **329**, 193 (1972).
[33] N. Rivier and D. Duffy, in: *Numerical Methods in the Study of Critical Phenomena*, eds. J. Della-Dora, J. Demongeot and B. Lacolle (Springer, Berlin, 1981).
[34] A. M. Kosevich, *Physical Mechanics of Real Crystals* (Naukova Dumka, Kiev, 1981).
[35] I. M. Lifshits, *Usp. Fiz. Nauk* **80**, 617 (1964).
[36] N. Rivier, *Philos. Mag. A* **40**, 859 (1979).
[37] J. C. Phillips, *J. Non-Cryst. Solids* **63**, 347 (1984).
[38] J. Ziman, *Models of Disorder* (Cambridge University Press, 1979).
[39] W. H. Zachariasen, *J. Am. Chem. Soc.* **54**, 3841 (1932).
[40] J. D. Bernal, *Proc. R. Soc. London, Ser. A* **280**, 299 (1964).
[41] J. Ziman, *Principles of the Theory of Solids* (Cambridge University Press, 1972).
[42] D. E. Polk, *J. Non-Cryst. Solids* **5**, 365 (1971).
[43] R. J. Bell and P. Dean, *Philos. Mag.* **25**, 1381 (1972).
[44] N. F. Mott, *Philos. Mag.* **19**, 835 (1969).
[45] L. D. Landau and E. M. Lifshitz, *Statistical Physics* (Pergamon, Oxford, 1973).
[46] J. L. Finney, *Proc. R. Soc. London Ser. A* **319**, 479 (1970).
[47] D. Nelson, *Phys. Rev. B* **28**, 5515 (1983); P. Steinhardt, D. Nelson and M. Ronchetti, *Phys. Rev. B* **28**, 784 (1983).
[48] F. C. Frank, *Proc. R. Soc. London, Ser. A* **215**, 43 (1952).
[49] M. Kleman and J. Sadoc, *J. Phys.* **40**, 569 (1979).
[50] J. Gaspard, R. Mosseri and J. Sadoc, *Philos. Mag. B* **50**, 554 (1984).
[51] G. Toulouse, *Phys. Rep.* **49**, 267 (1979).
[52] M. H. Cohen, in *Topological Disorder in Condensed Matter*, eds. F. Yonezawa and T. Ninomiya (Springer, Berlin, 1983).

[53] J. Phillips, *Phys. Rev. B* **32**, 5356, 6972 (1985).
[54] M. I. Klinger, *Phys. Rep.* **94**, 183 (1983).
[55] E. A. Porai-Koshits, in *Phase Separation in Glass*, eds. O. V. Mazurin and E. A. Porai-Koshits (North-Holland, Amsterdam, 1984).
[56] S. F. Edwards and P. W. Anderson, *J. Phys. F* **5**, 965 (1975).
[57] K. Binder and A. Young, *Rev. Mod. Phys.* **58**, 132 (1986).
[58] J. C. Phillips and M. F. Thorpe, *Solid State Commun.* **53**,699 (1986).
[59] S. Feng, P. Sen, B. I. Halperin and C. Lob, *Phys. Rev. B* **30**, 5386 (1984).
[60] J. W. Essam, *Rep. Prog. Phys.* **43**, 833 (1980).
[61] R. Blachnik and A. Hoppe, *J. Non-cryst. Solids* **34**, 191 (1979).
[62] K. Tanaka, *Solid State Commun.* **54**, 867 (1985); **60**, 295 (1986).
[63] N. Rivier, in *Topological Disorder in Condensed Matter*, eds. F. Yonezawa, T. Ninomiya (Springer, Berlin, 1983).
[64] J. Hubbard and J. Beeby, *J. Phys. C* **5**, 1277 (1969).
[65] S.-K. Ma, *J. Stat. Phys.* **26**, 221 (1981).
[66] M. Cyrot, *Phys. Lett. A* **83**, 275 (1981).
[67] A. Doolittle, *J. Appl. Phys.* **22**, 147 (1951).
[68] M. H. Cohen and D. Turnbull, *J. Chem. Phys.* **31**, 1164 (1959).
[69] A. Barlow, J. Lamb and A. Matheson, *Proc. R. Soc. London Ser. A* **292**, 322 (1966).
[70] W. Kauzmann, *Chem. Rev.* **43**, 219 (1948).
[71] G. Adams and J. Gibbs, *J. Chem. Phys.* **43**, 139 (1965).
[72] F. Simon and F. Lange, *Z. Phys.* **39**, 227 (1926).
[73] J. Gibbs and E. Di Marzio, *J. Chem. Phys.* **28**, 373, 807 (1958).
[74] M. H. Cohen and G. S. Grest, *Phys. Rev. B* **20**, 1077 (1979).
[75] Y. J. Frenkel, *Kinetic Theory of Liquids* (Clarendon, Oxford, 1946).
[76] J. Hirschfelder, D. Stevenson and H. Eyring, *J. Chem. Phys.* **5**, 896 (1935).
[77] J. Kirkwood, *J. Chem. Phys.* **18**, 380 (1950).
[78] H. Kanno, *J. Non-cryst. Solids* **37**, 203 (1980).
[79] N. F. Mott and E. A. Davis, *Electronic Processes in Noncrystalline Materials* (Clarendon, Oxford, 1979).
[80] E. Abrahams, P. W. Anderson, D. C. Licciardello and T. V. Ramakrishnan, *Phys. Rev. Lett.* **42** 673 (1979).
[81] S. F. Edwards and T. Vilgis, in *Physics of Disordered Materials*, eds. D. Adler and H. Fritzsche (Plenum, New York, 1985).
[82] G. S. Grest and M. H.Cohen, *Phys. Rev. B* **21**, 4413 (1980).
[83] E. J. Donth, *J. Non-cryst. Solids* **53**, 325 (1982).
[84] P. D. Gujrati and M. Goldstein, *J. Chem. Phys.* **74**, 259 (1981).
[85] K. P. Bordewijk, *Chem. Phys. Lett.* **32**, 592 (1975).
[86] S. A. Glarum, *J. Chem. Phys.* **33**, 639 (1960).
[87] J. Kosterlitz and D. Thouless, *J. Phys. C* **6**, 1181 (1973).
[88] V. L. Berezinskii, *Zh. Exp. Teor. Fiz.* **61**, 1144 (1971).

[89] J. Bosse, W. Goetze and M. Luecke, *Phys. Rev.* A **17**, 434, 447 (1978); E. Leutheusser, *Phys. Rev.* A **29**, 2765 (1984); J. Horbach, W. Kob and K. Binder, *Eur. Phys. J. B* **19**, 531 (2001).
[90] A. Bengtzelius, W. Goetze and A. Sjoelander, *J. Phys.* C **17**, 5917 (1984).
[91] S. P. Das and C. F. Mazenko, *Phys . Rev. Lett.* **54**, 118 (1985); *Rev. Mod. Phys.* **76**, 785 (2004).
[92] D. Forster, *Hydrodynamic Fluctuations, Broken Symmetry and Correlation Functions* (Benjamin, Reading, Massachusets, 1975).
[93] R. Kubo, *Rep. Prog. Phys.* **29**, 255 (1966).
[94] R. Gilmore, *Catastrophe Theory* (Wiley, New York, 1981).
[95] R. Hill and L. Dissado, *J. Phys.* C **18**, 3829 (1985).
[96] R. G. Palmer, D. L. Stein, E. Abrahams and P. W. Anderson, *Phys. Rev. Lett.* **53**, 958 (1984); **54**, 365 (1985).
[97] A. Blumen, G. Zumofen and J. Klafter, *Phys. Rev. B* **30**, 5379 (1984).
[98] M. F. Shlesinger, *J. Stat. Phys.* **39**, 639 (1984).
[99] J. Bendler, *J. Stat. Phys.* **39**, 625 (1984).
[100] S. A. Brawer, *J. Chem. Phys.* **72**, 4264 (1980).
[101] S. Chandrasekhar, *Rev. Mod. Phys.* **15**, 1 (1943).
[102] G. Pfister and N. Scher, *Adv. Phys.* **27**, 747 (1978).
[103] M. H. Cohen and G. C. Grest, *Phys. Rev.* B **24**, 4091(1981).
[104] R. C. Zeller and R. O. Pohl, *Phys. Rev.* B **4**, 2029 (1971).
[105] W. A. Phillips (ed.), *Amorphous Solids: Low-Temperature Properties* (Springer, Berlin, 1981).
[106] S. Hunklinger and W. Arnold, in *Physical Acoustics*, Vol. XII, eds. W. P. Mason and R. N. Thurnton (Academic Press, NY, 1976), p. 155; S. Hunklinger, M. V. Schickfus, in: *Amorphous Solids: Low-Temperature Properties*, ed. W. A. Phillips (Springer, Berlin, 1981), p. 81; S. Hunklinger and A. K. Raychaudhari, in *Progress in Low Temperature Physics*, Vol. IX, ed. D. F. Brewer (Elsevier, Amsterdam, 1986), p. 265.
[107] R. O. Pohl, in *Amorphous Solids: Low-Temperature Properties*, ed. W. A. Phillips (Springer, Berlin, 1981), p. 27.
[108] R. B. Stephens, *Phys. Rev.* B **8**, 2896 (1973); N. Bilir and W. A. Phillips, *Philos. Mag.* **32**, 113 (1975).
[109] X. Liu, B. E. White Jr., R. O. Pohl, E. Ivaniczko, K. M. Jones, A. H. Mahan, B. N. Nelson, R. S. Crandall and S. Veprek, *Phys. Rev. Lett.* **78**, 4418 (1997).
[110] B. Golding and J. E. Graebner, in *Amorphous Solids: Low-Temperature Properties*, ed. W. A. Phillips (Springer, Berlin, 1981), p. 107.
[111] A. C. Anderson, in: *Amorphous Solids: Low-Temperature Properties*, ed. W. A. Phillips (Springer, Berlin, 1981), p. 65.
[112] G. Ruocco and F. Sette, *J. Phys.*, CM **13**, 9141 (2001).

[113] R. Peierls, *Quantum Theory of Solids* (Clarendon, Oxford, 1955); P. G. Klemens, in *Physics of Non-Crystalline Solids*, ed. J. A. Prins (North-Holland, Amsterdam, 1965).
[114] H. S. Guentherodt and H. Beck (eds.), *Topics in Applied Physics*, Vol. 46, *Glassy Metals I, and* Vol. 53, *Glassy Metals II* (Springer, Berlin, 1983).
[115] E. M. Lifshits and L. P. Pitaevskii, *Statistical Physics, Part* 2 [*Theory of Condensed State*] (Pergamon, Oxford, 1980).
[116] P. W. Anderson, B. I. Halperin and C. M. Varma, *Philos. Mag.* **25**, 1 (1972).
[117] W. A. Phillips, *J. Low Temp. Phys.* **7**, 351 (1972).
[118] U. Buchenau, M. Prager, N. Nuecker, A. J. Dianoux, N. Ahmad and W. A. Phillips, *Phys. Rev. B* **34**, 5665 (1986); N. Ahmad, K. W. Hutt and W. A. Phillips, *J. Phys.* C **19**, 3765 (1986).
[119] M. I. Klinger, *Problems of Modern Physics* (100th Anniversary of A. F. Ioffe), ed. A. P. Aleksandrov (Nauka, Leningrad, 1980), p. 291.
[120] C. C. Yu and A. J. Leggett, *Comments. Condens. Matter. Phys.* **14**, 231 (1988).
[121] R. Orbach, *Science* **231**, 814 (1986).
[122] A. L. Burin, D. Natelson, D. Osheroff and Yu. Kagan, in *Tunneling Systems in Amorphous and Crystalline Materials* (Springer, Berlin, 1998), p. 223.
[123] M. I. Klinger and V. G. Karpov, *Pis'ma Zh. Tech. Fiz.* **6**, 1473, 1478 (1980); *Sol. State Commun.* **37**, 975 (1981); *Sov. Phys. JETP* **55**, 976 (1982); V. G. Karpov, M. I. Klinger and F. N. Ignatiev, *Sov. Phys. JETP* **84**, 760 (1983); *Sol. State Commun.* **44**, 333 (1982); F. N. Ignatiev, V. G. Karpov and M. I. Klinger, *Dokl. Akad. Nauk SSSR* **269**, 1341 (1983); *Sov. Phys. Sol. State* **25**, 727 (1983).
[124] M. I. Klinger, *Phys. Rep.* **94**, 183 (1983); *Philos. Mag. B* **81**, 1273 (2001).
[125] M. I. Klinger, *Phys. Rep.* **165**, 275 (1988); *Phys. Rep.* **492**, 111 (2010).
[126] M. I. Klinger, *Comments Condens. Matter Phys.* **16**, 137 (1992).
[127] Y. M. Galperin, V. G. Karpov and V. I. Kozub, *Adv. Phys.* **38**, 669 (1989).
[128] U. Buchenau and M. Ramos, in *Tunneling Systems in Amorphous and Crystalline Materials* (Springer, Berlin, 1998), p. 527.
[129] P. W. Anderson, *Concept in Solids* (Benjamin, New York, 1964); *Basic Notions of Condensed Matter Physics* (Benjamin-Cummings, London, 1984).
[130] C. K. Jones, P. G. Clemens and J. A. Rayne, *Phys. Lett.* **1**, 31 (1964).
[131] J. J. Freeman and A. C. Anderson, *Phys. Rev. B* **34**, 5684 (1986).
[132] R. O. Pohl, X. Liu and E. Thompson, *Rev. Mod. Phys.* **74**, 991 (2002).
[133] A. M. Stoneham, *Theory of Defects in Solids* (Clarendon, Oxford, 1975).
[134] S. Hunklinger, *Physica A* **261**, 26 (1998).
[135] H. A. Kramers, *Atti Congr. Inter. Fis.* (*Como*) **2**, 545 (1927).
[136] R. Kronig, *J. Opt. Soc. Amer.* **12**, 547 (1927).
[137] J. E. Graebner and B. Golding, *Phys. Rev. B* **19**, 964 (1979).

[138] J. L. Black and B. I. Halperin, *Phys. Rev. B* **16**, 2879 (1977); P. Hu and L. R. Walker, *Sol. State Commun.* **24**, 813 (1977).
[139] D. D. Osheroff, S. Rogge and D. Natelson, *Czech. J. Phys.* **46**, 3295 (1996); S. Rogge, D. Natelson and D. D. Osheroff, *Czech. J. Phys.* **46**, 2263 (1996).
[140] J. Jaeckle, in *Amorphous Solids: Low-Temperature Properties*, ed. W. A. Phillips (Springer, Berlin, 1981), p. 135.
[141] S. N. Taraskin and S. R. Elliott, *Phys. Rev. B* **59**, 8572 (1999).
[142] R. Shuker and R. W. Gammon, *Phys. Rev. Lett.* **25**, 222 (1970).
[143] N. V. Surovtsev and A. V. Sokolov, *Phys. Rev. B* **66**, 054205 (2002).
[144] D. Engberg, A. Wischnewski, U. Buchenau, L. Borjesson, A. Dianoux, A. Sokolov and L. Torell, *Phys. Rev. B* **59**, 4053 (1999); A. P. Sokolov, E. Roessler, A. Kisliuk and D. Quitmann, *Phys. Rev. Lett.* **71**, 2062 (1993).
[145] N. V. Surovtsev, J. A. H. Wiedersich, V. N. Novikov, E. Roessler and A. P. Sokolov, *Phys. Rev. B* **58**, 14888 (1998), *Phys. Rev. B* **64**, 064207 (2001); V. N. Novikov, A. P. Sokolov, B. Strube, N. V. Surovtsev, E. Duval and A. Mermet, *J. Chem. Phys.* **107**, 1057 (1997); T. Achibat, A. Boukenter and E. Duval, *J. Chem. Phys.* **99**, 2046 (1993).
[146] F. Sette, M. Krisch, C. Masciovecchio, G. Ruocco and G. Monaco, *Science* **280**, 1550 (1988).
[147] M. Foret, E. Courtens, R. Vacher and J. B. Suck, *Phys. Rev. Lett.* **77**, 3831 (1996); O. Pilla, A. Consulo, A. Fontana, C. Masciovecchio, G. Monaco, M. Montagna, G. Ruocco, T. Scopigno and F. Sette, *Phys. Rev. Lett.* **85**, 2136 (2000); E. Courtens, M. Foret, B. Hehlen and R. Vacher, *Sol. State Commun.* **117**, 187 (2001).
[148] M. Foret, B. Hehlen, G. Taillades, E Courtens, R. Vacher, H.Casalta and B. Domer, *Phys. Rev. Lett.* **81**, 2100 (1998); Kojima and M. Kodama, *Physica B* **263–264**, 336 (1999); Y. Inamura, M. Arai, O. Yamamuro, A. Inaba, N. Kitamura, T. Otomo, T. Matsuo, S. M. Bennington and A. C. Hannon, *Physica B* **263–264**, 299 (1999); A. Sokolov, U. Buchenau, D. Richter, A. Mermet, C. Masciovecchio, F. Sette, D. Fioretto, G. Ruocco, L. Willner and B. Frick, *Phys. Rev. E* **63**, 2464 (2001); M. Nakamura, M. Arai, Y. Inamura, T. Otomo and S. M. Bennington, *Phys. Rev. B* **67**, 064204 (2003).
[149] C. Masciovecchio, G. Ruocco, F. Sette, M. Krisch, R. Verbeni, U. Bergmann and M. Soltwisch, *Phys. Rev. Lett.* **76**, 3356 (1996); P. Benassi, M. Krisch, C. Masciovecchio, V. Mazzacurati, G. Monaco, G. Ruocco, F. Sette and R. Verbeni, *Phys. Rev. Lett.* **77**, 3835 (1996).
[150] A. Matic, L. Borjesson, G. Ruocco, C. Masciovecchio, A. Mermet, F. Sette and R. Verbeni, *Euro. Phys. Lett.* **54**, 77 (2001).
[151] A. Matic, D. Engberg, C. Masciovecchio and L. Borjesson, *Phys. Rev. Lett.* **86**, 3803 (2001).
[152] G. Ruocco,F. Sette, R. Di Leonardo, D. Fioretto, M. Krisch, M. Lorenzen, C. Masciovecchio, G. Monaco, F. Pignon and T. Scopigno, *Phys. Rev. Lett.* **83**, 5583 (1999).

[153] C. Masciovecchio, G. Baldi, S. Caponi, L. Comez, S. Di Fonzo, D. Fioretto, A. Fontana, A. Gessini, A. Santucci, F. Sette, G. Villiani, P. Vilmercati and G. Ruocco, *Phys. Rev. Lett.* **97**, 035501 (2006).

[154] A. I. Chumakov, I. Sergueev, U. van Buerck, W. Shirmacher, T. Asthalter, R. Rueffer, O. Leupold and W. Petry, *Phys. Rev. Lett.* **92**, 245508 (2004).

[155] A. K. Raychaudhury and R. O. Pohl, *Phys. Rev. B* **25**, 1310 (1982).

[156] W. Feller, *An Introduction to Probability Theory and Its Applications*, Vol. I (John Wiley, London, 1967).

[157] F. N. Ignatiev, V. G. Karpov and M. I. Klinger, *J. Non-Cryst. Sol.* **55**, 307 (1983).

[158] M. A. Krivoglaz, *Proc. Inst. Phys. Estonian Acad. Sci.* **59**, 31 (1986)

[159] M. A. Il'in, V. G. Karpov and D. A. Parshin, *Sov. Phys. JETP* **92**, 291 (1987).

[160] V. V. Fleurov and L. I. Trachtenberg, *Sol. State Commun.* **55**, 537 (1985).

[161] U. Buchenau, *Sol. State Commun.* **56**, 889 (1985); X. Wu, H. Chen, *J. Phys. C* **19**, 5957 (1986); X. Wu, H. Chen and J. Fang, *Sol. State Commun.* **60**, 373 (1986).

[162] M. H. Cohen and G. S. Grest, *Phys. Rev. Lett.* **45**, 1271 (1980); *Sol. State Commun.* **39**, 143 (1981).

[163] Ya. I. Frenkel, *Acta Physicochim. USSR* **3**, 633 (1935).

[164] H. R. Schober and B. B. Laird, *Phys. Rev. B* **44**, 6746 (1881), H. R. Schober and C. Oligschleger, *Phys. Rev. B* **53**, 11469 (1996); C. Oligschleger and H. R. Schober, *Phys. Rev. B* **59**, 811 (1999).

[165] H. R. Schober, *Physica A* **201**, 391 (1993); W. Jin, P. Vashishta, R. P. Kalia and J. R. Rino, *Phys. Rev. B* **48**, 9359 (1993).

[166] K. Trachenko and M. Turlakov, *Phys. Rev. B* **73**, 012203 (2006).

[167] J. A. Sussmann, *J. Phys. Chem. Soc.* **28**, 1643 (1967); M. I. Klinger, in *Physics of Disordered Materials*, eds. D. Adler and H. Fritzsche (Plenum, New York, 1985), p. 617.

[168] V. G. Karpov and D. A. Parshin, *Sov. Phys. JETP* **61**, 1308 (1985).

[169] M. A. Krivoglaz, A. L. Pinkevich, *Fiz. Tverd. Tela* (Russian), **11**, 96 (1969); M .A. Krivoglaz, *Zh. Exp. Teor. Fiz.* **88**, 2171 (1985).

[170] U. Buchenau, Y. M. Galperin, V. L. Gurevich and H. R. Schober, *Phys. Rev. B* **43**, 5040 (1991).

[171] M. Meissner and K. Spitzmann, *Phys. Rev. Lett.* **46**, 265 (1981); M. T. Loponen, R. C. Dynes, V. Narayanamurti and J. P. Garno, *Phys. Rev. B* **25**, 1161 (1982).

[172] L. D. Landau and E. M. Lifshits, *Quantum Mechanics* (Pergamon, Oxford, 1966).

[173] K. S. Gilroy and W. A. Phillips, *Philos. Mag.* **43**, 735 (1981).

[174] H. Froehlich, *Theory of Dielectrics* (Clarendon, Oxford, 1958).

[175] R. Tipping and J. Ogilvie, *Phys. Rev. A* **27**, 95 (1983).

[176] A. V. Turbiner, *Usp. Fiz. Nauk* (Russian) **144**, 35 (1984).

[177] L. van Hove, *Phys. Rev.* **89**, 1189 (1953).
[178] V. G. Karpov and D. A. Parshin, *Sov. Phys. JETP Lett.* **38**, 648 (1983).
[179] M. I. Klinger, *Sol. State Commun.* **51**, 503 (1984).
[180] J. E. Graebner, L. C. Allen and B. Golding and A. B Kane, *Phys. Rev. B* **27**, 3697 (1983); M. Meissner and P. Strehlow, *Czech J. Phys.* **46**, 2233 (1996).
[181] J. Jaeckle, L. Piche, W. Arnold and S. Hunklinger, *J. Non-Cryst. Sol.* **20**, 365 (1976).
[182] P. Esquinazi, R. Koenig and F. Pobell, *Z. Phys. B* **87**, 305 (1992); J. Classen, C. Enss, C. Bechinger, G. Weiss and S. Hunklinger, *Ann. Phys.* **3**, 315 (1994); J. E. Van Cleve, A. K. Raychaudhuri and R. O. Pohl, *Z. Phys. B* **93**, 479 (1994); C. Enss and S. Hunklinger, *Phys. Rev. Lett.* **79**, 2831 (1997).
[183] S. Hunklinger, W. Arnold and S. Stein, *Phys. Lett.* A **45**, 311 (1973); W. Arnold and S. Hunklinger, *Sol. State Commun.* **17**, 883 (1975).
[184] L. Bernard, L. Piche, G. S. Schumacher and J. Jofrin, *J. Low Temp. Phys.* **35**, 41 (1979); G. Baier and M. V. Schickfus, *Phys. Rev.* B **38**, 9952 (1988).
[185] C. C. Yu, *Phys. Rev. Lett.* **63**, 1160 (1989); E. R. Grannan, M. Randeria and J. P. Sethna, *Phys. Rev. B* **41**, 7784 (1990); S. N. Coppersmith, *Phys. Rev. Lett.* **67**, 2315 (1991).
[186] Abragam, in *The Principle of Nuclear Magnetism*, eds. N. F. Mott, E. C. Bullard and D. H. Wilkinson (Oxford University Press, London,1961).
[187] A. L. Burin and Y. M. Kagan, *Sov. Phys. JETP* **80**, 761 (1995).
[188] A. L. Burin and Y. M. Kagan, *Sov. Phys. JETP* **79**, 347 (1994).
[189] C. C. Yu, *Phys. Rev. B* **32**, 4220 (1985); S. V. Maleev, *Sov. Phys. JETP* **67**, 157 (1989).
[190] B. I. Shklovsky and A. L. Efros, *Electronic Properties of Doped Semiconductors* (Springer, Berlin, 1984).
[191] A. L. Burin and Y. M. Kagan, in *Tunneling Systems in Amorphous and Crystalline Materials* (Springer, Berlin, 1998).
[192] C. A. Angell, *J. Phys.* CM **12**, 6463 (2000).
[193] W. Goetze and M. R. Mayr, *Phys. Rev. E* **61**, 587 (2000).
[194] M. I. Klinger and A. M. Kosevich, *Phys. Lett.* A **280**, 365 (2001); **295**, 311 (2002); *J. Non-Cryst. Sol.* **345–346**, 242 (2004).
[195] M. I. Klinger and L. Vatova, *Phys. Rev. B* **72**, 134206 (2005).
[196] T. S. Grigera, V. Martin-Mayor, G. Parisi and P. Verrocchio, *Philos. Mag. B* **82**, 637 (2002).
[197] G. Parisi, *J. Phys.*, CM **15**, S765 (2003).
[198] A. F. Ioffe and A. R. Regel, *Progress in Semiconductors* **4**, 237 (1960).
[199] R. Brout and W. Wisscher, *Phys. Rev. Lett.* **9**, 54 (1962).
[200] Y. Kagan and Y. Iosilevskii, *Sov. Phys. JETP* **15**, 182 (1962).
[201] S. R. Elliott, *Euro. Phys. Lett.* **19**, 201 (1992).
[202] P. Sheng, M. Zhou and Z.-Q. Zhang, *Phys. Rev. Lett.* **72**, 263 (1994).

[203] A. M. Kossevich, *The Crystal Lattice: Phonons, Solitons, Dislocations* (Wiley-VCH, Berlin, 1999).
[204] I. M. Lifshits, S. A. Gredescul and L. A. Pastur, *Introduction to Theory of Disordered Systems* (Wiley, New York, 1988).
[205] J. Fabian and P. B. Allen, *Phys. Rev. Lett.* **82**, 1478 (1999); *Phys. Stat. Sol.* (c) **1**, 2860 (2004).
[206] V. L. Gurevich, D. A. Parshin and H. R. Schober, *JETP Lett.* **76**, 553 (2002); *Phys. Rev. B* **67**, 094203 (2003).
[207] D. A. Parshin, H. R. Schober and V. L. Gurevich, *Phys. Rev. B* **76**, 064206 (2007); *J. Phys., Conf. Ser.* **92**, 012131 (2007); D. A. Parshin and C. Laermans, *Physica B* **316–317**, 542 and 549 (2002); D. A. Parshin, *Phys. Stat. Sol.* (c) **1**, 2860 (2004).
[208] B. Ruffle, D. A. Parshin, E. Courtens and R. Vacher, *Phys. Rev. Lett.* **100**, 015501 (2008).
[209] V. L. Gurevich, D. A. Parshin, J. Pelous and H. R. Schober, *Phys. Rev. B* **48**, 16318 (1993); D. A. Parshin and C. Laermans, *Phys. Rev. B* **63**, 132203 (2001).
[210] M. I. Klinger, *Phys. Lett. A* **373**, 3563 (2009).
[211] M. I. Klinger, *Phys. Lett. A* **373**, 1782 (2009); *Theor. Phys. Math. Phys* **154**, 64 (2008).
[212] U. Buchenau, Y. Galperin, V. L. Gurevich, D. A. Parshin and H. R. Schober, *Phys. Rev. B* **46**, 2798 (1992).
[213] M. I. Klinger and V. Halpern, *Phys. Lett. A* **313**, 448 (2003).
[214] C. C. Yu and J. J. Freeman, *Phys. Rev. B* **36**, 7620 (1987); M. Randeria and J. P. Sethna, *Phys. Rev. B* **38**, 12607 (1988).
[215] A. L. Burin and Y. M. Kagan, *Sov. Phys. JETP* **82**, 159 (1996); *Phys. Lett. A* **215**, 191 (1996).
[216] L. S. Levitov, *Phys. Rev. Lett.* **64**, 547 (1990).
[217] W. Heitler, *The Quantum Theory of Radiation* (Clarendon, Oxford, 1954).
[218] N. J. Tao, G. Li, X. Chen, W. M. Du and H. Z. Cummins, *Phys. Rev. A* **44**, 6665 (1991).
[219] T. S. Grigera, V. Martin-Mayor and G. Parisi, *Philos. Mag.* **82**, 637 (2002).
[220] S. N. Taraskin and S. R. Elliott, *Phys. Rev. B* **61**, 12017 (2000).
[221] E. Maurer and W. Schirmacher, *J. Low Temp. Phys.* **137**, 453 (2004).
[222] J. W. Kantelhardt, S. Russ and A. Bunde, *Phys. Rev. B* **63**, 064302 (2001).
[223] S. N. Taraskin, J. J. Ludlam, G. Natarajan and S. R. Elliott, *Philos. Mag. B* **82**, 197 (2002).
[224] S. N. Taraskin and S. R. Elliott, *J. Phys., CM* **14**, 3143 (2002).
[225] W. Schirmacher, *Euro. Phys. Lett.* **73**, 892 (2006).
[226] W. Schirmacher, G. Ruocco and T. Scopigno, *Phys. Rev. Lett.* **98**, 025501 (2007).
[227] C. Ganter and W. Schirmacher, *Phys. Rev. B* **82**, 094205 (2010).

[228] T. Nakayama, *Rep. Prog. Phys.* **65**, 1195 (2002).
[229] T. Nakayama, *J. Phys. Jap.* **68**, 3540 (1999).
[230] K. Seeger, *Semiconductor Physics* (Springer, New York, 1973).
[231] P. W. Anderson, *Phys. Rev. Lett.* **34**, 953 (1975).
[232] T. Holstein, *Ann. Phys.* (NY) **8**, 325, 343 (1979).
[233] M. I. Klinger, *Problems of Electron (Polaron) Transport Theory in Semiconductors* (Pergamon, Oxford, 1979).
[234] M. I. Klinger, *Sov. Phys., Adv. Phys. Sci. (Russia)* **147**, 528 (1985).
[235] J. Hubbard, *Proc. R. Soc.* A **276**, 238 (1963).
[236] W. A. Phillips, *Philos. Mag. B* **34**, 983 (1984).
[237] V. Karpov, *Fiz. Tech. Poluprovod.* **19**, 123 (1985).
[238] R. A. Street and N. F. Mott, *Phys. Rev. Lett.* **35**, 1293 (1975).
[239] N. F. Mott, E. A. Davis and R. A. Street, *Philos. Mag. B* **32**, 961 (1975).
[240] M. Kastner, D. Adler and H. Fritzsche, *Phys. Rev. Lett.* **37**, 1504 (1976).
[241] N. F. Mott, *J. Phys.* C **13**, 5433 (1980).
[242] D. Adler, *J. Non-Cryst. Sol.* **35–36**, 819 (1980).
[243] H. Fritzsche and M. Kastner, *Philos. Mag. B* **37**, 285 (1978).
[244] M. I. Klinger, *Sov. Phys., Adv. Phys. Sci. (Russia)* **152**, 623 (1987).
[245] S. I. Pekar, *Sov. Phys., Adv. Us. Fiz.* **44**, 156 (1956).
[246] M. I. Klinger and I. I. Yaskovets, *J. Phys.* C **17**, 949 (1984).
[247] M. I. Klinger, *Sol. State Commun.* **45**, 949 (1983); *Proc. Acad. Sci. USSR* **279**, 91 (1884).
[248] V. Karpov, *Zh. Eksp. Teor. Fiz.* **85**, 883 (1983).
[249] M. I. Klinger and S. N. Taraskin, *Phys. Rev. B* **52**, 2557 (1995).
[250] M. I. Klinger and S. N. Taraskin, *Phys. Rev. B* **47**, 10235 (1993).
[251] F. D. Haldane and P. W. Anderson, *Phys. Rev. B* **13**, 2553 (1978).
[252] W. Fowler and R. J. Elliott, *Phys. Rev. B* **34**, 5525 (1986).
[253] K. Shimakawa, A. V. Kolobov and S. R. Elliott, *Adv. Phys.* **44**, 475 (1995).
[254] A. V. Kolobov, B. T. Kolomiets, O. V. Konstantinov and V. M. Lyubin, *J. Non-Cryst. Sol.* **45**, 335 (1981).
[255] V. L. Averyanov, A. V. Kolobov, B. T. Kolomiets and V. M. Lyubin, *J. Non-Cryst. Sol.* **45**, 34r, y3 (1981).
[256] M. I. Klinger, *Phys. Lett.* A **327**, 210 (2004).
[257] M. I. Klinger, in *Photo-Induced Metastability in Amorphous Semiconductors*, ed. A. V. Kolobov (Wiley-VCH, Berlin, 2003).

Index